Accelerating the Discovery of New Dielectric Properties in Polymer Insulation

Boxue Du
Tianjin University, China

A volume in the Advances in Computer and Electrical Engineering (ACEE) Book Series

IGI Global
DISSEMINATOR OF KNOWLEDGE
www.igi-global.com

Published in the United States of America by
IGI Global
Engineering Science Reference (an imprint of IGI Global)
701 E. Chocolate Avenue
Hershey PA, USA 17033
Tel: 717-533-8845
Fax: 717-533-8661
E-mail: cust@igi-global.com
Web site: http://www.igi-global.com

Library of Congress Cataloging-in-Publication Data

Names: Du, Boxue, 1961- author.
Title: Accelerating the discovery of new dielectric properties in polymer insulation / by Boxue Du.
Description: Hershey, PA : Engineering Science Reference, [2017] | Includes bibliographical references.
Identifiers: LCCN 2016057815| ISBN 9781522523093 (hardcover) | ISBN 9781522523109 (ebook)
Subjects: LCSH: Electric insulators and insulation--Polymers. | Dielectric devices.
Classification: LCC TK3441.P58 D823 2017 | DDC 621.319/20284--dc23 LC record available at https://lccn.loc.gov/2016057815

This book is published in the IGI Global book series Advances in Computer and Electrical Engineering (ACEE) (ISSN: 2327-039X; eISSN: 2327-0403)

British Cataloguing in Publication Data
A Cataloguing in Publication record for this book is available from the British Library.

All work contributed to this book is new, previously-unpublished material.
The views expressed in this book are those of the authors, but not necessarily of the publisher.

For electronic access to this publication, please contact: eresources@igi-global.com.

Advances in Computer and Electrical Engineering (ACEE) Book Series

ISSN:2327-039X
EISSN:2327-0403

MISSION

The fields of computer engineering and electrical engineering encompass a broad range of interdisciplinary topics allowing for expansive research developments across multiple fields. Research in these areas continues to develop and become increasingly important as computer and electrical systems have become an integral part of everyday life.

The **Advances in Computer and Electrical Engineering (ACEE) Book Series** aims to publish research on diverse topics pertaining to computer engineering and electrical engineering. **ACEE** encourages scholarly discourse on the latest applications, tools, and methodologies being implemented in the field for the design and development of computer and electrical systems.

COVERAGE

- Digital Electronics
- Programming
- Qualitative Methods
- Computer science
- Electrical Power Conversion
- Chip Design
- VLSI Fabrication
- Optical Electronics
- Power Electronics
- Computer Architecture

IGI Global is currently accepting manuscripts for publication within this series. To submit a proposal for a volume in this series, please contact our Acquisition Editors at Acquisitions@igi-global.com or visit: http://www.igi-global.com/publish/.

The Advances in Computer and Electrical Engineering (ACEE) Book Series (ISSN 2327-039X) is published by IGI Global, 701 E. Chocolate Avenue, Hershey, PA 17033-1240, USA, www.igi-global.com. This series is composed of titles available for purchase individually; each title is edited to be contextually exclusive from any other title within the series. For pricing and ordering information please visit http://www.igi-global.com/book-series/advances-computer-electrical-engineering/73675. Postmaster: Send all address changes to above address.

Titles in this Series

For a list of additional titles in this series, please visit:
http://www.igi-global.com/book-series/advances-computer-electrical-engineering/73675

Handbook of Research on Nanoelectronic Sensor Modeling and Applications
Mohammad Taghi Ahmadi (Urmia University, Iran) Razali Ismail (Universiti Teknologi Malaysia, Malaysia) and Sohail Anwar (Penn State University, USA)
Engineering Science Reference • ©2017 • 579pp • H/C (ISBN: 9781522507369) • US $245.00

Field-Programmable Gate Array (FPGA) Technologies for High Performance Instrumentation
Julio Daniel Dondo Gazzano (University of Castilla-La Mancha, Spain) Maria Liz Crespo (International Centre for Theoretical Physics, Italy) Andres Cicuttin (International Centre for Theoretical Physics, Italy) and Fernando Rincon Calle (University of Castilla-La Mancha, Spain)
Engineering Science Reference • ©2016 • 306pp • H/C (ISBN: 9781522502999) • US $185.00

Design and Modeling of Low Power VLSI Systems
Manoj Sharma (BVC, India) Ruchi Gautam (MyResearch Labs, Gr Noida, India) and Mohammad Ayoub Khan (Sharda University, India)
Engineering Science Reference • ©2016 • 386pp • H/C (ISBN: 9781522501909) • US $205.00

Reliability in Power Electronics and Electrical Machines Industrial Applications ...
Shahriyar Kaboli (Sharif University of Technology, Iran) and Hashem Oraee (Sharif University of Technology, Iran)
Engineering Science Reference • ©2016 • 481pp • H/C (ISBN: 9781466694293) • US $255.00

Handbook of Research on Emerging Technologies for Electrical Power Planning, Analysis, ...
Smita Shandilya (Sagar Institute of Research Technology & Science, India) Shishir Shandilya (Bansal Institute of Research & Technology, India) Tripta Thakur (Maulana Azad National Institute of Technology, India) and Atulya K. Nagar (Liverpool Hope University, UK)
Engineering Science Reference • ©2016 • 410pp • H/C (ISBN: 9781466699113) • US $310.00

For an enitre list of titles in this series, please visit:
http://www.igi-global.com/book-series/advances-computer-electrical-engineering/73675

IGI GLOBAL
DISSEMINATOR OF KNOWLEDGE

www.igi-global.com

701 East Chocolate Avenue, Hershey, PA 17033, USA
Tel: 717-533-8845 x100 • Fax: 717-533-8661
E-Mail: cust@igi-global.com • www.igi-global.com

Table of Contents

Preface

Modern power system is more complex than before and new technologies are integrated in even faster. We can see power systems with HVDC, wind generator and other components came into practice around the world. Obviously new insulation materials matching new standards are now in urgent need. Polymer dielectrics have been used in power system for long. As a fast-moving technology, the old principles of polymer dielectrics have been revised in recent decades due to the marvelous development of power system.

There are several important properties that matters a lot in polymer insulation such as interface property, space charge and treeing. Researches on these properties have always been a significant issue especially when the concept of nanocomposites came into our view. This provide a prospective path to new properties in discovery new properties of polymer dielectrics. However, polymer dielectrics is really an interdisciplinary area concerning chemistry, physics and mechanics. Many principle and theory are not confirmed or even unknown for both academic researchers and industrialists. To this end, we write this book to provide some valuable researches, hoping to inspire readers with new ideas of polymer properties and new research methods.

This book is a collection of our previous researches. People with some basic background knowledge may find it with abundant content covering many hot topics in recent years. In Chapter 1 and 8, we focused on the insulation in new power energy. Polymer properties in wind generator winding and the effect of radiation in nuclear power station on polymer were discussed. In Chapter 2, we provided an interesting study on insulators in high speed railway system which required new dielectric properties different from traditional power system. In Chapter 3, 4, and 5, polymer insulation exposed to multi-field in HVDC power system provided clues for the discovery of new polymer materials. We also paid attention to traditional equipment like GIS in chapter 6. Besides, the edge of new technology is our first concern. In Chapter 7, 9, and 10, polymer dielectrics with better thermal behavior and insulation applied in superconducting environment were involved. Some of this technology has just come into practice in power system for few years and we are aiming to provide data and experiments support for practice with our current research.

Accelerating the Discovery of New Dielectric Properties in Polymer Insulation started for me when I worked as the leader of the High voltage lab of Tianjin University, China. This book includes many aspects of power system insulation, some of which are quite different from each other. It is really a team effort to finish this book. It is my pleasure to acknowledge the numerous contributions made by my team, among which the data provided by Dr. Yong Liu in Chapter 2 and the proofreading done by Mr. Zhuoran Yang are highly recognized.

Chapter 1
Introduction for Topics in Polymer Insulation

ABSTRACT

Polymer dielectrics have been used as insulation for long. Many Properties have drawn researches' attention in different occasions. Generally, these topics are usually interdisciplinary while requiring necessary professional knowledge of electrical system. Potential readers may find some ideas and methods mentioned in this book hard to understand if they are not quite into the modern high voltage power system. People may want to learn some basic principles and concepts in the area of polymer insulation. To this end, this chapter provides the introduction of methods and concepts which concern with the content of this book. It is hard to be all-sided, but the introduction is sure to be helpful to understand these topics better.

MODIFICATION OF POLYMER

Modern power system is developing fast, becoming large-scaled, ultra-high voltage, distributed, flexible and hybrid. The insulation materials are now required to be multifunctional to match the operation of smart grid. Traditional materials such as silicon rubber, epoxy resin and low-density polyethylene are of great insulation, but the defects are also quite obvious, especially when the working condition or the ambient influences are quite extreme. For example, insulation in transformer usually suffers from temperature much higher than cables while super-conduction system works in cryogenic

DOI: 10.4018/978-1-5225-2309-3.ch001

environment. Hence the insulation materials are thought to be not only insulating, but also adaptive to different external factors. However it is usually a costly and long-term project to develop a brand new material from initial study to actual use. Considering this, researchers prefer to modify traditional insulation materials instead of invented a new one from zero. The following two methods are widely applied in both industry and research. They are also included in this book.

Nanocomposite

Nanocomposite has become a very hot topic since the concept was first put forward by T. J. Lewis (1994). Nanocomposite usually consist with polymer matrix and nano-scale filler. It has been proved that nanocomposite shows better resistance to corona, electric erosion, particle discharge and breakdown. Recent researches have also presented its property of space charge and electrical treeing suppression. The ingredient of nanocomposites is quite flexible since most polymer insulation material can be applied as matrix.

The properties of nanocomposites are greatly influenced the filler. To summarize, there are three kind of filler:

1. Clay particles
2. Inorganic oxides
3. Nonlinear particles and some other materials.

In this book, different kinds of fillers are applied to achieve different function. In chapter 8, BN nano-particle is mixed with matrix to improve the thermal conductivity. In chapter 2, nano Al_2O_3 is used to improve the breakdown property of insulation. In occasion such as the cable accessary, space charge accumulation is a serious problem. In Figure 1, recent research has also presented the effect of nonlinear SiC particle in silicon rubber contributing to nonlinear conductivity which accelerates the dissipation of space charge (Du et al., 2016).

The dispersion of filler is another factor that matters. A widely used compounding method is called melt compounding. Nanocomposites in this book are most produced in this way. To examine the dispersion of filler, the scanning electron microscope (SEM) is routinely used (in Figure 2). There are also other advanced method such as transmission electron microscopy (TEM) and a Quadrat-based method put forward by Kim in 2007 (Kim et al., 2007).

Figure 1. Relationship between the conductivity and the electric field

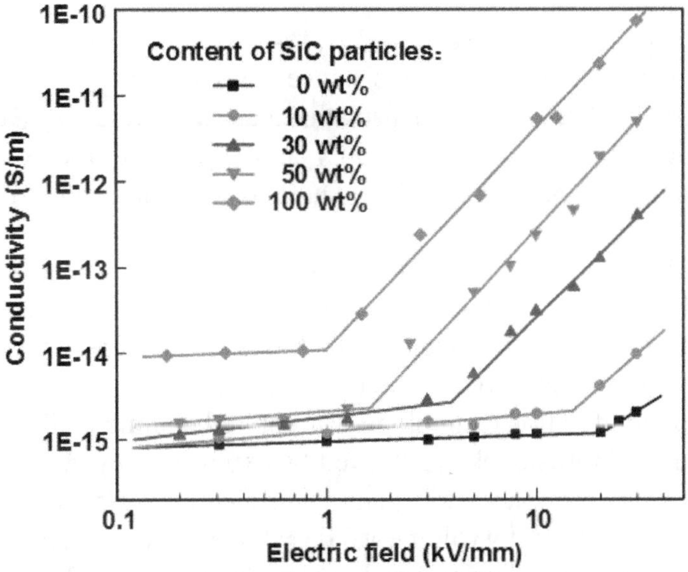

Figure 2. Eaxample: SEM observation of SiR/SiC composites, (a) 10 wt%, (b) 30 wt%, (c) 50 wt%, and (d) 100 wt%

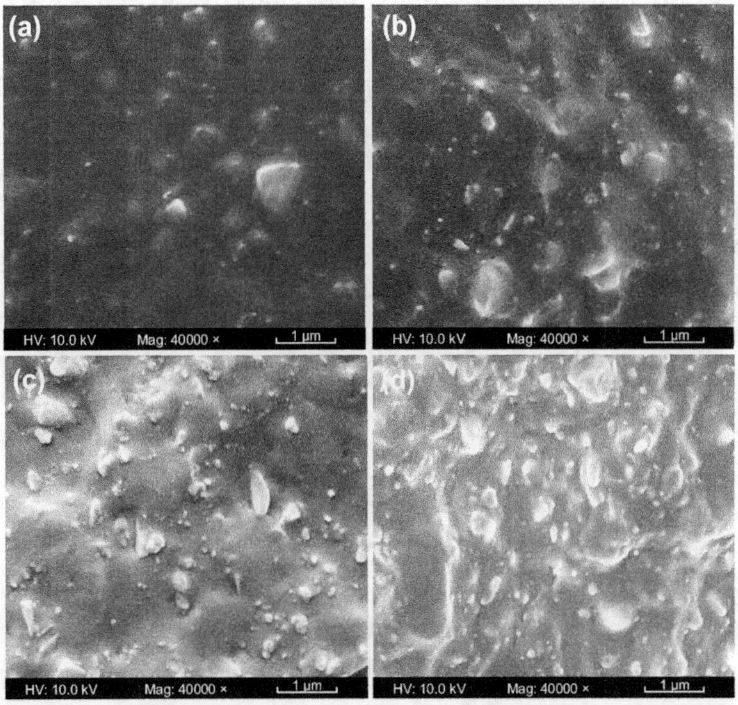

Fluorination

Composites with fluorine are believed to present excellent electrical, chemical and thermal property such as the resistance of high temperature and erosion, insulation and inertia. This kind of insulation material has drawn great attention since polytetrafluoroethylene (PTFE) and polytrifluorochloroethylene (PTFCE) were first produced by E. I. Du Pont Company and 3M Company in 1940s.

The surface molecular structure modification based on fluorination started in 1930s when Ruff and Keim proved that the carbon chain can exist in fluorocarbon on certain condition. In 1940s, L. A. Bigelow was the first to expound the principles of fluorination systematically. It was in 1947 that this surface modification technology was applied commercially.

Fluorination in this book is achieved by exposing polymer insulation to fluorine gas. The samples are usually fluorinated at 25 °C using a F2/N2 mixture gas with 20% F2 by volume at 0.05 MPa. A typical schematic diagram is shown in Figure 3. Fluorination can change the chemical component of surface molecule. Recent researches present that the fluorination level with the thickness of only several microns can improve the resistance of corona and reduce the depth of traps which contributes to faster charge dissipation rate. Energy dispersive spectroscopy (EDS) analysis was employed to obtain

Figure 3. Schematic diagram of fluorination system

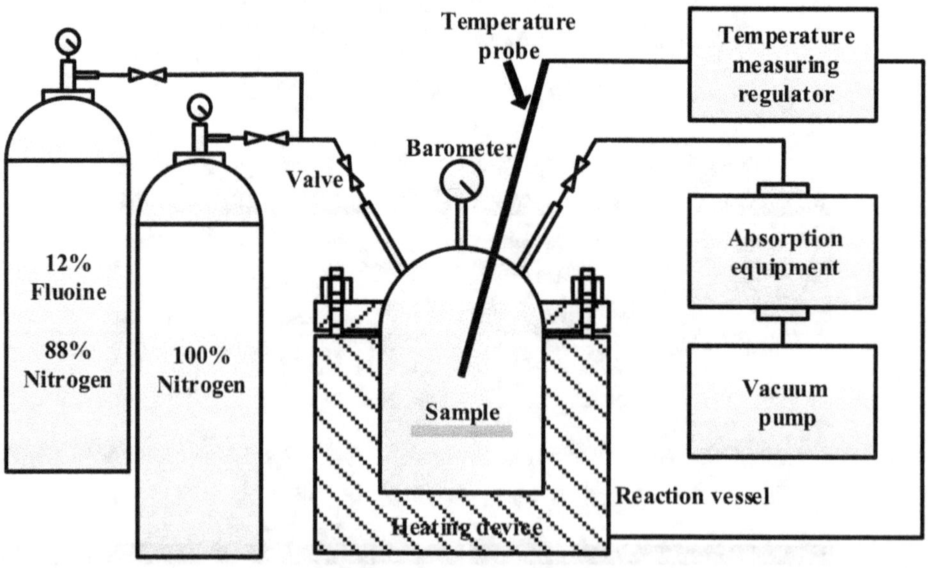

the quantitative distributions of elements. Fluorination level can be observed clearly by EDS. Besides, ATR-IR spectra is also widely applied to detect the change of chemical component of surface molecule in fluorination research. A detail example for ATR-IR spectra analysis is shown in chapter 5.

PROPERTIES THAT MATTERS IN POWER SYSTEM

As for a branch science of electrical engineering, polymer insulation materials are always required to process brilliant electrical properties. Regardless of the properties in semplice mechanology and chemistry, researches care more about insulating, charge behavior, discharge and property which has significant concern with electric behavior of polymer insulation material such as hydrophobicity. Modern power system is considerable complex, making the devices suffering from different problems. For instance, the cable accessories in HVDC need to reduce the quantity of accumulated space charge; The outdoor insulator always require good hydrophobicity and pollution flashover resistance; The main insulation of cable is thought to have resistance to the origination of electrical tree; For the rest problems such as partial discharge is so pervasive that might be detected in most power system insulation. All these problems along with different kinds of insulation material makes the topic of polymer insulation an abundant and complicated science. In this section, some highly focused properties are introduced and it is intended to make the cause of book clear for readers who are not in the area of electrical engineering.

Surface Charge

Surface charge is a basic property that always be mentioned in previous researches. It describes the charges accumulated on the surface of insulation when it is exposed to high voltage or electric field. A widely recognized surface charge measurement system is shown in Figure 4.

Generally, surface charge is considered in the research on external insulation such as outdoor insulator. The charge on surface can affect the discharge and surface flashover. Besides, fluorination can usually give rise to a significant change in surface charge property. Another purpose of measuring surface charge is that it can be used to calculate surface trap distribution and carrier mobility. The method can be drawn from the theory put forward by Simmons (1973). Relative research can be found in chapter 2.

Figure 4. Surface charge measurement system

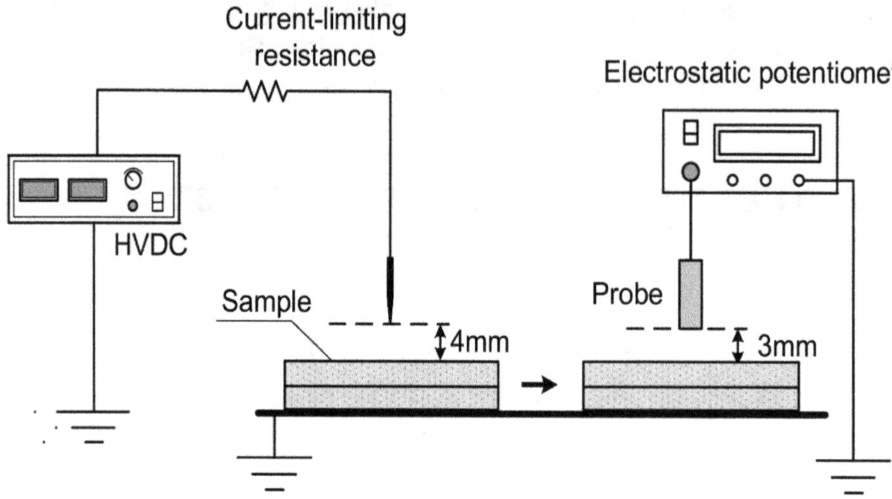

Space Charge

Space charge is basically caused by the same reason with surface charge. Space charge can be to some degree counteracted in AC system due to the alternating voltage. This problem becomes rather serious in HVDC system since the insulation is required to suffer single polarity voltage. For example, in the HVDC cable accessories with complex constructs, the distribution of electric field tends to be non-uniform, resulting in the formation of space charge in the region where the field is enhanced. The space charges in accessory insulation would further distort the local electric field, accelerate the early aging of insulation and initiate partial discharge, electrical treeing or eventual breakdown of accessory insulation, which is considered to be a significant issue with respect to life expectation (Montanari, 2011).

There are some different methods to measure space charge, among which the most widely applied one is named as Pulsed-electro-acoustic (PEA) Method. A PEA system is shown in Figure 5. The pulse voltage drives the space charge in sample to vibrate. The vibration is conducted as mechanical wave to the ground which is then detected and transformed to electrical signal.

Space charge property is another important path to the analysis of trap and carrier mobility. Carrier mobility is a key factor of space charge accumulation and dissipation. Especially, in the dissipation process, the dissipation rate is

Figure 5. Schematic illustration of the experimental setup for PEA

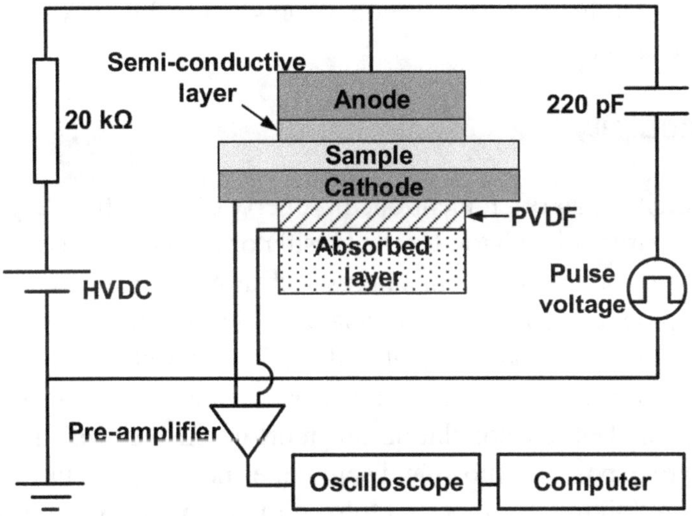

greatly influenced by the carrier mobility. A mathematic method to calculate it based on PEA method can be found in Montanari's research (Montanari, 2000).

Partial Discharge

Partial discharge describes the phenomenon that a discharge caused by high voltage or strong electric field arises in partial region of insulation, but not yet throughout the insulation, forming a conducting path which is recognized as breakdown. Usually, partial discharge in a tiny region cannot affect the breakdown voltage of the insulation. However, the by-produce such as electron and ion retained in insulation can destruct and decompose polymer in the long-term. Besides, partial discharge also results in high temperature which is also quite adverse to insulation.

The research on partial discharge has a history of nearly 80 years. The initial voltage of partial discharge is the most significant feature. Current researches aims to increase the initial voltages in order to avoid partial discharge. Partial discharge happens in micro holes in solid insulation, bubble in liquid insulation and interface of different polymer insulation materials. It is an obvious feature and cause of the degradation of insulation. Partial discharge is influenced by the form of voltage, insulation material, local electric field distribution and space charge generated by discharge. Since partial discharge is so pervasive, the measurement system is relatively flexible. It depends on the material and the research a scholar wants to conduct. Examples can be referred to chapter 3.

Hydrophobicity

Hydrophobicity is an often mentioned property of insulation, especially for the study of outdoor insulator. Current insulator can be classified as ceramic and polymeric. Other kinds of insulators such as glass insulator are gradually replaced in practice. Among the abundant species of polymer, silicon rubber is widely applied as insulator material for its remarkable hydrophobicity. High voltage power system insulator usually runs in coastal region since the industry and urban population are more intensive, which means that insulators are exposed to moist and sometimes polluted air. In China, many power system failures are reported to result from pollution flashover every year. Researches pointed out that hydrophobicity is a key factor to prevent pollution flashover (Wang et al., 1999).

It is considered that migration of hydrophobicity is the most significant property that matters in power system. The migration of hydrophobicity means that the pollution covering the surface of insulator, such as a silicon rubber insulator, will also become hydrophobe. High temperature is proved to accelerate this process. Besides, the molecular mass of polymer, degree of crosslinking and the thickness of silicon rubber coating can also have effect on hydrophobicity. An intuitionistic experiment named droplet test are usually applied in the research of hydrophobicity (in chapter 3).

Treeing

Electrical tree is another important reason for insulation failure. Treeing is observed to initiate at points where gas voids, impurities, mechanical defects or projections of conductions. Electric field is greatly distorted and concentrated in these regions. Some literature points out that treeing is caused by embedded electrode which injects and extracts charges into or out of polymer (Hussin., 2011). These charges are recognized as space charge in insulation. Other reason such as mechanical fatigue is also considered. According to this theory, mechanical fatigue might result in cracks in polymer due to Maxwell compressive stress under AC voltage. Then it is possible for an electrical tree to initial because of gas discharge (Tanaka et al., 1978). Even though many concept and process of treeing is still uncertain, an acceptable research points out that electrical treeing is labyrinthine structures of narrow gas-filled tubules which created by localized partial discharge activity (Auckland., 1995). The growth of electrical tree is characterized as three phases which are inception phase, propagation phase and completion phase.

Typical category of electrical tree can be summarized as branch, bush-branched and bush-shaped structures (in Figure 6). This is classified by fractal dimension (in chapter 4 and 10). Since electrical tree is generally less than 2 mm, a digital microscope imaging system is also quite necessary in research. The inception voltage and accumulated damage of electrical treeing is thought to the most important features. In this book, these characteristics are discussed.

Thermal Conductivity

As mentioned above, temperature affect insulation behaviors obviously in power system. In most occasion, engineers and researches wish to keep

Figure 6. Different types of trees

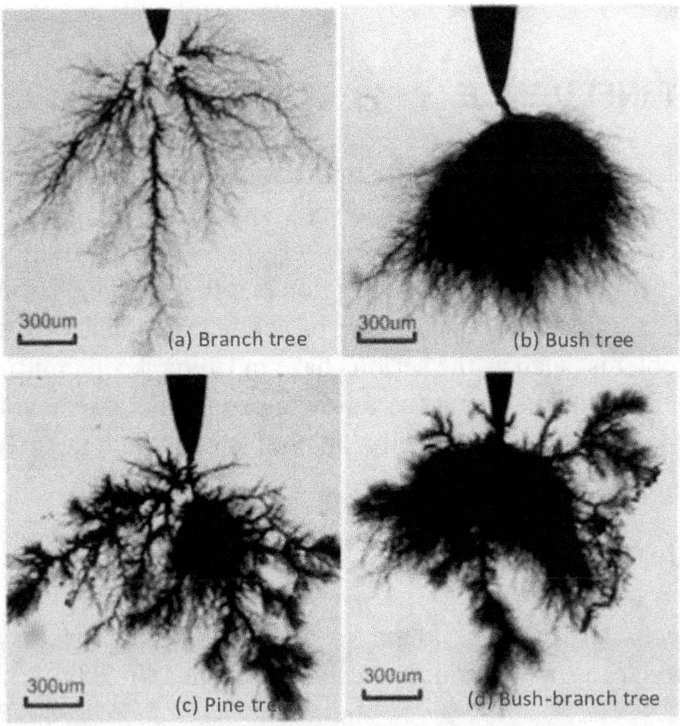

insulation temperature in an acceptable range. The running temperature of HVDC cable is around 90 °C while that of transformers is about 85-105 °C. For a bushing connected with a converter valve, the running temperature is relative higher, but still no more than 110 °C. The temperature fluctuates a lot when there is a failure in power system. A short circuit current is extremely dangerous since large current can bring temperature to a high level in a short time. Considering polymer material, high temperature accelerates the aging process, resulting in premature failure of insulation. Moreover, the existing flaws in insulation may also be aggravated by high temperature.

To this end, the polymer with high thermal conductivity is highly appreciated. This could be achieved by high thermal conductivity fillers. Actually this could also be seen as an application of nanocomposites. There are sever options of the fillers. Oxide particles is a widely considered formula, such as aluminum oxide (Kozako et al., 2010). The disadvantage is that aluminum oxide is of high dielectric constant hence silicon dioxide is another option in many occasions (Procter et al., 1991). Nitride particles is an effective path to high thermal conductivity as well. In chapter 8, we presented some relative researches of boron nitride. Graphene oxide and graphene oxide are both under study in this field. There is also an interesting choice of filler, diamond, which is obviously too costly for industrial manufacture.

AMBIENT INFLUENCE

Ambient factors are quite diverse and complex. These factors becomes increasingly important for modern power system. The wet air and micro salt particles in coastal region, the low atmospheric pressure in high altitude region, the low temperature in high latitude region and the sand and strong wind in gobi desert. Besides, these factors are basically combined with each other and new problems, such as the serious haze in China, keep emitting. It is so difficult to list all of them that the following content is intended to provide some basic knowledge of the main ambient factors which are mentioned in this book.

Cryogenic Temperature

Superconductivity was first found in 1911. Modern power system has already introduced superconducting cable to practice. In 2004, China built a

33.5m long superconducting cable transmission line in Yunnan and it is still in operation (Xin et al., 2004). This technology is of great advances since superconductivity of cable can get rid of about 15% power loss happens on normal cable. Superconductivity is also applied in fault current limiter in order to reduce the current greatly. It could be predicted that superconducting material will play an increasingly important role in future.

The cryogenic temperature is the necessary condition of superconductivity. This is normally achieved by the vaporization temperature of liquid nitrogen, which is -196 °C. Even through new materials present superconductivity in higher temperature, which is also known as high temperature superconducting, the temperature is still low enough to change the property of polymer insulation. For instance, cross linked polyethylene (XLPE) presents improvement in electric properties compared with the properties in room temperature. A research done in Japan shows that the AC dielectric strength of XLPE improved by 60%, which is over 40kV (Kosaki, 1996). However the brittle temperature is another factor which should be consider at cryogenic temperature. Cracks are also reported in polyethylene at the temperature of 170 K which obviously destroy the insulation (Kosaki, 1998). Relative research can be found in chapter 10.

Mechanical Factors

Mechanical factors, usually stretching, are inevitable in power system. Polymer insulation is stretched, constringent, bended and expanded since the first second when it is manufactured. The transportation, installation and operation of insulation are all effected by mechanical factors. Figure 7 shows a typical example of cable accessory. The insulation is always stretched in two directions, namely normal and tangent during the whole running period. Mechanical factors play a role which must not be ignored. Some data collected in power system proves that for overhead lines, the mechanical failure occupies 65% of the failure causes while the electrical failure shares only 3%.

Mechanical factors changes the property of insulation through the microstructure. Polymer materials are normally either in the form of chain (such as PE) or in the form of net (such as XLPE). Mechanical forces can stretch the chain or net micro structure in polymer. This definitely changes the free volume and potential energy distribution which further changes the electrical properties. Some of these changes are temporary, but if the forces break the crystal region into small pieces or even crack the insulation macroscopically, an insulation breakdown will then be very predictable.

Figure 7. Stretching of cable accessories

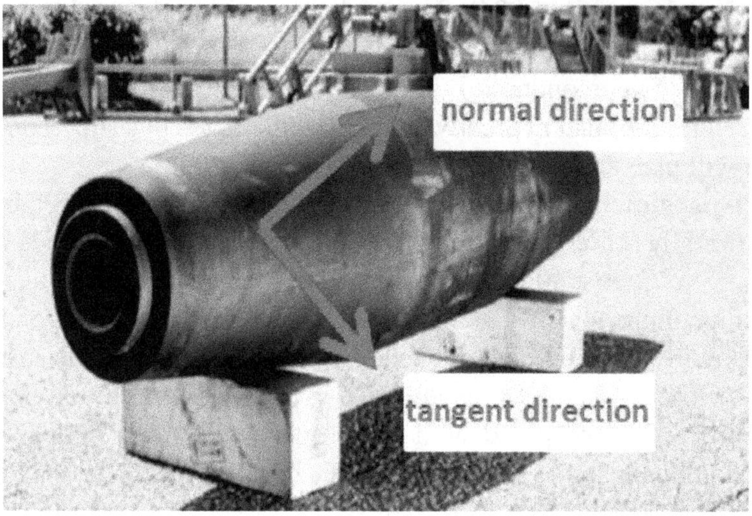

Radiation

Nuclear power is usually the first that readers might the come up with where radiation is so obvious to influence the insulation. Actually in some small scale power system, such as a space station, insulation is also exposed to a mass of radiation. Relative research can be found in chapter 9. The following are some basic theory to explain the reason why the radiation matters in polymer insulation.

Similar to mechanical forces, radiation can changes polymer micro structure as well. However these changes are usually chemical. Radiation crosslinking is a basic form of reaction. In this process, different linear molecules are combined with chemical bond, forming space polymers or reticular polymer. Another reaction is named as radiation degradation reaction. This is an opposite process of radiation crosslinking. The main chain of polymer breaks and the molecule mass decreases. These two reactions normally happens synchronously and the final product depends on which one is more dominating. This dynamic process is relative to the structure of polymer. For those polymer which tends to the degradation reaction, the breaking of molecule chain makes it no more polymer gradually along with the failure of mechanical and electrical properties. Unfortunately, this process happens in most polymer if the radiation intensity is large enough.

REFERENCES

Auckland, D. (1995). Electrical treeing in solid polymeric insulation. *Engineering Science and Education Journal*, *4*(1), 11–16. doi:10.1049/esej:19950105

Du, B. X., Li, Z. L., & Yang, Z. R. (2016). Field-dependent conductivity and space charge behavior of silicone rubber/SiC composites. *IEEE Transactions on Dielectrics and Electrical Insulation*, *23*(5), 3108–3116. doi:10.1109/TDEI.2016.7736876

Hussin, N. (2011). *The Effects of Crosslinking Byproducts on the Electrical Properties of Low Density Polyethylene* (Doctoral Thesis). University of Southampton.

Kim, C. H., Lim, H. O., Chung, I., Lee, I. W., Chun, H. H., & Jo, N.-J. (2007). Actuation behavior of waterborne polyurethane/conductive filler nanocomposite electrode. *Composite Interfaces*, *14*(5–6), 477–491. doi:10.1163/156855407781291209

Kosaki, M. (1996). *Research and development of electrical insulation of superconducting cables by extruded polymers. IEEE Electrical Insulation Magazine.*

Kosaki, M. (1998). Development and test s of extruded polyethylene insulated superconducting cable.*Second International Conference on Properties and Applications.*

Kozako, M., Okazaki, Y., Hikita, M., & Tanaka, T. (2010). Preparation and evaluation of epoxy composite insulating materials toward high thermal conductivity.*10th IEEE International Conference on Solid Dielectrics*, 1–4 doi:10.1109/ICSD.2010.5568250

Lewis, T. J. (1994). Nanometric dielectrics. *Trans IEEE*, *DEI-1*, 12–825.

Montanari, G. C. (2011). Bringing an Insulation to Failure: The Role of Space Charge. *IEEE Transactions on Dielectrics and Electrical Insulation*, *18*(2), 339–364. doi:10.1109/TDEI.2011.5739438

Montanari, G. C., Mazzanti, G., & Palmieri, F. (2000). Mobility Evaluation from Space Charge Measurements Performed by the Pulsed Electroacoustic Technique.*International Conference on Properties and Applications of Dielectric Materials*, 38~41. doi:10.1109/ICPADM.2000.875624

Procter, P., & Solc, J. (1991). Improved thermal-conductivity in microelectronic encapsulants. *IEEE Trans. Compon. Hybrids Manuf. Technol.*, *14*(4), 708–713. doi:10.1109/33.105121

Simmons, J. G., & Tam, M. C. (1973). Theory of Isothermal Currents and the Direct Determination of Trap Parameters in Semiconductors and Insulators Containing Arbitrary Trap Distributions. *Physical Review B: Condensed Matter and Materials Physics*, *7*(8), 3706–3713. doi:10.1103/PhysRevB.7.3706

Tanaka, T., & Greenwood, A. (1978). Effects of Charge Injection and Extraction on Tree Initiation in Polyethylene. *IEEE Transactions on Power Apparatus and Systems*, *PAS-97*(5), 1749–1759. doi:10.1109/TPAS.1978.354668

Wang, S. W., Liang, X. D., & Huang, L. C. (1999). *Laboratory Study on Hudrophobicity and Pollution Flashover Properties of Polymer Insulators*. Beijing: Electrical Engineering Dept, Tsinghua University.

Xin, Y., Hou, B., Bi, Y. F., Cao, K. N., Zhang, Y., Wu, S. T., & Han, Z. H. et al. (2004). Chinas 30m 35kV/2kA ac HTS Power Cable Project. *Superconductor Science and Technology*, *17*(5), 332–335. doi:10.1088/0953-2048/17/5/048

Chapter 2
Polyimide Film for Generator Winding

ABSTRACT

Polyimide film is widely used in different areas due to its brilliant chemical and electrical properties. This chapter discusses effect of direct fluorination on surface charge and surface partial discharge of polyimide films. The influences of different fluorination time and voltage frequency are studied. The trap properties of polyimide film are analyzed. It is clear that the fluorination has changed the micro structure of the sample so that the trap property and the dissipation time are different. Further research focuses on the effects of water absorption on surface charge and dielectric breakdown of nanocomposite. The polyimide/Al_2O_3 film is selected as sample and the relation between water absorption and surface charge and breakdown voltage is discussed.

INTRODUCTION

Polyimdie is one type of universal engineering materials. Owing to its excellent chemical and physical properties, polyimide films have been widely used in turn to turn insulation and turn to ground insulation in wind turbine generators. PI has been used as an important insulation material (Du,B.X.,2013;Xin,H.,2005). Researches show that PI exhibits excellent combination properties: good dielectric property and heat-resistance, excellent chemical stability, high flame retardancy and mechanical strength, as well as radiation and weather resistance(Du,B.X.,2013).

DOI: 10.4018/978-1-5225-2309-3.ch002

However, the pulse voltage from the converter, which is imposed on the double-fed induction generator insulation, has a fast rising time, high repetition frequency and amplitude compared with the common sinusoidal voltage(Sun,P.,2012). It often causes overvoltage and localized high electrical field in the insulation system, leading to the premature failure of the insulations, partial discharge and acceleration of insulation degradation (Wheeler, J. C. G.,2005). Partial discharge will occur in insulation systems of high voltage power equipment. Surface discharge can occur in liquid, vacuum or gas near to the surface of solid dielectric (Preetha, P.,2011). Insulating system usually contains various types of defects. These defects will be the root of partial discharge(Florkowska, B.,2012). Due to the discharge plasma close to the surface of solid dielectric, the surface of insulation material will be degraded gradually. The longtime of partial discharge can cause the breakdown of electrical insulation system which will result in apparatus failures. Because of the existence of carbonyl and amine groups, the PI surface is hydrophilic which results in the absorption of water. Its insulating properties are vulnerable to be degraded after absorbing water, thus leading to the increasing risk of electrical breakdown and the appearance of malfunction or irremediable damage in the operating system.

Many papers have investigated insulating properties of polyimide films and other materials under pulse voltage. Serdyuk et al investigated charging of polymer surfaces by means of impulse corona discharges and concluded that charge deposition was different in case of glow corona and burst corona(Kumara, S.,2009). Wu et al studied the structure change of polyimide film aging under pulse voltage and found that the degradation of structure in polyimide propagates from surface to interior with continuous aging(Wu, G. N.,2010). Some researchers had investigated the insulation aging and failure mechanism of polyimide film. Kaufhold et al had reported that the partial discharge were the main cause of insulation failure(Kaufhold, M.,2000). Bellomo et al had found that the life of insulation materials used in variable-frequency motors was mainly affected by the partial discharge(Bellomo, J. R.,1999). Kimura et al had investigated the surface discharge of polyimide film and reported that surface charges of polyimide had an effect on the partial discharge(Kimura, K.,2002). Study on the partial discharge mechanisms of PI is of great significance. However, little work has been reported on the chemical modification of polyimide film surface layer to modulate its surface electrical property so as to suppress the partial discharge. Some researchers have added nano particles in PI for the purpose of improving the dielectric properties. The incorporation of inorganic nano particles (TiO_2, Al_2O_3 and SiO_2 etc.)

with PI greatly improves its dielectric property, such as the corona resistance and thermal conductive performance, which makes a more broad application prospects for PI (Du, B .X.,2014). Wu et al. showed that when 25 wt% nano Al_2O_3 particles were added in PI, the corona-resistant property was improved and the corona-resistant lifetime was prolonged a lot than that in original PI film (Luo, Y.,2014). Our previous research showed that the surface charge was restrained by the nano particles in PI/Al_2O_3 nanocomposite films (Du, B. X.,2013). As water absorption in moisture condition is a main reason for breakdown in PI insulation, many papers have investigated dielectric properties of PI films after absorbing water (Wang, F.,2004). Kaneko et al reported that the charge injection and the carrier mobility of PI were improved by the water absorption (Kaneko, K.,2005). Takada et al found that the band gap of PI chain was narrowed because of a dipole moment of water molecules, thus increasing the hetero charge and thermal breakdown occurred because of the increasing conductivity by water (Takada, T.,2013).

As a method of surface modification, direct fluorination using fluorine gas could make the properties of polymer surface similar to the fluoropolymer, without changing the bulk characteristics of the original polymer (Du, B. X.,2013). It has been known that the fluorination can change the chemical component in the surface layer of polymers, thus influencing charge injection and electrical properties of the surface layer when the polymers are used as an insulator (Du, B. X.,2013). However, most researches on fluorination are focused on improving the properties concerning wettability, adhesion, chemical stability and barrier properties (Liu, Y. Q.,2012). In fact, the change in chemical component in the polymers surface layer by fluorination can cause the corresponding change in electrical property of the surface layer, thus influence the charge injection from electrodes and suppress the occurrence of the partial discharge when they are used as insulator (Jiang, Y.,2010).

In this chapter, researches were carried out to investigate the effect of surface fluorination on surface charge behaviors of polyimide films under positive pulse voltage, the effects of direct fluorination on surface charge coupling behaviors of the PI films under the combination of dc and pulse voltage, and the influence of water absorption on the breakdown strength and surface charge of the PI/Al_2O_3 composite films. The results showed that the surface charge behaviors were affected by the fluorination, corona discharging time, pulse voltage amplitude and pulse frequency and polarities of the dc and pulse voltage combination. The product of the derivative of the surface potential and time tdV/dt was used to analyze the differences between the unfluorinated and fluorinated films. Fluorination can effectively suppress the occurrence of partial discharge and degradation of the polyimide film. The water absorption of PI films is enhanced by the nano particles, which causes the sharp decrease of breakdown strength. The decay process of surface charge is greatly affected by the immersion time.

EFFECT OF DIRECT FLUORINATION ON SURFACE CHARGE OF POLYIMIDE FILMS USING REPETITIVE PULSE POWER

Effect of Fluorination

The polyimide film surface is charged by pulse corona discharging. It has been studied in many papers that the amplitude of the charging voltage and charging time by using DC power are two important factors affecting the accumulation and decay of the surface charge (Zhuang, Y.,2012; Du, B. X.,2012; Plopeanu, M. C.,2011). Considering the repetitive pulse power used in this research, different charging time corresponds to different pulse corona discharging times. The research in this chapter took into consideration how the surface charge behavior was influenced by the pulse voltage amplitude, corona discharging times and pulse frequency, together with the direct fluorination of the films.

Figure 1 shows the surface charge decay curves under different pulse amplitude. It can be found that the charge density decays fast at the initial time. The initial value depends on the trapped charge density and the effective capacitance of the film, while the decay could be associated with the release of charges from trapping states in the film (Watson, P. K.,1990). It is the detraping of these charges that leads to the decline of the surface charge density. The trapped charges in shallow traps are easy to escape, which causes the rapid decay at the initial time. As shown in Figure 1a and 1b, the initial surface charge density of the unfluorinated and fluorinated samples increases with increasing the pulse amplitude. The charge density of the unfluorinated sample increases from ~650 to ~3000 pC/mm^2 with changing the applied pulse from 1.5 to 3.5 kV. Thus more charges could be injected into the film under higher pulse voltage. It can also be found from Figures 1a and1b that the surface charge density of the fluorinated sample is quite small after 200 s, while the density of the unfluorinated one is still much higher even after 1800 s. Therefore, surface charge of the fluorinated sample decreases much faster compared with the unfluorinated one.

Figure 1. Relationship between the surface charge density and the decay time under different pulse voltage, (a) was the unfluorinated sample; (b) was fluorinated for 30 min

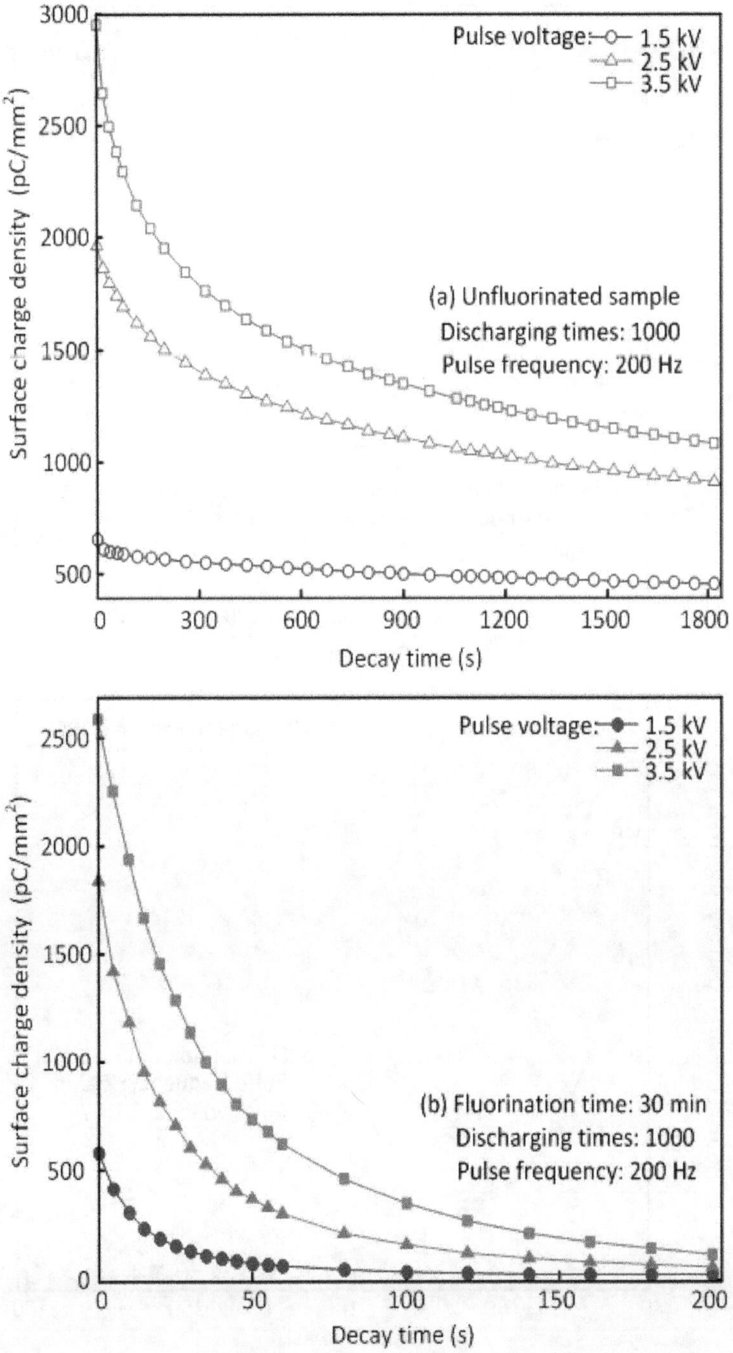

Figure 2. Relationship between the surface charge density and the decay time with different discharging time, (a) was the unfluorinated sample; (b) was fluorinated for 45 min

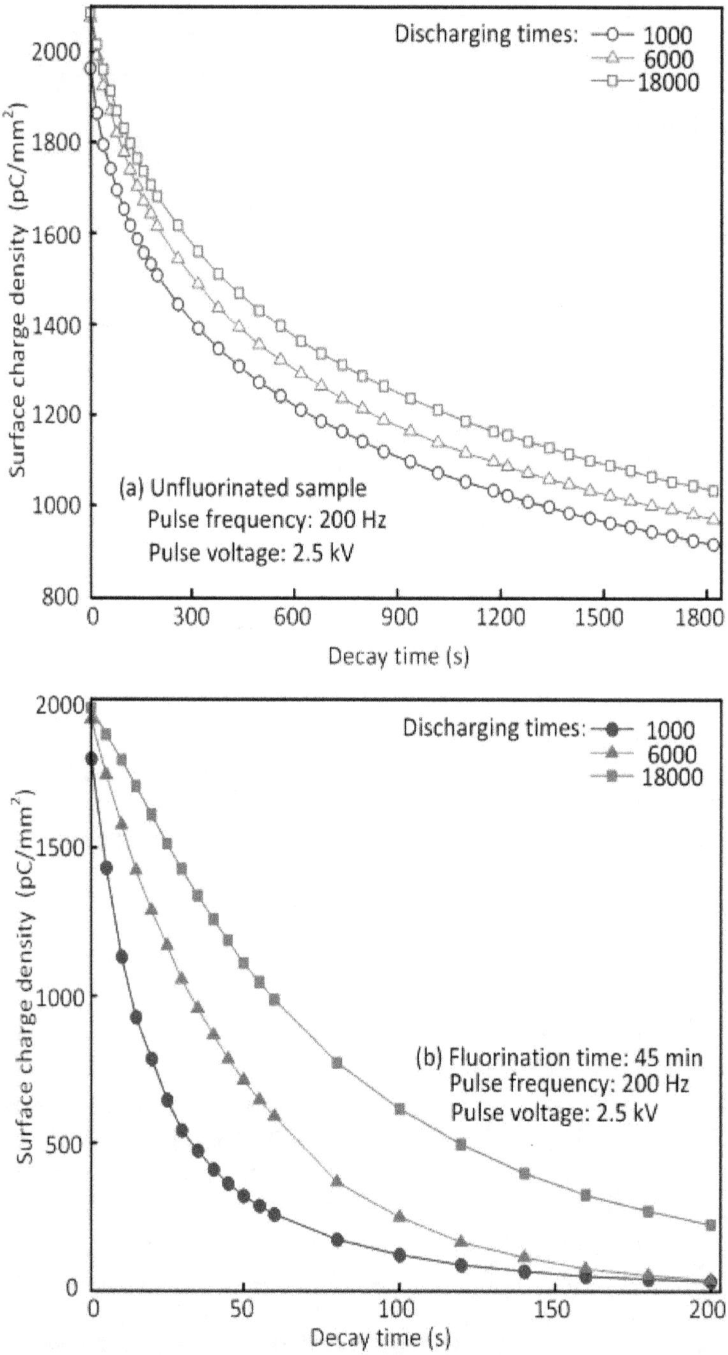

The curves in Figure 2 present the surface charge density decay with different corona discharging times. For the unfluorinated sample, the surface charge density increases a little with increasing the discharging time from 1000 to 18000, as is shown in Figure 2a. Figure 2b shows that the charge density of the samples with the 6000 and 18000 discharging times are higher compared with the 1000 times. It can be found that the curves of the unfluorinated sample have similar decreasing tendency for different discharging times, especially when the decay time is over 300 s. For the fluorinated sample, however, the charge density decays slower under the 6000 and 18000 discharging times than that under the 1000 discharging times. While the discharging time increases, more charges tend to accumulate in the deep traps, the result of which is that the surface charge decays a little slower under the 6000 and 18000 than that under the 1000 discharging times.

Figure 3 presents the relationship between the surface charge density and the decay time with different pulse frequency. It can be found that the charge density tends to increase as the frequency changes from 400 to 1000 Hz. For the unfluorinated sample, the initial charge density is ~2200 pC/mm^2 under 400 Hz, while it is ~2400 pC/mm^2 under 1000 Hz. It can be seen from Figure 3b that the initial charge density of the fluorinated sample increases from ~1900 to ~2100 pC/mm^2 with increasing the frequency from 400 to 1000 Hz. As is shown in Figure 3a, the curves of the unfluorinated sample have a similar decreasing tendency while the decay time is over 200 s. Figure 3b shows that a similar decreasing tendency appears under 600 and 1000 Hz, while the surface charge decays faster under 400 Hz. After the pulse voltage is applied to the needle electrode, the air is ionized to produce ions, some of which would be driven to the sample surface. Some ions produced during the previous pulse would be able to exist until the subsequent pulse comes, which is defined as the accumulation effect (Zhang, C.,2013). As the frequency increases, the time interval between two pulses becomes smaller and the residual ion density before the subsequent pulse is applied could probably be higher (Zhang, C.,2013). Therefore, more charges tend to be driven to the sample surface under higher frequency.

Figure 4 shows the relationship between the surface charge density and the decay time for different samples. It is presented that the initial surface charge density of the unfluorinated sample is ~2150 pC/mm^2, while it is ~2050 pC/mm^2 for the 60 min fluorinated sample and ~1900 pC/mm^2 for the 30 min fluorinated sample. Under the same conditions, the surface charge density of the fluorinated samples is lower than that of the unfluorinated one. The results indicate that the fluorination can suppress the accumulation of charge

Figure 3. Relationship between the surface charge density and the decay time with different pulse frequency, (a) was the unfluorinated sample; (b) was fluorinated for 15 min

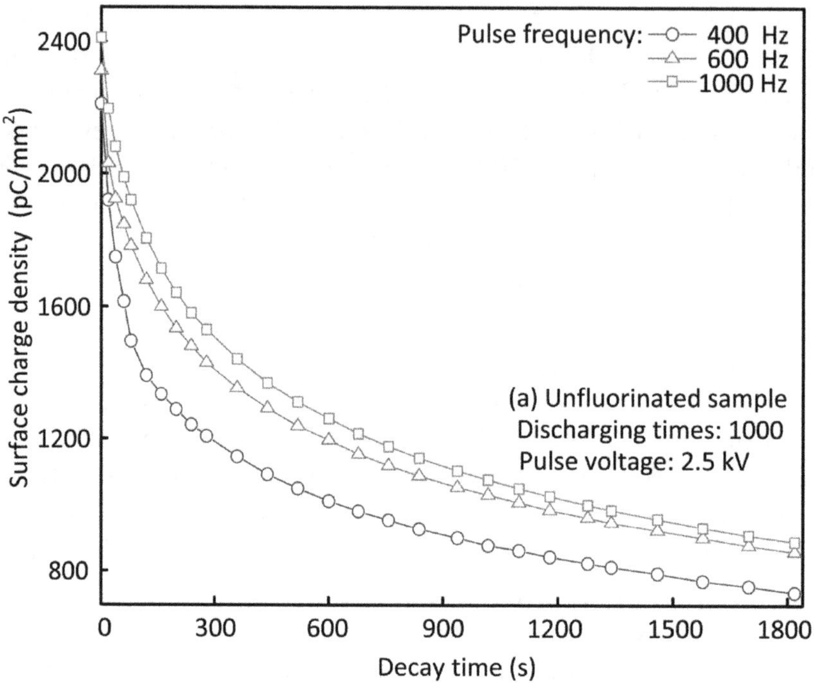

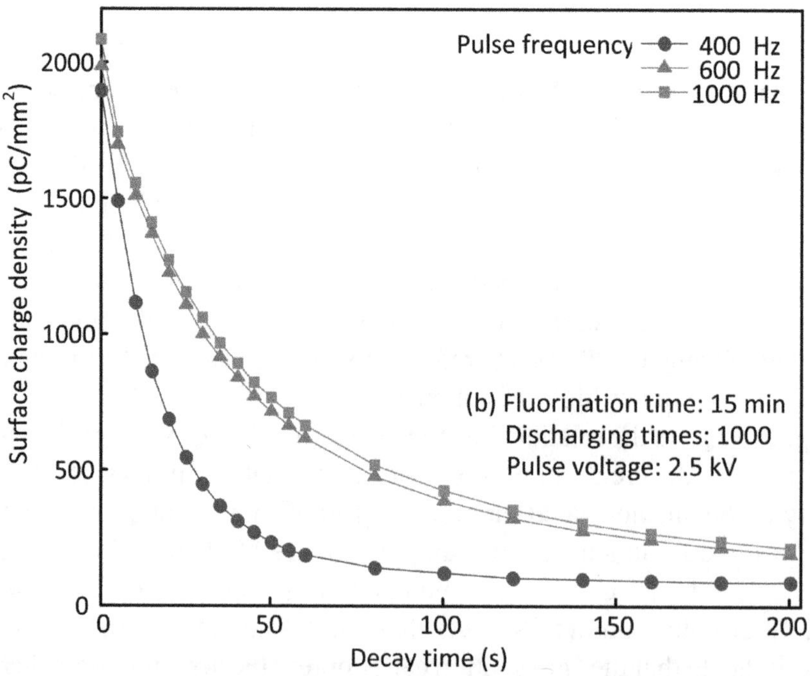

Figure 4. Relationship between the surface charge density and the decay time for different samples with the fluorination time of 0, 30 and 60 min

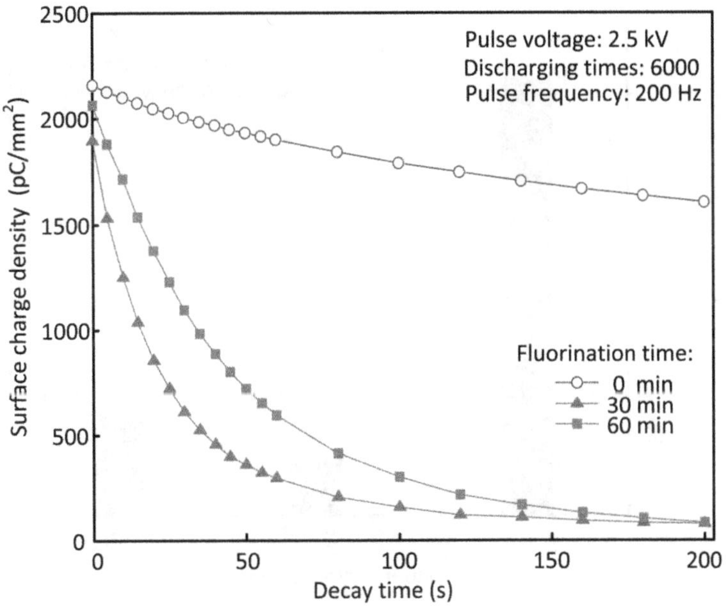

on the sample surface. As a result of the fluorination, there will be a formation of fluorinated layer on the surface of the sample. It could be assumed that the fluorinated layer is helpful to capture charges and form a shielding layer, which shallows the deep traps (Du, B. X.,2013). It can be observed that the surface charge density of the fluorinated samples becomes quite small after 200 s, while it is still ~1600 pC/mm^2 for the unfluorinated sample. It can also be noticed that the sample with 30 min fluorination time shows a faster decay tendency compared with the 60 min one.

Surface fluorination could break the C-H bond of polymer and the CF, CF_2 and CF_3 groupings are generated, thus forming a fluorinated layer. The layer depth depends on the polymer nature and the treatment conditions (pressure of the reactive mixture, fluorine concentration, treatment time and temperature). The ATR-IR spectra show that the C=C, C=O, C–O–C and phenyl–H bonds are significantly reduced in strength for the fluorinated film. And a very broad strong absorption band appears in the 950–1340 cm^{-1} region, which is ascribed to CF, CF_2 and CF_3 groupings (Du, B. X.,2013).

It has been shown that the relative permittivity of the fluorinated samples is lower than that of the unfluorinated sample (Du, B. X.,2013). The fluorine

Figure 5. Relationship between the dissipation time and the fluorination time

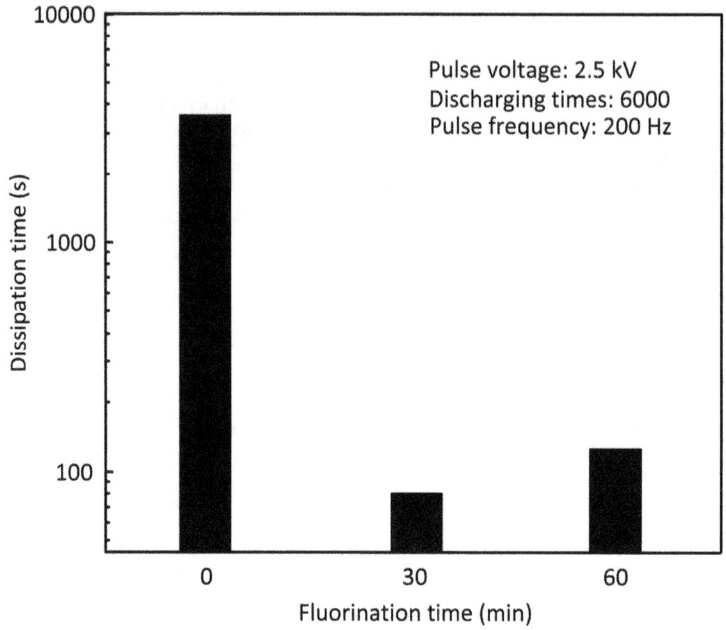

atom is the smallest atom in all the elements except the hydrogen atom. The electrons in 2s and 2p orbit are very close to the nucleus, which leads to the low electron polarization of F element (Du, B. X.,2013). Therefore, the permittivity becomes smaller after the fluorination.

The definition of the dissipation time is the time after which the surface charge decreases to 10% of the initial value (Sjöstedt, H.,2009). A longer dissipation time means that the surface charge decays slower. The relationship between the dissipation time and the fluorination time is shown in Figure 5. It could be found that the dissipation time of the fluorinated samples are around 100 s. However, for the unfluorinated sample, the dissipation time is over 3000 s, which is much longer than that of the fluorinated samples. As the fluorination time changes from 30 to 60 min, the dissipation time increases a little. The result indicates that the fluorination can accelerate the surface charge decay.

Figure 6 shows the relationship between the dissipation time and the pulse frequency. With increasing the frequency, the dissipation time of both samples increases. For the fluorinated sample, the dissipation time is ~100 s under 400 Hz and ~200 s under 600 and 1000 Hz. The dissipation time of the unfluorinated sample is several thousand seconds, which is much greater

Figure 6. Relationship between the dissipation time and the pulse frequency under 2.5 kV pulse and 1000 discharging times

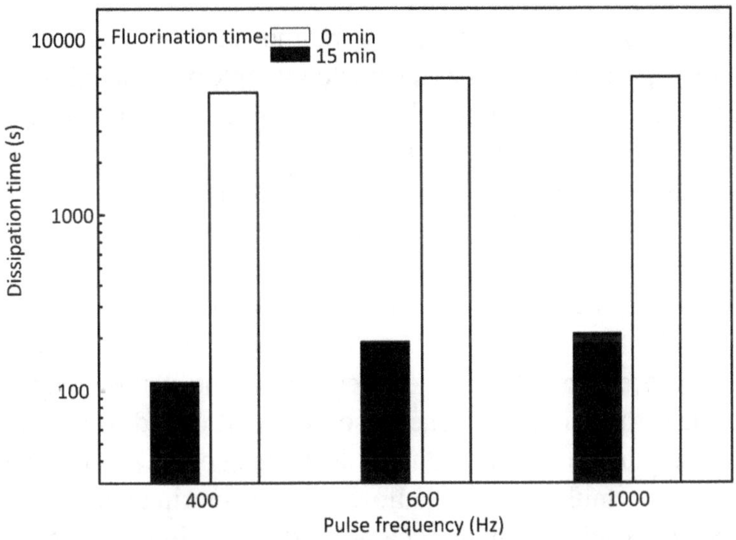

than that of the fluorinated sample, indicating that the surface charge of the unfluorinated sample decays much slower.

The tdV/dt Characteristics

Charges can be transferred by contact, corona or surface states, which are usually associated with traps. The energy of the traps is related to time by considering demarcation energy, indicating the border between emptied and occupied traps (Sjöstedt, H.,2009). The product of the derivative of the surface potential and time tdV/dt has been used by Watson for the study of charge trapping (Watson, P. K.,1990). The tdV/dt curve enables the representation of slow phenomena that would be not clear in the surface potential V vs t curve. Molinié and Llovera have studied how the peaks appearing on the tdV/dt curve correspond to physical phenomena with different characteristic time (Zhuang, Y., 2012; Llovera, P.,2004). In this chapter, the tdV/dt curve is used to analyze the surface charge behavior of polyimide films under different conditions.

The tdV/dt curves of the unfluorinated and fluorinated samples under different pulse voltage are shown in Figure 7. The characteristic time is defined as the time when a peak appears in the curve. For the unfluorinated sample, the curves present two peaks with different amplitude and characteristic time, as shown in Figure 7a. It is observed that the amplitude of the peak increases with increasing the pulse from 1.5 to 3.5 kV. The characteristic time of the first peak is ~200 s while it reaches several thousand seconds for the second peak. However, the curves of the fluorinated sample present only one obvious peak and the characteristic time is between 10 and 40 s, as is shown in Figure 7b. The amplitude of the peak is directly related to the charging voltage and becomes higher with increasing the pulse amplitude.

A double exponential decay equation has been used to analyze the surface potential decay curve and it was assumed that there were two types of decay process (Zhuang, Y., 2012). For the tdV/dt curves in Figure 7, it can be inferred that the first peak with a smaller characteristic time is related to a faster decay process that the charges escape from the shallow trap, while the second peak with a greater characteristic time is related to a slower decay process that the charges escape from the deep trap. For the fluorinated sample, most charges accumulate in the shallow traps or fluorinate layer and there is one obvious peak with a smaller characteristic time in the curve. However, for the unfluorinated sample, charges accumulate in both shallow and deep traps and the curve presents two peaks.

The characteristic time and peak in the tdV/dt curve could be related to the decay speed of the V vs t curve, which has been discussed from a mathematical viewpoint in (Llovera, P.,2004). While the peaks of two curves have similar amplitude, a smaller characteristic time corresponds to a faster decay. In the case that the characteristic time is similar, a higher peak corresponds to a faster decay. Therefore, the characteristic time and peak could provide information about the V vs t curve from different aspects. In Figure 7a, for the case of 3.5 kV, the peak is higher and the characteristic time is smaller, which reflects a faster decay. In Figure 7b, the characteristic time presents a little difference and the amplitude of peaks differ greatly. Therefore, the amplitude of the peak in Figure 7b is the main factor to reflect the decay speed and the decay under 3.5 kV is faster.

Figure 8 shows the tdV/dt curves with different discharging times. For the unfluorinated sample, the curves in Figure 8a show similar shapes with two peaks. With changing the discharging time, the characteristic time of two peaks are around 300 and 4000 s. In Figure 8b, the curves of the fluorinated sample present only one peak, the amplitude of which increases with increasing the discharging time. The characteristic time of the fluorinated sample is much smaller than that of the unfluorinated one. With changing the discharging time from 1000 to 18000, the characteristic time increases from 20 to 100

Figure 7. tdV/dt curves under different pulse voltage, (a) unfluorinated sample; (b) sample fluorinated for 30 min

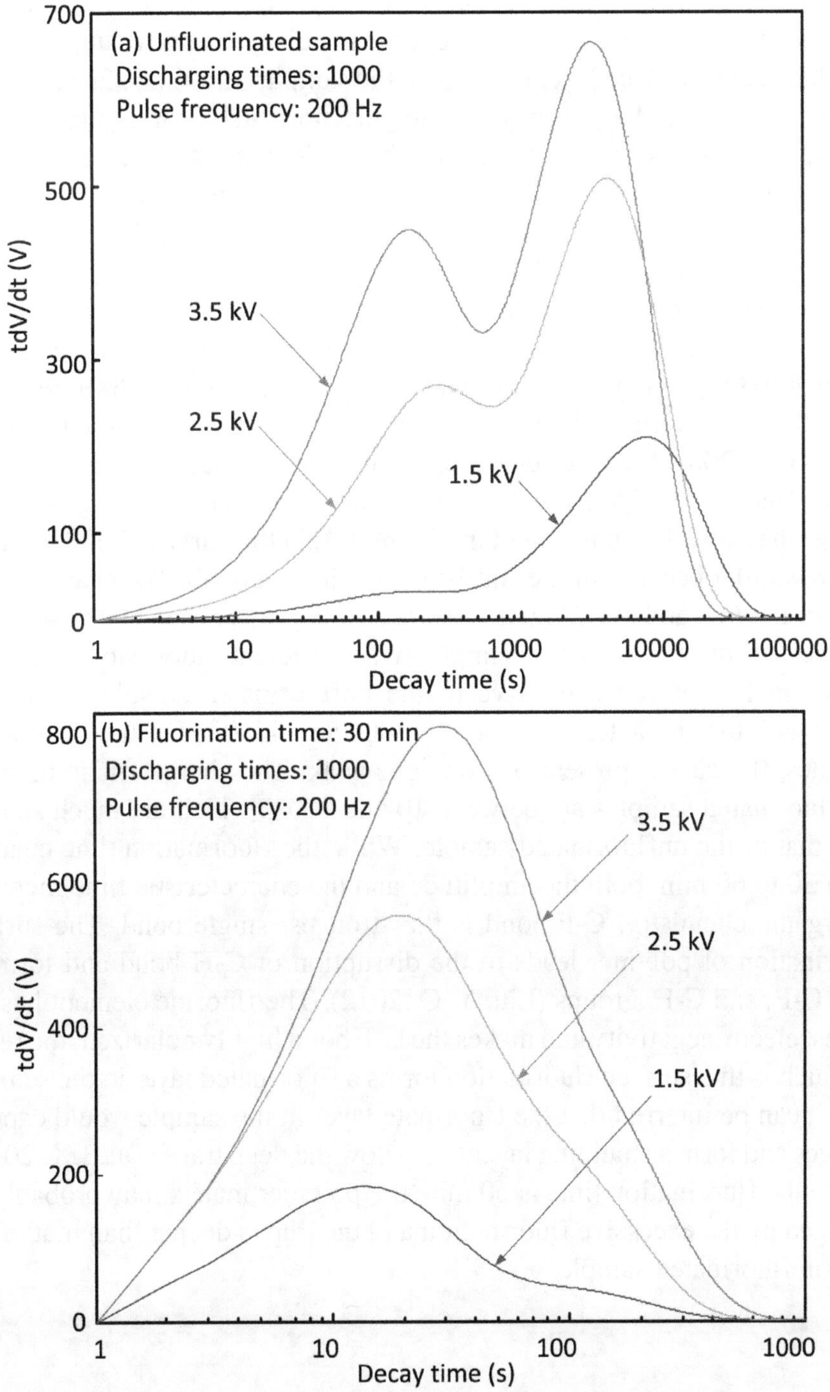

s, indicating that the surface charge decays slower. With increasing the discharging time, more charges tend to accumulate on the sample surface in the force of electric field, leading to the increase of surface charges. For the fluorinated sample, charge injection can be significantly suppressed by the fluorinated surface layer existing in the sample (Liu, Y. Q.,2012). Under 1000 times, more charges may probably accumulate in the shallower traps and decay a little faster compared with that under the 6000 and 18000 times.

Figure 9 shows the tdV/dt curves under different frequency. For the unfluorinated sample, the characteristic time of the first peak under 400 Hz is smaller than that under 600 and 1000 Hz, while the second peaks under different frequency present small difference in characteristic time as is shown in Figure 9a. Combined with the discussions in Figure 3a, the surface charge under 400 Hz shows a faster decay as the decay time is from 0 to 200 s, resulting a smaller characteristic time in the first peak in Figure 9a. While the decay time is over 200 s, the three curves in Figure 3a present a similar decreasing speed. Hence, the characteristic time of the second peaks in Figure 9a are almost the same. It can be found in Figure 9b that the curves of 600 and1000 Hz are similar because of the similar curves in Figure 3b. The characteristic time of the 600 and 1000 Hz curves are a little greater than that of 400 Hz.

The tdV/dt curves of the samples with different fluorination time are shown in Figure 10. The curve of the unfluorinated sample shows two peaks with the characteristic time of ~300 and ~4000 s. For the fluorinated samples, the curves present one obvious peak. The characteristic times of the fluorinated samples are between 10 and 100 s, which are much smaller than that of the unfluorinated sample. While the fluorination time changes from 30 to 60 min, both the amplitude and the characteristic time increase. In organic chemistry, C-F bond is the strongest single bond. The surface fluorination of polymer leads to the disruption of C-H bond and forming C-F, C-F_2 and C-F_3 groups (Liu, Y. Q.,2012). The fluorine element has the largest electronegativity and makes the C-F bond highly polarized, the result of which is that surface fluorination forms a fluorinated layer in the sample. And it can be inferred that the fluorinate layer in the sample would capture charges and form a shielding layer to shallow the deep traps (Du,B.X.,2013). While the fluorination time is 60 min, the polymer matrix may probably be changed by the excessive fluorination and the trap is deeper than that of the 30 min fluorinated sample.

Figure 8. tdV/dt curves with different corona discharging time, (a) unfluorinated sample; (b) sample fluorinated for 45 min

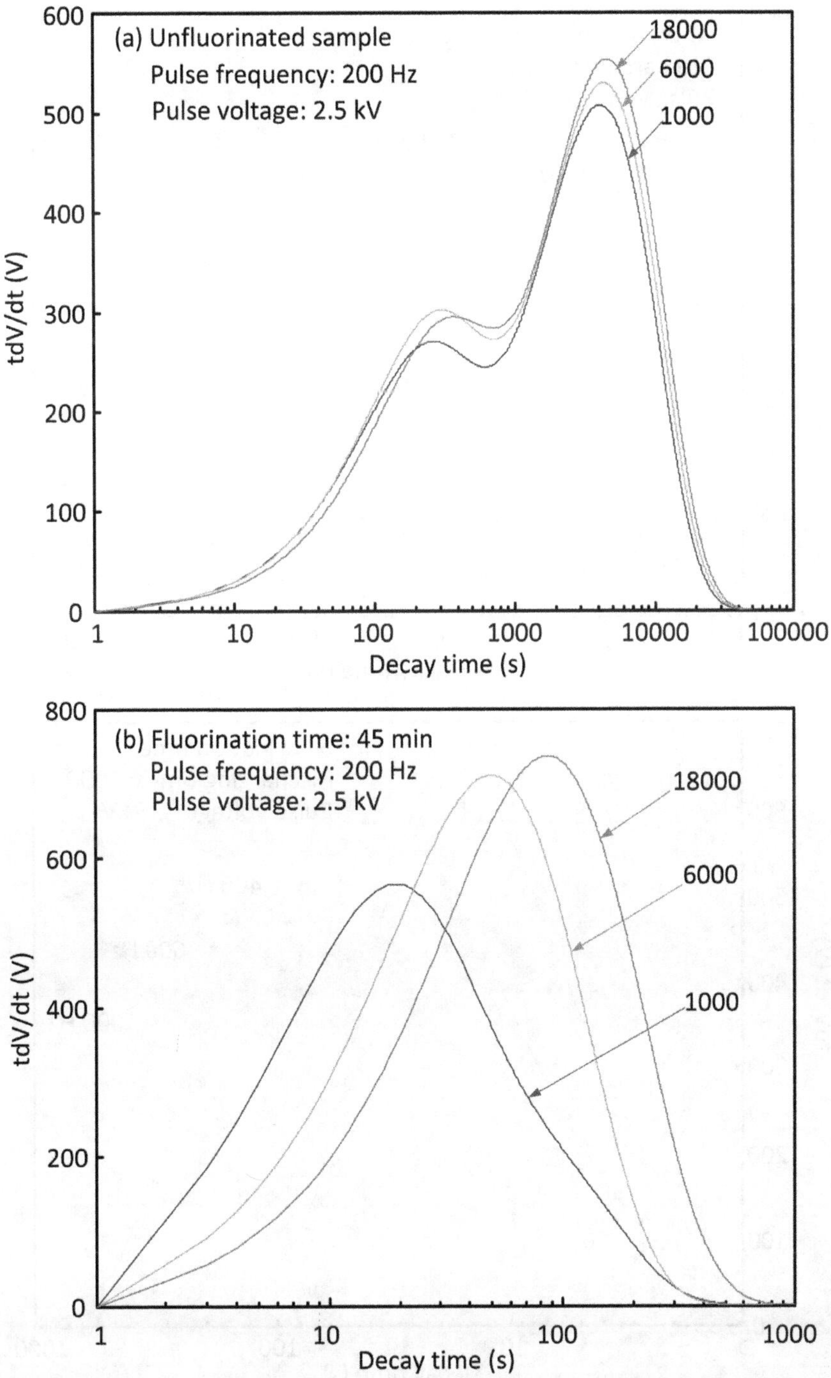

Figure 9. tdV/dt curves with different pulse frequency, (a) was unfluorinated sample; (b) was fluorinated for 15 min

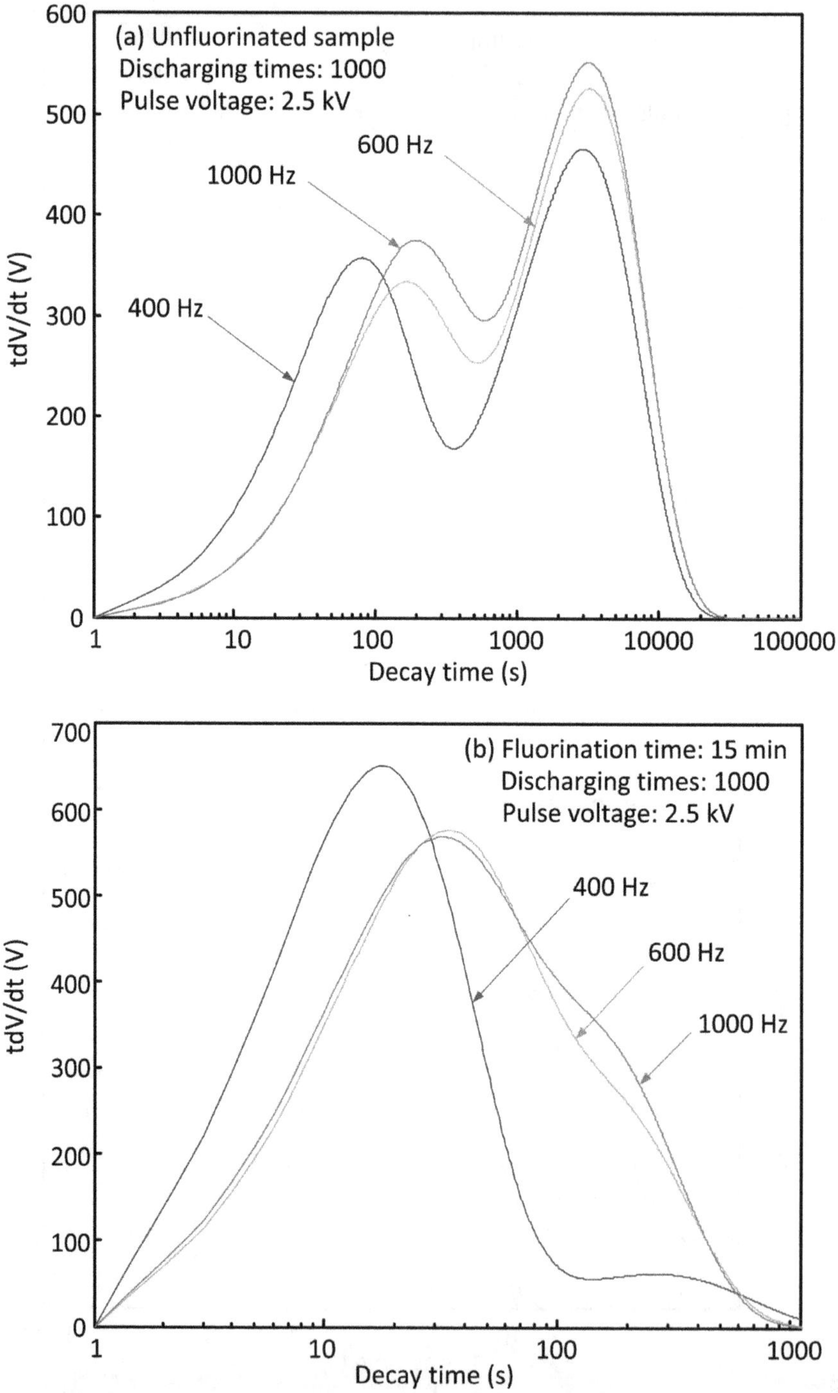

Figure 10. tdV/dt curves of different samples with the fluorination time of 0, 30 and 60 min

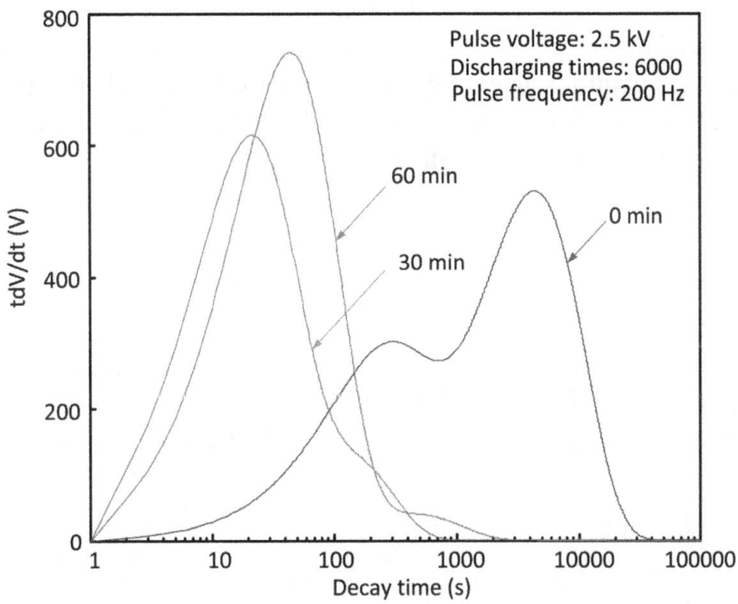

EFFECT OF DIRECT FLUORINATION ON SURFACE PARTIAL DISCHARGE CHARACTERISTICS OF POLYIMIDE FILMS

Surface Charge Accumulation and Decay with Different Voltages

In order to investigate the effects of voltage application on the surface charge dynamic characteristics, the results of unfluorinated PI film will be introduced in detail here considering the similar trend of samples with different fluorination time. Relationship between the surface charge density and the decay time after applying ±3 kV dc and positive pulse voltage with different amplitude combination is shown in Figure 11. Figure 11a demonstrates the combination effects of positive dc voltage and positive pulse voltage. The charge density decreases with the increase of time. The decay is fast at the beginning, and then becomes slower with time. Figure 11b shows the relationship between the surface charge density and the decay time after applying -3 kV dc voltage and positive pulse voltage with different amplitude combination. The unexpected results of surface charge decay occur. When the positive pulse voltage is 2 kV. The surface charge density is negative, increasing from -557 to -775 pC/mm^2 and then becoming smaller. While the pulse voltage increases to 4 kV, the initial surface charge density is positive, decreasing faster from 799 to 0 pC/mm^2 and then becoming negative. The 6 kV case has a higher initial charge density and shows a faster decay than the 4 kV case. The charge density of the 6 kV case is nearly positive during the decay process.

Figure 11. Relationship between the surface charge density and the decay time of the unfluorinated sample after applying (a) +3 kV dc and positive pulse voltage; (b) -3 kV dc and positive pulse voltage

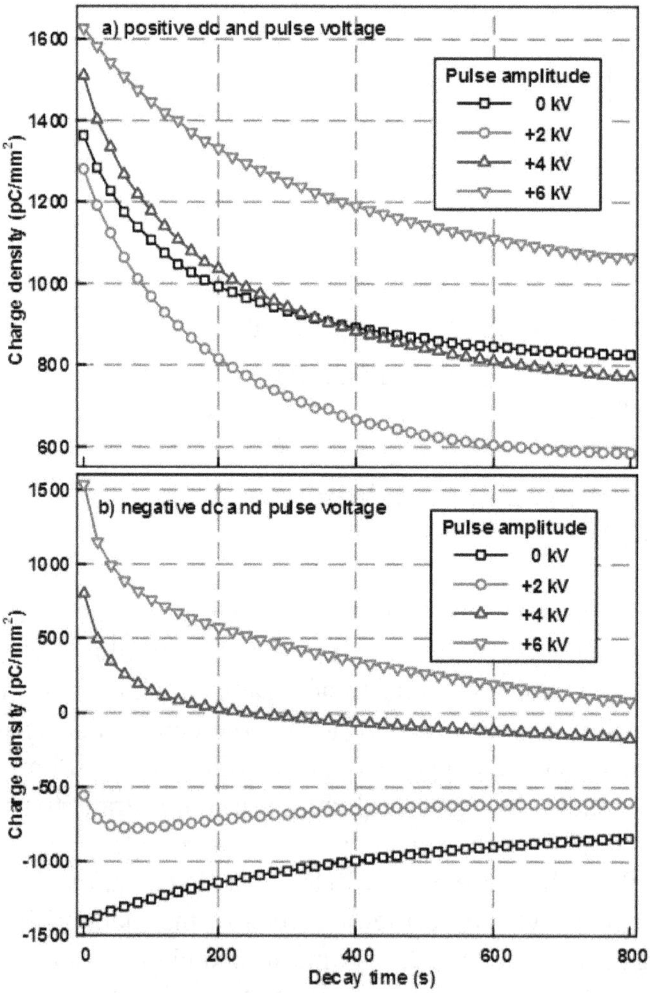

Relationship between the initial surface charge density and the pulse voltage amplitudes of the unfluorinated sample after applying ±3 kV dc and pulse voltage combination is shown in Figure 12. It can be seen that the +2 kV pulse addition decreases the initial charge density of independent positive dc corona charging. When the amplitude of positive pulse voltage rises from 2 to 6 kV, the initial charge density increases from 1281 to 1627 pC/mm^2. For the negative pulse addition, the initial charge density decreases from 1363 to -1620 pC/mm^2 with the pulse voltage ranging from 0 to -6 kV. As these two situations are the same trend with the effect of negative dc and pulse voltage combination.

The surface charge density distribution of the unfluorinated sample after applying ±3 kV dc and positive pulse voltage with different amplitude combination is demonstrated in Figure 13. For the positive dc and positive pulse voltage combination, the surface charge density shows an exponential decrease from the sample center to the edge. Specifically, the +2 kV pulse addition reduces the local charge density in the center compared with the effect of independent dc corona charging, while close to the edge, the charge density is nearly the same. As the pulse amplitude increases from 2 to 6 kV, the entire surface charge density distribution is enhanced. For the combination of

Figure 12. Relationship between the initial surface charge density and the pulse voltage amplitudes of the unfluorinated samples

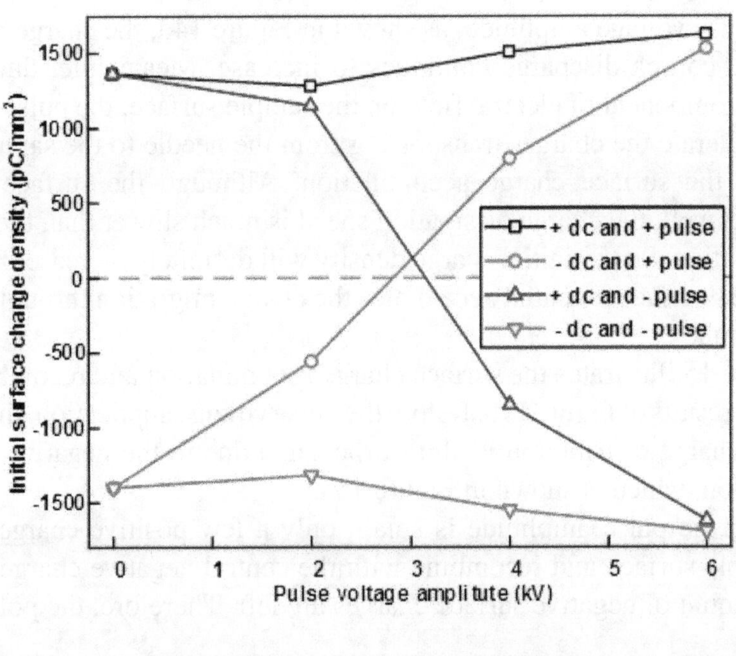

dc and pulse with different polarities, the surface charge distribution changes dramatically. One may clearly distinguish differences in the shapes of the distributions. Thus, the charge density of dc and 2 kV pulse corona charged sample has a local minimum at the center and its peak is located at ~0.8 cm away from the central point that yields a bell-shaped profile. In the case of the 4 kV pulse, the peak of the surface charge density is at ~1.5 cm and a positive peak appears in the center. A higher peak and nearly whole positive charges distribute along the sample surface for the 6 kV case.

Figure 14 illustrates the surface charge accumulation and recombination from the results of Figure 11a. Before the pulse voltage application, the initial surface charge distribution is almost consentaneous because of the same dc voltage corona charging, which is shown in Figure 14a. The followings are the situations after the pulse voltage application with different amplitudes. In Figure 14b, the accumulation of homopolar surface charges makes the charges from small pulse voltage corona difficult to migrate onto the PI film surface. Besides, the central charges trapped in the shallow traps will detrap and move to the surrounding areas by the repulsive force produced by the pulse voltage, whose direction is indicated by the arrows. Simultaneously, the small pulse voltage has little effects on the surface charge on the marginal region.

Figure 14c shows more charges generated from corona will move towards the sample surface and migrate from the center to the marginal area, when the pulse voltage amplitude increases. The surface charge traping and detraping procedure is strengthened by the higher surface potential. With the increasing of the pulse voltage amplitude, as shown in Figure 14d, the charge quantity from the corona discharge continues to increase. Meanwhile, due to the vertical component of electric field on the sample surface, the pulse voltage will accelerate the charges transporting from the needle to the sample, and increase the surface charge accumulation. Although the surface charge decays as well, the charge dissipation speed is much slower than the charge accumulation. So the initial charge density will definitely increase. Besides, the higher surface potential accelerates the charge migration throughout the sample bulk.

Figure 15 illustrates the surface charge accumulation and recombination from the results of Figure 11b. Before the pulse voltage application, the initial surface charge distribution is almost the same due to the negative voltage application, which is shown in Figure 15a.

When the pulse amplitude is small, only a few positive charges reach the sample surface and recombine with the central negative charges, but a large amount of negative surface charges are left. Therefore, the polarity of

Figure 13. Surface charge density distribution of the unfluorinated sample after applying (a) +3 kV dc and positive pulse voltage; (b) -3 kV dc and positive pulse voltage

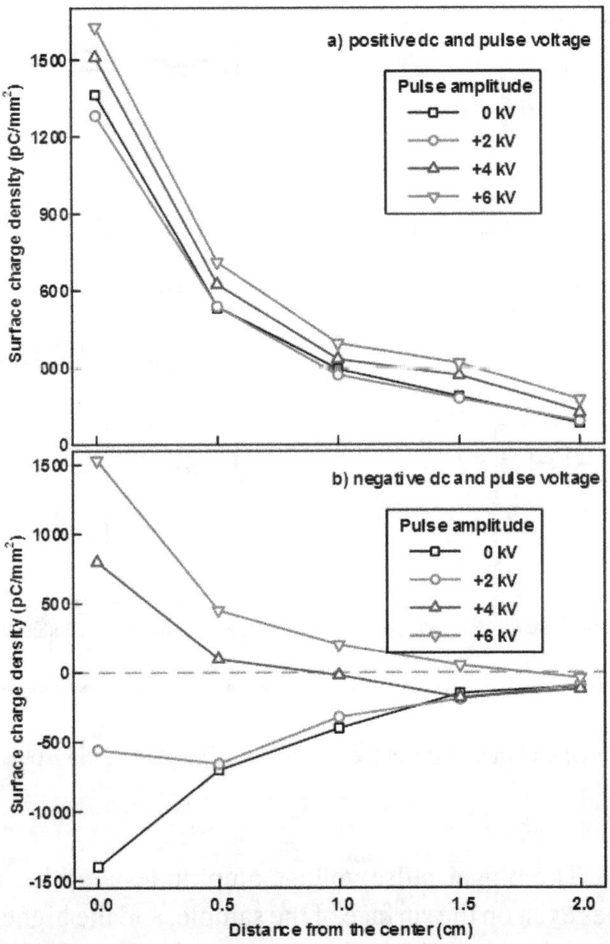

the surface charge is still negative. After removing the pulse voltage, the surrounding charges will surge towards the sample center, ending with an increase of negative charge density. For the medium pulse voltage, more positive charges are pushed onto the surface and recombine with the negative charges. Once the amount of positive charges exceeds the negative ones in the center, the initial surface charge density is positive. As the surface potential is lower than the surrounding areas, the polarity of the central charge turns to negative after a period of time with the horizontal electrical field. When the pulse amplitude continues increasing, more positive charges will accumulate

Figure 14. Illustration of surface charge accumulation after applying (a) +dc voltage; (b) +dc and small +pulse voltage; (c) +dc and medium +pulse voltage; (d) +dc and big +pulse voltage

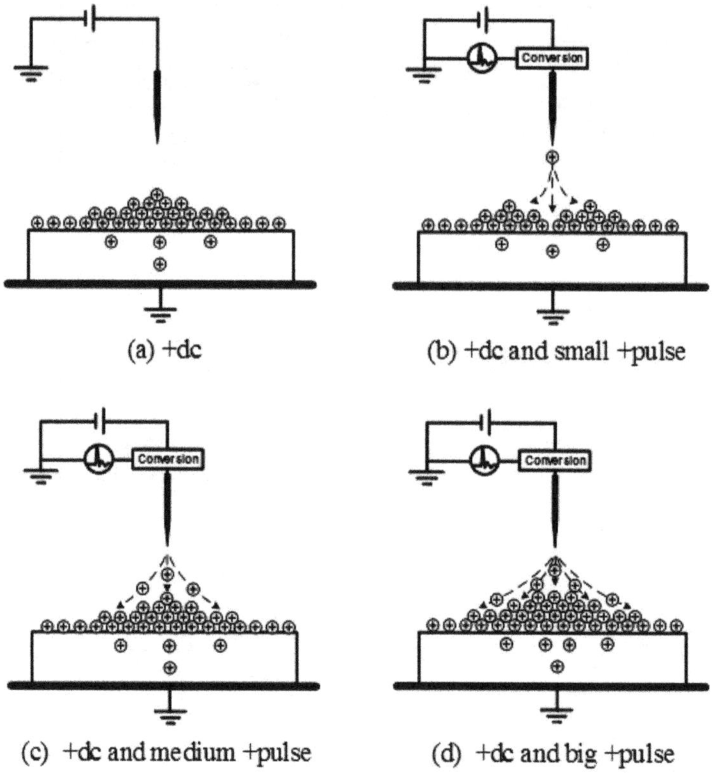

on the surface. The bigger pulse voltage amplitudes are, the more residual positive charges exist on the surface of the sample, and the higher initial value becomes. At the same time, the high positive surface potential can accelerate the charge neutralization in the sample bulk.

Effect of Direct Fluorination

In order to investigate the trend for the surface charge distribution and dissipation of samples with different fluorination time, a certain voltage application is selected here.

Figure 16 illustrates the relation between the surface charge density and the decay time with different fluorination time after applying ±3 kV dc and 4 kV negative pulse voltage combination. It is similar with the results in Figure

Figure 15. Illustration of surface charge accumulation after applying (a)–dc voltage; (b) –dc and small +pulse voltage; (c)-dc and medium +pulse voltage; (d) –dc and big +pulse voltage

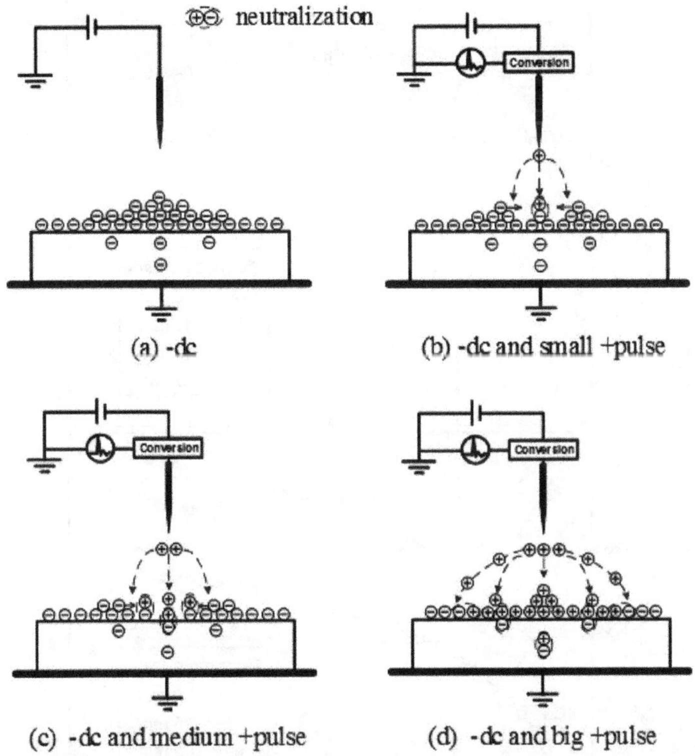

11b for the combination of dc and pulse voltage with different polarities. It can be found that the surface charge of the unfluorinated sample decays first and then increases after crossing zero point. For the 15 min fluorinated sample, the surface charge decays faster than that of the unfluorinated one but tends to be saturated after reducing to zero instead of continuous rise. Clearly, the decay beginning period of the 45 min fluorinated sample is close to the above results while it increases to a peak point and falls to zero again after decaying to zero. As for the combination of negative dc and negative pulse voltage, the effect of fluorination time is clarified in Figure 16b. With the lapse of fluorination time from 0 to 45 min, the surface charge decays faster. Because the surface fluorinated layer changes the surface state and accelerates the decay process. But 60 min fluorination produces an opposite

Figure 16. Relationship between the surface charge density and the decay time with different fluorination time after applying (a) +3 kV dc and -4 kV pulse voltage; (b) -3 kV dc and -4 kV pulse voltage

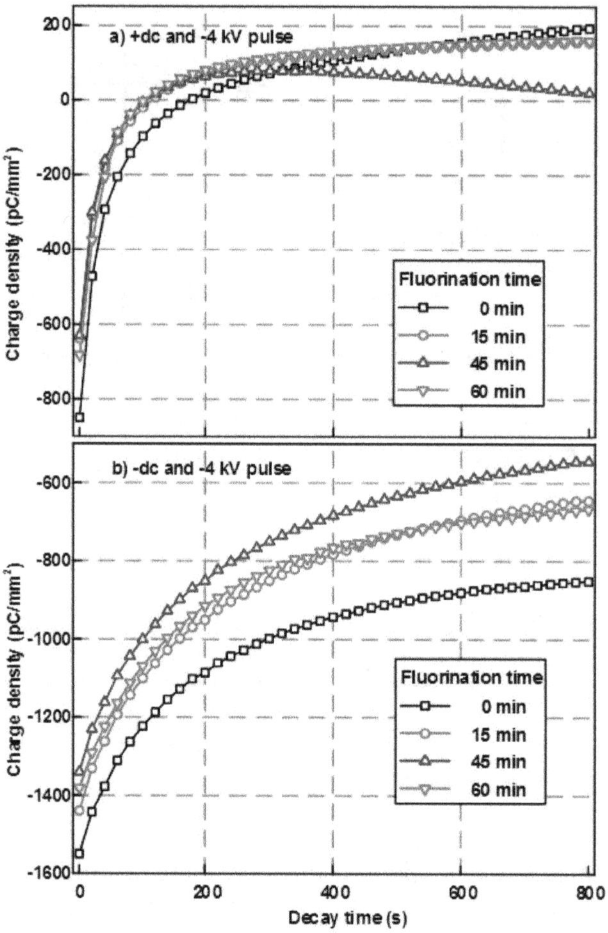

effect, meaning that the decay tends to be slower, which may be caused by the chain break in the polymer bulk due to excessive fluorination (Du, B. X.,2015)

The relationship between the initial surface charge density and the fluorination time is shown in Figures 17. It can be seen that direct fluorination decreases the initial charge density for the four corona charging methods. When the fluorination time rises from 0 to 45 min, the initial charge density decreases except for the 60 min fluorinated sample. Besides, the initial charge density of the corona charging by dc and pulse voltage with different polarities is lower than that of voltages with the sample polarity.

Figure 17. Relationship between the initial surface charge density and the pulse voltage amplitudes of the fluorinated samples

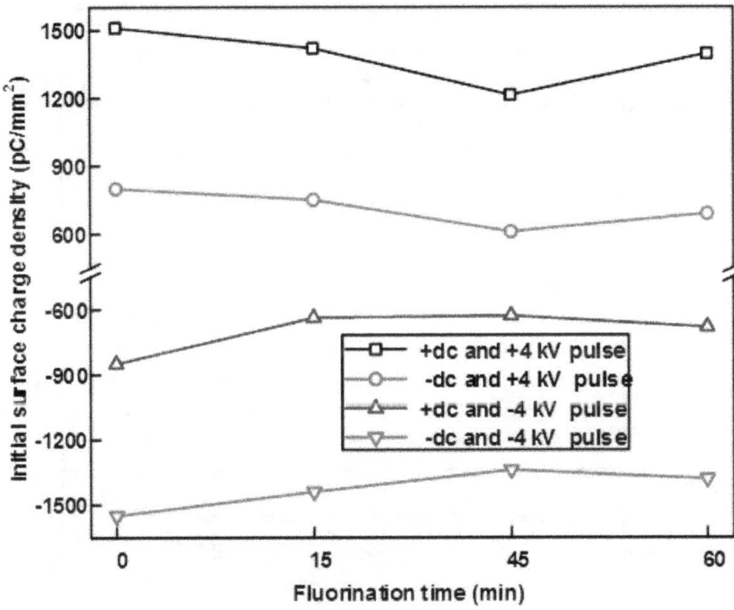

The surface charge density distribution of the fluorinated sample after applying ±3 kV dc and -4 kV pulse voltage combination is demonstrated in Figure 18. Like the results in Figure 13a, the surface charge density shows an exponential decrease from the sample center to the edge after corona charging with the sample polarity. Specifically, the 45 min fluorination reduces larger charge density as a whole, compared with other samples. For the combination of dc and pulse with different polarities, one may clearly distinguish the peak of the positive surface charge accumulation at 1.5 cm and a negative peak appears in the center. Lower peaks and less charges distribute along the sample surface with increasing the fluorination time from 0 to 45 min. But 60 min fluorination makes more surface charge accumulate on the PI sample surface.

Trap Distribution

Trap distribution of the samples with different fluorination time after applying dc and pulse voltage with different amplitude is shown in Figure 19. The energy of the traps is related to the decay time by considering energy levels (Sjöstedt, H.,2009). In this chapter, the trap distribution is calculated to analyze the surface charge dynamic behaviors of PI films under combined dc and pulse voltages.

Figure 18. Surface charge density distribution of the fluorinated sample after applying (a) +3 kV dc and -4 kV pulse voltage; (b) -3kV dc and -4 kV pulse voltage

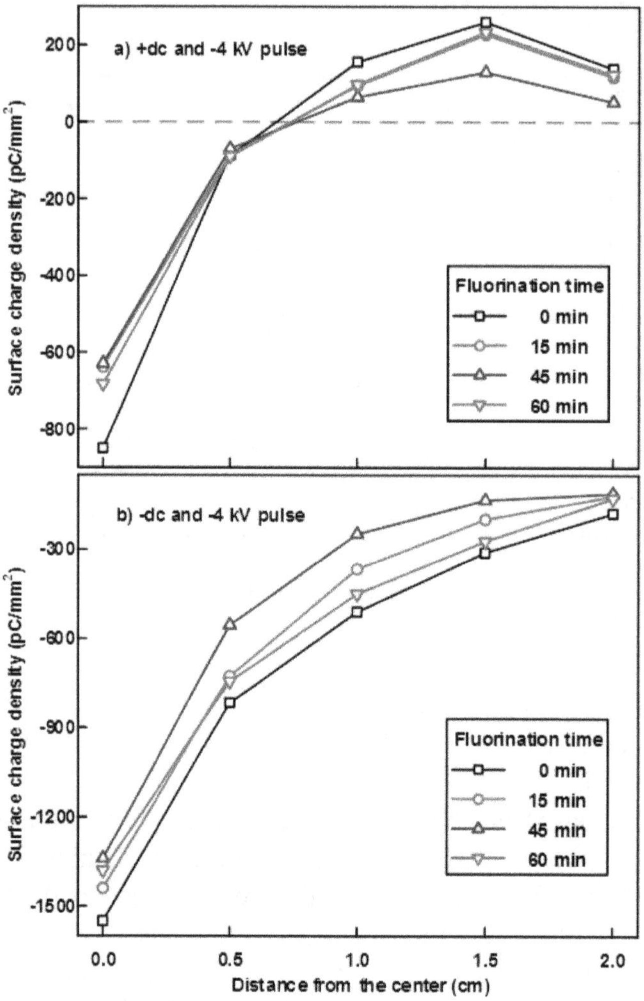

In Figure 19a, for the case of independent corona charging, there is a lower peak at 0.82 eV and a higher peak at 0.93 eV, corresponding to the shallow and deep trap region. The +2 kV pulse addition reduces the higher peak and raises the lower peak, which seems to distribute more uniformly. The repulsive force produced by small pulse voltage will make the charges located in shallow traps detrap, resulting in a faster decay of surface charge. In the cases of 4 and 6 kV pulse, the lower peak alters a little while the

Figure 19. Trap distribution of the samples with different fluorination time after applying dc and pulse voltage with different amplitude

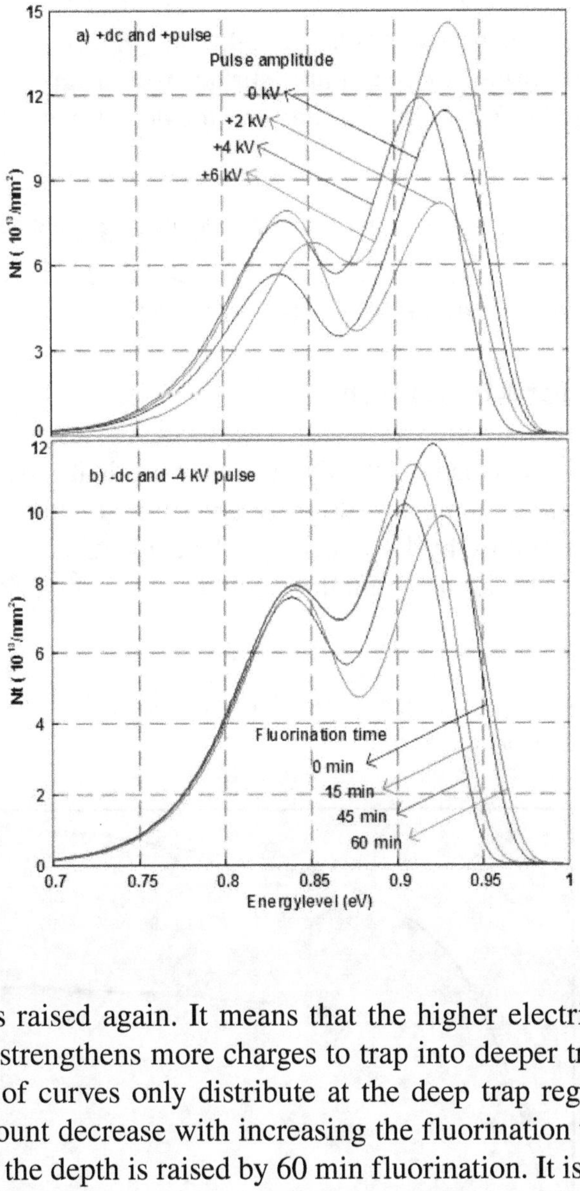

higher peak is raised again. It means that the higher electric field by the pulse voltage strengthens more charges to trap into deeper traps. In Figure 19b, changes of curves only distribute at the deep trap regions. The trap depth and amount decrease with increasing the fluorination time from 0 to 45 min, while the depth is raised by 60 min fluorination. It is noted that the 45min fluorinated sample has a lower value of the second peak, which is consistent with the results in Figure 16b that the surface charge of the 45 min fluorinated sample decays the fastest. The excessive direct fluorination produces polymer molecular chains breaking and deepens the traps as newly introduced defects. Meanwhile, previous results show that more C-H bonds are replaced by C-F bonds gradually with the increase of fluorination degree (Du,B.X.,2013). A much denser fluorinated layer is formed on the surface of PI samples, because of more stable C-F bonds. The fluorine element with the largest electronegativity makes the C-F bond highly polarized, and the fluorinated layer formed on the sample surface would capture electrons and form a shielding layer, which also shallows the deep traps.

EFFECTS OF WATER ABSORPTION ON SURFACE CHARGE AND DIELECTRIC BREAKDOWN OF POLYIMIDE/Al_2O_3 NANOCOMPOSITE FILMS

Effect of Water Absorption

The relationship between the water absorption of original and nanocomposite films with different immersion time is shown in Figure 20. The water absorption is determined by the percentage of weight gain when the specimen is immersed in water for a certain time (Shimizu, N.,2001; Zou, C.,2008).

Figure 20. Relationship between the water absorption of original and nanocomposite films with different immersion time

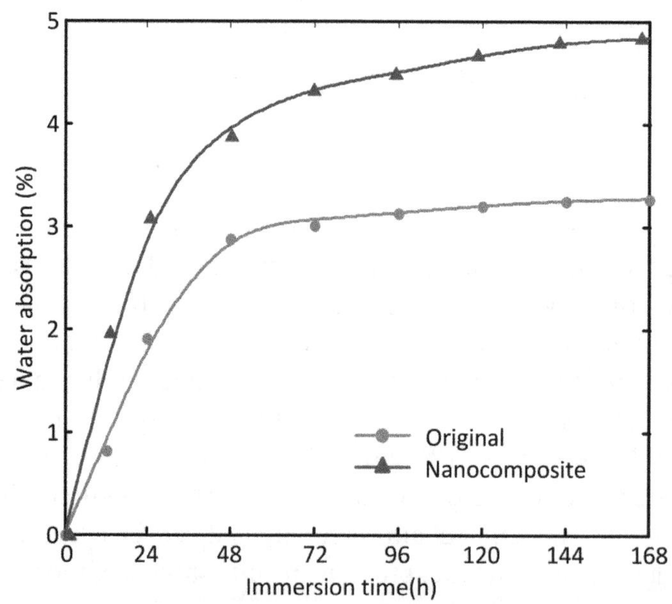

Since the water absorption of the nanocomposite film showed no apparent increase when the film was immersed for more than 168 h, the absorption behavior more than 168 h was not shown in the Figure. With increasing the immersion time from 0 to 168 h, the water absorption increases for all specimens. It can also be observed that the original film is saturated and the water absorption is stabilized at about 3% in about 48 h. However, for the nanocomposite specimens, the value of water absorption gradually builds up to about 5% with increasing the immersion time to 168 h. For the same immersion time, the nanocomposite specimens absorb more water than the original ones.

PI film exhibits high water absorption capacity because of its imide bonds. The molecular structure of PI contains hydrophilic groups such as carbonyl and amine groups. The hydrolysis of these groups to polyamic acid will break the molecular chain (Brewer, A. K.,1998). Therefore, the PI is hydrophilic and will absorb water in wet conditions. Simultaneously, according to the "water shell model" (Zou, C.,2008), when nano particles are incorporated in polymer matrix, a nano particle plays as a role of "core". The core is surrounded by a water layer with the thickness of several water molecules. Thus the water absorption is increased in nanocomposite films.

Effect on Surface Charge

Surface charge is injected in PI specimens by the method of corona discharge. The charges are encouraged by the electric field and injected in the upper surface of specimen through air. The addition of the grid electrode can ensure the surface potential at different positions to be identical. After the corona discharge, the injected charges in the specimen are concentrated in the region of micro size thin layer below the surface (Torpey, P. A.,1984; Torpey, P. A.,1986). Taking the dimension of the back electrode and the specimen into consideration, the electric field generated by the injected charges between the upper surface of the specimen and the back electrode is higher than that along the upper surface. Thus, the main way of charge decay transporting is through the bulk. The charges are firstly transferred from the surface to the bulk, and then it undergoes trapping and de-trapping processes, leading to the decay of surface charge. The different decay curves show the effects of different immersion time on the surface charge of original and nanocomposite films.

To show variation features of the surface potential at different conditions more clearly, the potential data from 0 to 1200 s are chosen in Figure 21 and

22. The relationship between the surface potential and the decay time with different immersion time is shown in Figure 21. It is shown from Figure 21a that for all the original specimens, the surface potential of immersed specimens decays faster than the dry ones. With increasing immersion time, the initial value of the surface potential decrease from ~2176 to ~ 1312 V for the original ones. This means the charge injection is affected by the absorbed water. The initial value of the surface potential for the nanocomposite specimens is higher than the original ones when the film contains no water. This may due to the introduction of more traps in the specimen caused by the incorporation of nanoparticles. However, the initial value drops rapidly to ~735 V and is lower than the original ones with increasing the immersion time to 168 h. Meanwhile, with increasing the decay time the nano specimen presents a faster descending speed. For example, when immersed for 168 h, the potential of nanocomposite specimens at 1200 s drops rapidly to ~ 20 V. While the potential of the original ones drops slower to ~ 600V.

The decay of surface potential is consistent with the release of electrons from trapping states. With increasing the time, it is the detraping of these electrons which leads to the decline of the surface potential. The trapped charges in shallow traps are easy to escape, which causes the rapid decay at the initial time. With increasing the time, the detraping of charges in deep traps occurs, thus the decay of surface potential slows down. When immersing the specimens into the water, the absorbed water will enhance the mobility of the charge carriers, which leads to the increase of the decay rate. The hydrolysis of the hydrophilic groups in PI to polyamic acid is also helpful to enhance the mobility of the charge carriers. In this way, the decay rate of all the specimens will be increased when they are immersed in water. It is shown in Figure 21 that the nanocomposite specimens immersed for 168 h have the biggest water absorption, which leads to the potential curve at 1200 s declining the largest to ~ 20 V. According to the "water shell model", when enough water is absorbed in the nanocomposite films, there is a high concentration water regions interconnected between the nano particles and the polymer matrix, which is the conductive layer 2 in the model (Zou, C.,2008). With a high content of nano particles in the nanocomposite film, the distance between the particles is short and the water layer surrounding the particles may make it easier for the charge to decay. Hence, the incorporation of the nano particles may alter the distribution of the water absorbed in the PI film, which may cause the decay of charge in nanocomposite films easier than the original ones.

Figure 21. Relationship between the normalized surface potential and the decay time with different immersion time of (a) original samples; (b) nanocomposites at 300 Hz

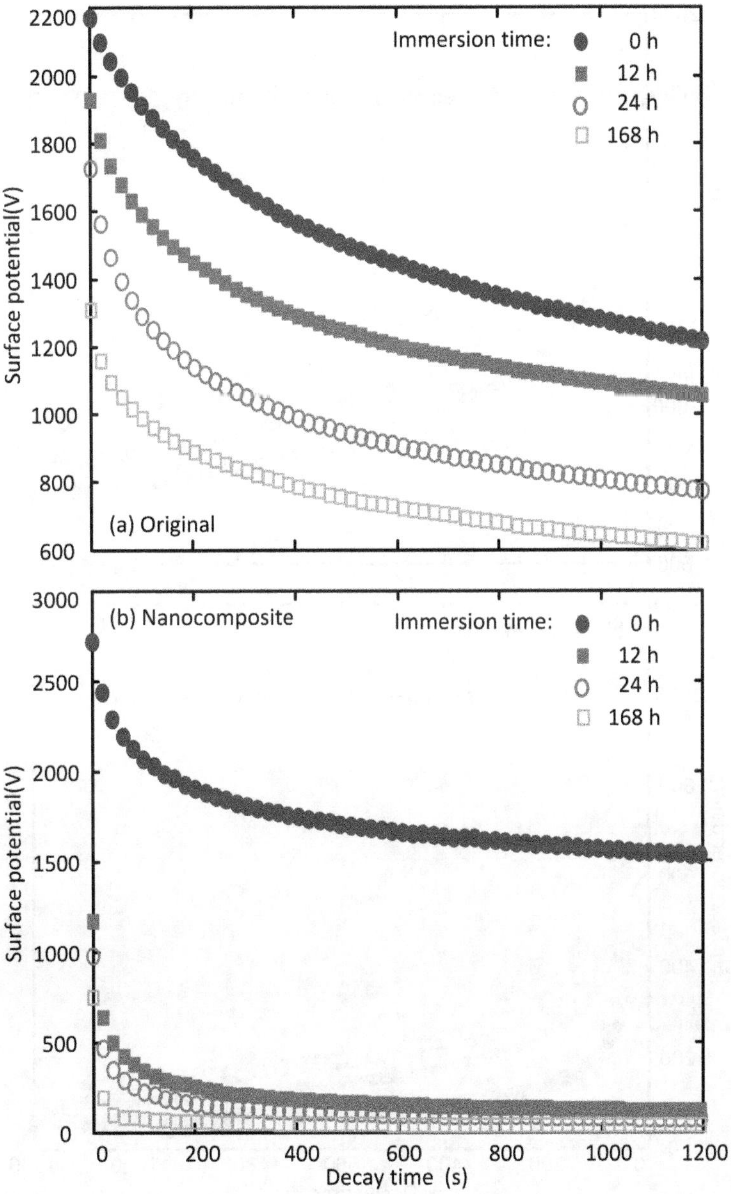

Figure 22. Relationship between the surface potential and the decay time with different pulse frequencies of (a) original samples; (b) nanocomposites immersed for 24 h

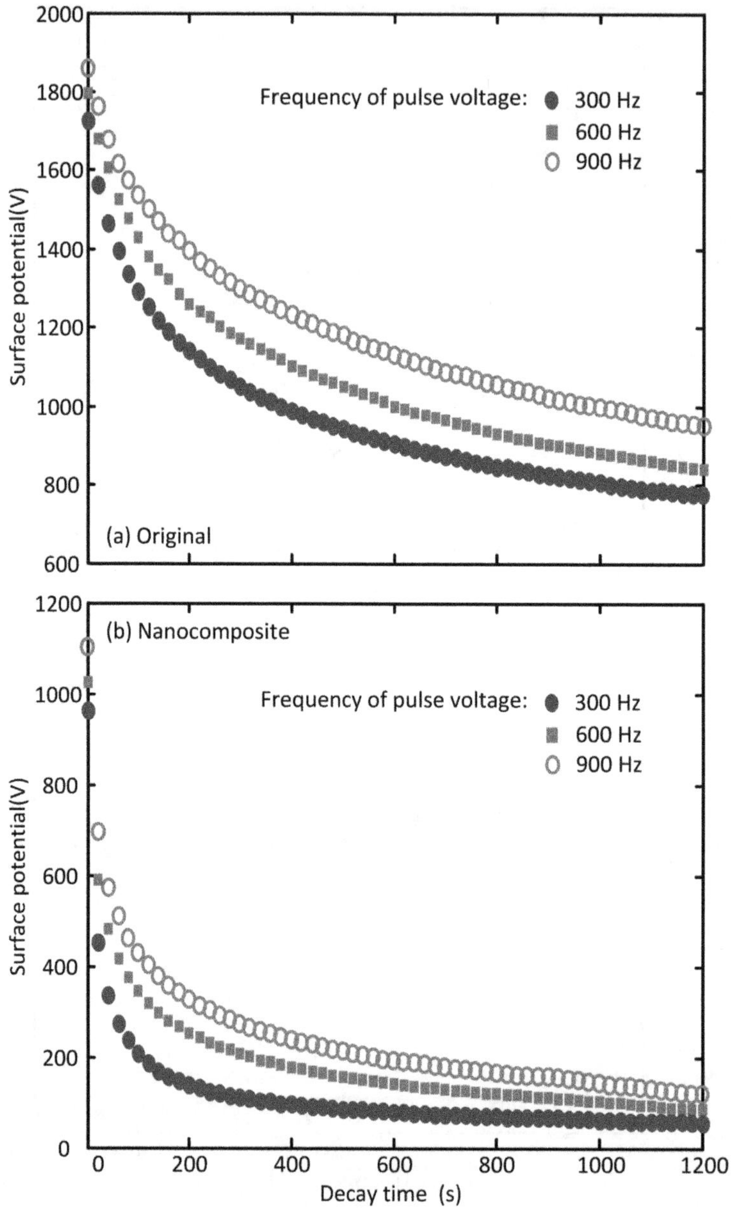

To reflect effects of applied pulse frequencies on the surface potential decay, Figure 22 shows the relationship between the surface potential and the decay time with different pulse frequencies of the specimens immersed for 24 h. It can be observed that with increasing the frequency, the surface potential increases. With increasing the pulse frequency from 300 to 900 Hz, the surface potential of original specimens at 1200 s increases from ~ 780 to ~ 970, while the nano ones increases from ~ 60 to ~ 120. It reflects that the immersion time impacts the nanocomposite specimens more than the original ones.

As time goes on, the time when the total number of charges returns to 10% of the original charged value can be defined as the dissipation time. A longer dissipation time indicates a slower corresponding decay process. With increasing the immersion time, the dissipation time of all kinds of specimens decreases. However, it decreases a lot for the nanocomposite specimens and decreases slower for the original specimens. The results show that the immersion time has a greater impact on the nanocomposite specimens, which are consistent with the water absorption shown in Figure 21.

Charge transportation is considered to be associated with carrier trapping and de-trapping. The decay of surface potential is caused by the charge transportation. The detraping process of charges is dependent upon time. With the lapse of time, the trapped charges in shallow traps are preferentially escaped as electrons are excited into the conduction levels and the detraping of charges in deeper traps will escape later. Simmons and Tam have defined a time-dependent demarcation energy E_m, which shows the border between the energy state which has been emptied of electrons or the electrons remain frozen at time t (Simmons, J. G.,1973). The current is determined by the rate of thermal excitation of electrons from traps and thus is proportional to dV/dt. Meanwhile, through the plotting of tdV/dt against t, the change of the charges detraping from the traps at time t can be estimated (Watson, P. K.,1990).

The change of charges over the time with various immersion time of original and nanocomposite specimens immersed for 0 to 168 h is shown in Figure 23. It can be observed from that for the all specimens, the time needed for all charges to escape from traps (the intersection between the curve and the X axis) increased in the order 168<24<12<0 h. For the dry specimens (0 h), the time of nano ones is longer than that of the original ones. However, when immersed in water, the time of nanocomposite specimens is much shorter than the original ones.

Figure 23. The change of the charges over the time with various immersion time of (a) original samples; (b) nanocomposite samples immersed for 0 to 168 h

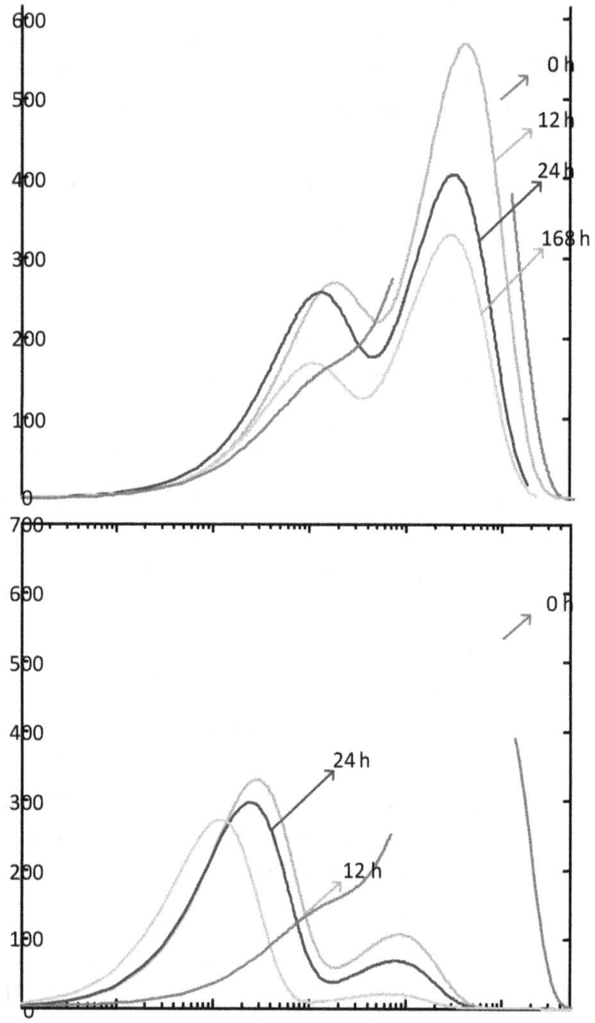

The results show that the duration of the charge decay process of different specimens is influenced by immersion time. And it is in line with the change of the water absorption for the original and nanocomposite films. Meanwhile, there is another interesting feature in the Figure, which is the peak of the curve. It is shown that there is only one peak in the curves of dry specimens, which represents there is only one changing peak of the charges in the corresponding trap depth. However, when the specimens are immersed in water, the original

Figure 24. The change of the charges over the time with different pulse frequencies of (a) original samples; (b) nanocomposite samples immersed for 24 h

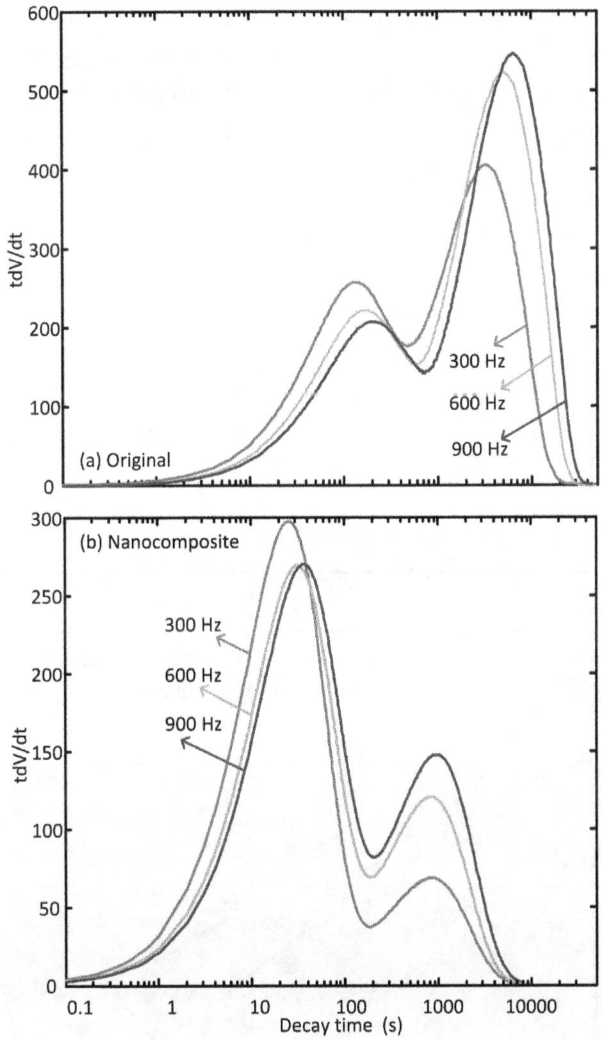

peak decreases with increasing the immersion time and a new peak occurs in the curve. It is assumed that the carrier mobility is enhanced by the absorbed water, which may cause the occurrence of the new peak. The hydrolysis of the hydrophilic groups in the specimens into polyamic acid will break the molecular chain, which may affect the number of the carriers and the trap depth in the specimen. The nanocomposite specimens exhibit higher water absorption and the water distribution is changed because of the incorporation

Figure 25. Relationship between (a) bulk resistivity and (b) surface resistivity of original and nanocomposite films with different immersion time

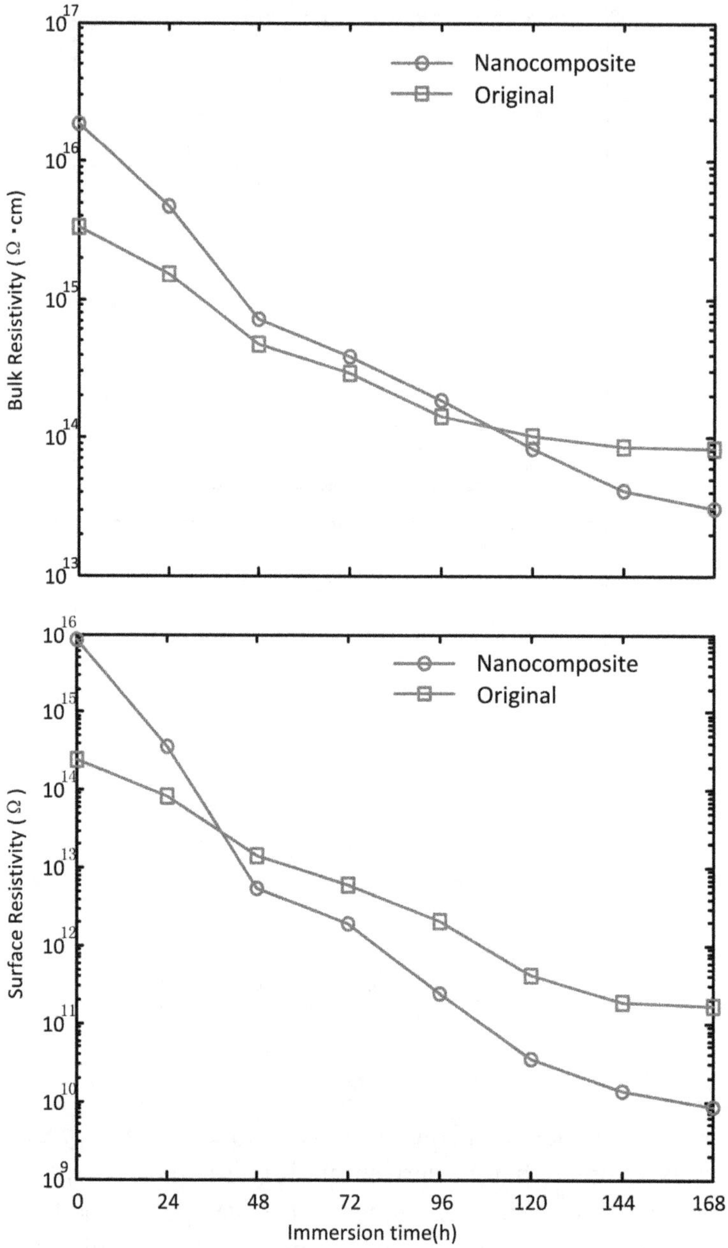

of the nano particles, thus may cause the new peak of the curve to appear earlier and the peak value to be higher.

Figure 24 shows the change of charges over the time with different pulse frequencies of specimens which are immersed for 24 h. It can be observed from Figure 24a that the time for all charges to escape from traps decreases in the order 900>600>300 Hz. The same tendency for the nanocomposite specimens is shown in Figure 24b. With increasing the applied frequencies, the value of the original peak increases. The results show that with increasing the frequency, the duration of decay process is extended. Meanwhile, with immersing the specimens into water, a compound peak also occurs in the curve, representing a compound peak of the detraping charges.

Effect on Dielectric Breakdown

Since the bulk and surface resistivity will influence the charge transportation and insulation properties. The relationship between the bulk and surface resistivity of original and nanocomposite with different immersion time are shown in Figure 25. When the immersion time is 0 h, the resistivity of nanocomposite film is higher than the original one. With increasing the immersion time, the resistivities of all specimens are decreased. Meanwhile, the resistivities of nanocomposite specimens drop faster than the original ones.

The relationship between the bulk breakdown and the surface flashover voltages of original and nanocomposite films with different immersion time are shown in Figure 26. It can be observed that for the dry specimens (0 h), both the surface flashover and bulk breakdown voltage of nanocomposite films are higher than those of the original ones.

Generally speaking, the local field strength can be enhanced by the surface charge (Foster, J.,2011), which means the breakdown results are opposite to the surface charge results mentioned above. Considering that a non-uniform electric field is generated by electrodes, this phenomenon may be the result of the additive nano particles, which cause the charge distribution to be changed and the field strength to be uniform. With immersing the specimens in water, the breakdown voltage decreases significantly with increasing the immersion time, which may be mainly caused by the increasing conductivity and dielectric loss. Although the surface charge of the PI film decays faster when absorbing water, which are beneficial to increase the breakdown voltage, however, the existence of water greatly reduced the breakdown strength of the film, which leads to the decrease of the breakdown voltage. The breakdown

Figure 26. Relationship between (a) breakdown voltage of original and nanocomposite films and (b) surface flashover of original and nanocomposite films with different immersion time

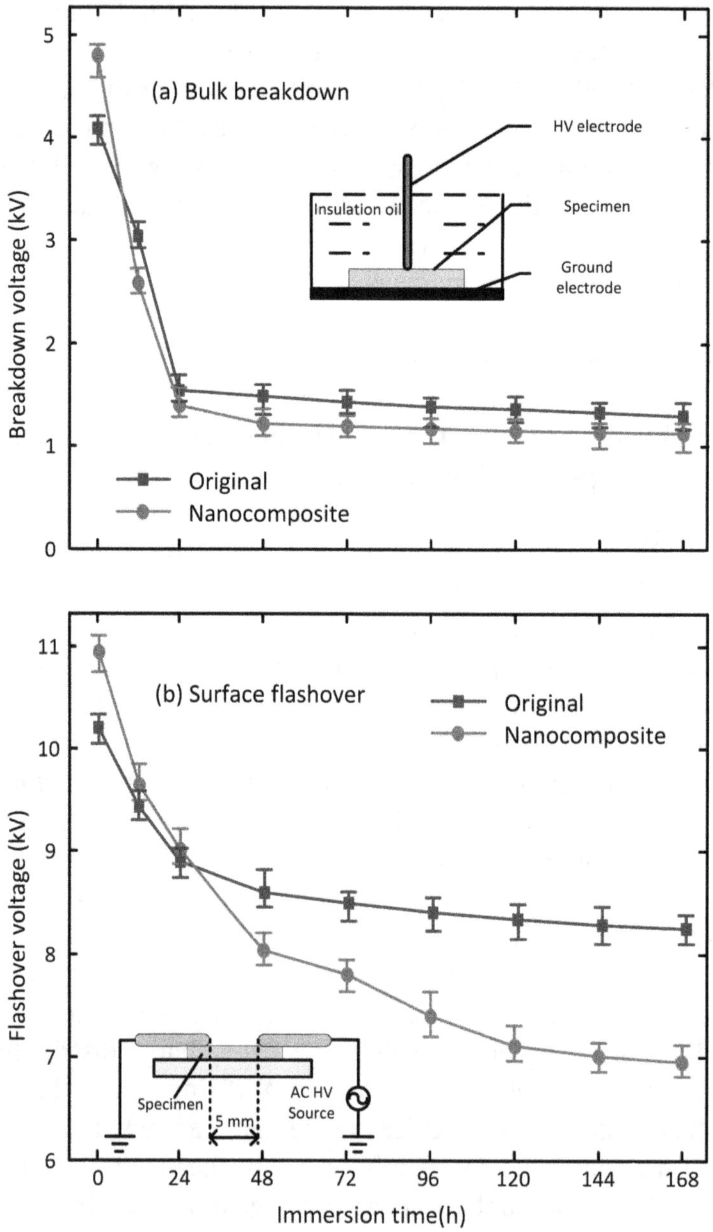

strength of nanocomposite specimens drops faster than the original ones. This may be the result of higher water absorption affected by the nano particles. The results show that the dielectric property of PI films is enhanced by the addition of nano particles. However, when the films work in contaminated and moist conditions, the breakdown property falls even faster for the nano composite films, which may cause the insulation of PI films to be degraded.

REFERENCES

Bellomo, J. R., Castelan, P., & Lebey, T. (1999). The Effect of Pulsed Voltages on Dielectric Material Properties. *IEEE Transactions on Dielectrics and Electrical Insulation*, *6*(1), 20–26. doi:10.1109/94.752005

Brewer, A. K., Orr, R. J., & Janicke, T. A. (1988). Hydrolytic Deterioration of Polyimide Insulation on Naval Aircraft Wiring. *IEEE Conf. Electr. Insul. Dielectr. Phenomena (CEIDP),* 180-188.

Du, B. X., & Guo, Y. G. (2015). Effects of surface fluorination on dielectric properties and surface charge behavior of water absorbed polyimide film. *IEEE Transactions on Dielectrics and Electrical Insulation*, *22*(3), 1369–1375. doi:10.1109/TDEI.2015.7116324

Du, B. X., Jie Li, , & Wei Du, . (2013). Dynamic Behavior of Surface Charge on Direct-fluorinated Polyimide Films. *IEEE Transactions on Dielectrics and Electrical Insulation*, *20*(3), 947–954. doi:10.1109/TDEI.2013.6518964

Du, B. X., Li, J., & Du, W. (2013). Surface Charge Accumulation and Decay on Direct-fluorinated Polyimide/Al_2O_3 Nanocomposites. *IEEE Transactions on Dielectrics and Electrical Insulation*, *20*(5), 1764–1771. doi:10.1109/TDEI.2013.6633707

Du, B. X., & Xiao, M. (2014). Effects of Thermally Conducting Particles on Resistance to Tracking Failure of Polyimide/BN Composites. *IEEE Transactions on Dielectrics and Electrical Insulation*, *21*(4), 1565–1572. doi:10.1109/TDEI.2014.004310

Du, B. X., Zhang, J. W., & Gao, Y. (2012). Effects of TiO_2 Particles on Surface Charge of Epoxy Nanocomposites. *IEEE Transactions on Dielectrics and Electrical Insulation*, *19*(3), 755–762. doi:10.1109/TDEI.2012.6215077

Florkowska, B., Roehrich, J., Zydron, P., Florkowski, M., & Rybak, A. (2012). Interaction of Conductor with Polymeric Materials (XLPE/EPR) at Partial Discharges. *IEEE Transactions on Dielectrics and Electrical Insulation, 19*(6), 2119–2127. doi:10.1109/TDEI.2012.6396972

Foster, J., Beeson, S., Thomas, M., Krile, J., Krompholz, H., & Neuber, A. (2011). Rapid Formation of Dielectric Surface Flashover due to Pulsed High Power Microwave Excitation. *IEEE Transactions on Dielectrics and Electrical Insulation, 18*(4), 964–970. doi:10.1109/TDEI.2011.5976082

Jiang, Y., An, Z. L., Liu, C., Zheng, F., & Zhang, Y. (2010). Influence of Oxyfluorination Time on Space Charge Behavior in Polyethylene. *IEEE Transactions on Dielectrics and Electrical Insulation, 17*(6), 1814–1823. doi:10.1109/TDEI.2010.5658233

Kaneko, K., Ozaki, T., Nakane, E., & Mizutani, T. (2005). Effects of Absorbed Water on Space Charge and Conduction Phenomena. *IEEE Conf. Electr. Insul. Dielectr. Phenomena (CEIDP)*, 657-660.

Kaufhold, M., Aninger, H., Berth, M., Speck, J., & Eberhardt, M. (2000). Electrical Stress and Failure Mechanism of the Winding Insulation in PWM-inverter-fed Low-voltage Induction Motors. *IEEE Transactions on Industrial Electronics, 47*(2), 396–402. doi:10.1109/41.836355

Kimura, K., Itaya, S., & Hikita, M. (2002). *Partial Discharge Behavior on PI Film with Micro Gap under Steep-front Step Voltage*. Boston, MA: IEEE Int'l. Sympos. Electr. Insul.

Kumara, S., Serdyuk, Y. V., & Gubanski, S. M. (2009). Charging of Polymeric Surfaces by Positive Impulse Corona. *IEEE Transactions on Dielectrics and Electrical Insulation, 16*(3), 726–733. doi:10.1109/TDEI.2009.5128512

Kurtz, S. R., & Anderson, R. A. (1986). Properties of The Metal-polymer Interface Observed with Space-charge Mapping Techniques. *Journal of Applied Physics, 60*(2), 681–687. doi:10.1063/1.337414

Liu, Y. Q., An, Z. L., Cang, J., Zhang, Y. W., & Zheng, F. H. (2012). Significant Suppression of Surface Charge Accumulation on Epoxy Resin by Direct Fluorination. *IEEE Transactions on Dielectrics and Electrical Insulation, 19*(4), 1143–1150. doi:10.1109/TDEI.2012.6259982

Llovera, P., & Molinie, P. (2004). New Methodology for Surface Potential Decay Measurements: Application to Study Charge Injection Dynamic on Polypropylene Films. *IEEE Transactions on Dielectrics and Electrical Insulation, 11*(6), 1049–1056. doi:10.1109/TDEI.2004.1387828

Luo, Y., Wu, G., Liu, J., Peng, J., Zhu, G., & Gao, G. (2014). Investigation of Temperature Rffects on Voltage Endurance for Polyimide/Al_2O_3 Nanodielectrics. *IEEE Transactions on Dielectrics and Electrical Insulation, 21*(4), 1824–1834. doi:10.1109/TDEI.2014.004305

Plopeanu, M. C., Notingher, P. V., Dumitran, L. M., Tabti, B., Antoniu, A., & Dascalescu, L. (2011). Surface Potential Decay Characterization of Non-woven Electret Filter Media. *IEEE Transactions on Dielectrics and Electrical Insulation, 18*(5), 1393–1400. doi:10.1109/TDEI.2011.6032807

Preetha, P., & Joy Thomas, M. (2011). Partial Discharge Resistant Characteristics of Epoxy Nanocomposites. *IEEE Transactions on Dielectrics and Electrical Insulation, 18*(1), 264–274. doi:10.1109/TDEI.2011.5704518

Shimizu, N., & Tanaka, H. (2001). Effect of Liquid Impregnation on Electrical Tree Initiation in XLPE. *IEEE Transactions on Dielectrics and Electrical Insulation, 8*(2), 239–243. doi:10.1109/94.919943

Simmons, J. G., & Tam, M. C. (1973). Theory of Isothermal Currents and the Direct Determination of Trap Parameters in Semiconductors and Insulators Containing Arbitrary Trap Distributions. *Physical Review B: Condensed Matter and Materials Physics, 7*(8), 3706–3713. doi:10.1103/PhysRevB.7.3706

Sjöstedt, H., Gubanski, S. M., & Serdyuk, Y. V. (2009). Charging Characteristics of EPDM and Silicone Rubbers Deduced from Surface Potential Measurements. *IEEE Transactions on Dielectrics and Electrical Insulation, 16*(3), 696–703. doi:10.1109/TDEI.2009.5128508

Sun, P., Li, J., & Yan, Y. (2012). Study on Aging Characteristics and Lifetime Model of Polyimide Film: A Review. *Inter. Conf. High Voltage Eng. (ICHVE)*, 259-264.

Takada, T., Miyake, H., Tanaka, Y., Yoshida, M., & Hayase, Y. (2013). Quantum Chemical Calculation Studies on Hetrero Charge Accumulation in Polyimide Film Containing Water. *IEEE Conf. Electr. Insul. Dielectr. Phenomena (CEIDP)*, 246-249.

Torpey, P. A. (1984). Double-carrier Injection and Recombination in Insulators Including Diffusion Effects. *Journal of Applied Physics*, *56*(8), 2284–2294. doi:10.1063/1.334262

Wang, F., Qiu, Y., Pfeiffer, W., & Kuffel, E. (2004). Insulator Surface Charge Accumulation under Impulse Voltage. *IEEE Transactions on Dielectrics and Electrical Insulation*, *11*(5), 847–854. doi:10.1109/TDEI.2004.1349790

Watson, P. K. (1990). The Energy Distribution of Localized States in Polystyrene, Based on Isothermal Discharge Measurements. *Journal of Physics. D, Applied Physics*, *23*(12), 1479–1484. doi:10.1088/0022-3727/23/12/002

Watson, P. K. (1990). The Energy Distribution of Localized States in Polystyrene, Based on Isothermal Discharge Measurements. *Journal of Physics. D, Applied Physics*, *23*(12), 1479–1484. doi:10.1088/0022-3727/23/12/002

Wheeler, J. C. G. (2005). Effects of Converter Pulses on the Electrical Insulation in Low and Medium Voltage Motors. *IEEE Electr. Insul. Mag.*, *21*(2), 22–29. doi:10.1109/MEI.2005.1412216

Wu, G. N., Wu, J. D., Zhou, L. R., Gao, B., Zhou, K., Guo, X. X., & Cao, K. J. (2010). Microscopic View of Aging Mechanism of Polyimide Film under Pulse Voltage in Presence of Partial Discharge. *IEEE Transactions on Dielectrics and Electrical Insulation*, *17*(1), 125–132. doi:10.1109/TDEI.2010.5412010

Xin, H., Fu, S. Y., Li, G. T., Guo, R. R., & Li, L. F. (2005). Novel Silica Tube/Polyimide Composite Films with Variable Low Dielectric Constant. *Advanced Materials*, *17*(8), 1056–1059. doi:10.1002/adma.200401330

Zhang, C., Shao, T., Ma, H., Zhang, D., Ren, C., Yan, P., & Schamiloglu, E. et al. (2013). Experimental Study on Conduction Current of Positive Nanosecond-pulse Diffuse Discharge at Atmospheric Pressure. *IEEE Transactions on Dielectrics and Electrical Insulation*, *20*(4), 1304–1314. doi:10.1109/TDEI.2013.6571449

Zhuang, Y., Chen, G., Chappell, P. H., & Rotaru, M. (2012). *Surface Potential Decay: Effect of Different Corona Charging Times. In Annu. Rep. Conf. Electr. Insul. and Dielectr. Phenomena* (pp. 620–623). CEIDP.

Zou, C., Fothergill, J. C., & Rowe, S. W. (2008). The Effect of Water Absorption on the Dielectric Properties of Epoxy Nanocomposites. *IEEE Transactions on Dielectrics and Electrical Insulation*, *15*(1), 106–117. doi:10.1109/TDEI.2008.4446741

Chapter 3
SiR Insulation for Outdoor Insulator

ABSTRACT

Silicone rubber (SiR) insulators were employed in power system due to its light weight, high electrical strength and excellent contamination performance. Much attention has been paid on the accurate evaluation of the insulator hydrophobicity. In this chapter, the acoustic characteristics of surface discharge were investigated to evaluate the hydrophobic properties of SiR insulator at different ageing degrees. The obtained results indicate that the acoustic characteristic of discharge is sensitive to the hydrophobic properties in the laboratory tests, which can be applied as a non-contact method for hydrophobicity evaluation of SiR insulator. Besides, the surface charge behaviors considering the hydrophobic properties of SiR insulator were discussed. The droplet experiment was carried out to investigate the maximum value, cumulative charge and energy eigenvector of discharge current. Finally, considering the strong airflow for SiR insulators employed in high-speed railway, the hydrophobic properties and corona discharge properties were investigated in this chapter.

INTRODUCTION

The ±800kV UHVDC transmission line is now in experimental operation in China (Shang, 2006; Shu et al., 2006). More than ninety percent of insulators used in the HVDC transmission lines are polymer insulators due

DOI: 10.4018/978-1-5225-2309-3.ch003

to advantages of light weight and surface hydrophobicity (Nazir et al., 2016; Matsuoka et al., 2002; Fu et al., 2011). In these lines, corona discharge could occur even on well-designed hardware and insulators, and the corona can damage the polymeric insulators (Du et al., 2009; Liu et al., 2016; Fernando et al., 2010; Du et al., 2009; Baker et al., 2009; Rajini et al., 2009). Under dc voltage, compared with ac voltage, the electrostatic force due to the constant electrostatic field makes more ions bump against the insulator surface. By causing some chemical changes, significant material degradation will reduce the insulator surface hydrophobicity (Kim et al., 1994; Khattak et al., 2016). When insulators are exposed to long-term humidity and severe contamination, with the decrease of hydrophobicity, the conductive layer will be easily accelerated to form on the surface. Thus, leakage current will be driven to pass through the layer and cause the dry-band discharges (Lopes et al., 2002). When there is sufficient heat energy, UV radiation, acids etc. produced by the discharges, more molecular chains of silicone rubber will be broken, which induces severe surface deterioration. Due to the hydrophobicity loss and the development of discharges, flashover may take place and affect operating reliability of power system (Chandrasekar et al., 2009). From the viewpoint of safety and reliability, to develop methods for hydrophobicity evaluation of SiR insulators is mainly the basis for the insulation design and the reliable operation of power system (Sabri et al., 2010; Liu et al., 2010).

The common methods for hydrophobicity evaluation are contact angle (CA) method and the spray method (HC method) (IEC/TS 62073-2003, 2003; Barsch et al., 1999; STRI Guide 92-1, 1992). Although these methods have been applied for the evaluation, they still have some limitations. For the CA method, it can accurately reflect the state of water droplet on the insulator surface, but it requires the critical measuring condition which is extensively applied in the laboratory tests (Barsch et al., 1999). For the spray method given in IEC 62073 (IEC/TS 62073-2003, 2003) that is also called the HC method recommended by Swedish Transmission Research Institute (STRI) (STRI Guide 92-1, 1992), although it is convenient in both laboratory test and engineering application, the data may have some dispersion to influence the evaluation results. Therefore, it is required to establish more accurate methods for the hydrophobicity evaluation. Acoustic discharge diagnosis techniques have been studied and developed for a long period, and a substantial number of studies have been accumulated (Kim et al., 2003; Boczar et al., 2004; Markalous et al., 2008; Sakoda et al., 2010; Sharkawy et al., 2008). L. E. Lundgaard has demonstrated several applications of acoustic discharge diagnosis, such as outdoor insulation, gas-insulated switchgear, capacitor, and etc. (Lundgaard,

1992; Lundgaard, 1992; Lundgaard et al., 1998). Acoustic emission (AE) has widely gained acceptance as a passive technique for observing the faults of equipment in power system. However, the characteristics of acoustic signal for hydrophobicity evaluation are still unexplored. In Section 2, we would focus on the acoustic characteristics of discharge for hydrophobicity evaluation on different aged SiR specimens by using the droplet test in the laboratory. Through analyzing the acoustic characteristics, the relationship among the surface discharges, the voltage range, the maximum slope, the pulses number and the acoustic signal energy with the surface hydrophobicity was established. The Weibull distribution parameter of the wave was employed to reflect the hydrophobicity condition. With the decrease of surface hydrophobicity, all these characteristics showed an obvious tendency. Therefore, the method based on the acoustic characteristics was proposed based on the laboratory tests for the hydrophobicity evaluation of SiR insulators. Besides, also based on droplet tests, surface discharges and flashover induced by discrete water droplets on polymer insulator specimens were carried out in the laboratory experiments. In Section 3, characteristics of discharge current including maximum discharge current, cumulative charge and energy eigenvector of surface discharges were investigated to reveal effects of discrete water droplets on surface discharges. The variation in droplet properties was droplet numbers, droplet location, volume and contamination. By using methods of wavelet transform and energy eigenvector of surface discharges, relationship between discharge characteristics and droplet parameters can be obtained, which is helpful to further investigation on the influence of water droplets in flashover process of polymer insulator.

Moreover, insulators are not only applied in the HV power system. In recent years, the high-speed railway is developing rapidly worldwide and SiR insulators are employed on railway roof. Corona usually occurs due to the high electric fields caused by poorly designed insulator hardware, manufacturing deficiencies or water drops on insulator surface (Moreno et al., 2001; Phillips et al., 1999). If allowed to occur continuously, corona discharges may induce serious flashover or breakdown accidents (Kim et al., 1994; Lopes et al., 2001). During corona process, the bombardments of the charged particles can break the molecular bonds of SiR, and the surface becomes hydrophilic because of the formation of hydroxyl and carboxyl caused by chemical reactions (Li et al., 2008; Zhu et al., 2006). A brittle, silica-like layer can be formed on the SiR surface due to oxidation during the corona process (Hillborg et al., 1999; Fritz et al., 1995). Most researchers focused on the degradation of SiR caused by corona in stationary air, but few paid attention to the hydrophobicity loss due to corona in high-velocity airflow. The airflow velocity around SiR insulators on railway roof is extremely high. It is urgent to investigate the corona performance of SiR insulators in high-velocity airflow.

The corona discharge characteristics in the presence of wind are different from that in stationary air. Avella et al. reported that the corona onset voltage was related to the wind speed. Besides, when the wind speed was increased, the corona current magnitude increased correspondingly (Avella et al., 2012). According to the results in a mini corona cage, the wind caused an increment in DC corona current of a conductor (Li et al., 2010). Wang et al. reported that the gas breakdown voltage was higher in the presence of airflow than that in the absence of airflow and the plasma channel on the surface was blown downstream by airflow (Wang et al., 2008). It can be concluded that the existence of airflow has much effect on corona discharge, so the investigation in the corona behavior of SiR in high-velocity airflow is of significant importance.

In order to research the effects of wind on corona behavior of SiR, pure HTV-SiR was treated with corona generated in a needle-plate electrode system in the absence or presence of wind. In Section 4, the variations in corona discharge intensity, discharge current and contact angles of specimens were measured and analyzed under different wind conditions. The hydrophobicity recovery process was also investigated under different wind conditions.

HYDROPHOBICITY EVALUATION OF SILICONE RUBBER INSULATOR USING DC DISCHARGE INDUCED ACOUSTIC WAVE

Specimens

The specimens were SiR slices removed from SiR insulators which were applied on transmission lines. To obtain the specimens with different degradation conditions, the corona-degradation experiments were carried out, as shown in figure 1. The corona discharge was generated by using a multi-needle to plate electric field configuration. As shown in figure 1a, the plate electrode was grounded, while the multi-needle electrodes were connected to the dc HV of $\pm$10.0 kV. After the corona degradation for 2h to 10h respectively, the hydrophobicity variation of the specimens were classified by the static

Figure 1. Schematic setup and hydrophobicity variation of the specimens

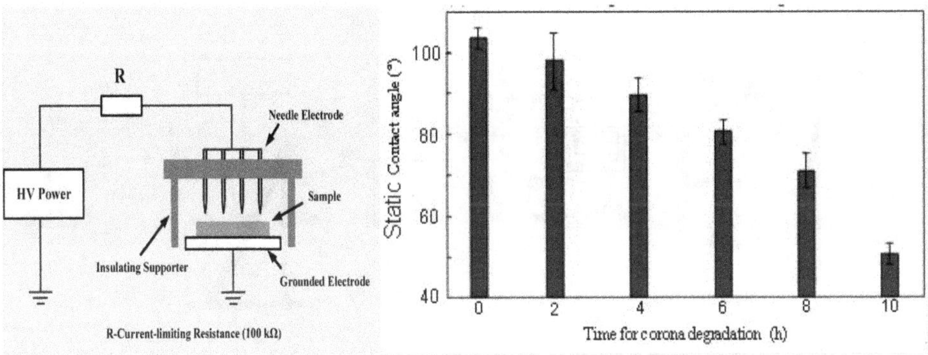

contact angle method published in IEC 62073, as shown in figure 1b. With the increase of corona-degradation time, the hydrophobicity of specimen surface decreases. And then, the experiments according to static droplet test and dynamic droplet test were carried out in a climate chamber, in which the environmental conditions were set at 1.01×105 Pa, 23 °C and 70% relative humidity.

Droplet Test

Figure 2a shows the arrangement of SDT experiment. The dimensions of specimens are 50 mm × 21 mm × 5 mm. After the corona ageing, two droplets were put on the specimen surface in an epoxy support frame in a short time, as more droplets may make it hard to distinguish the flashover voltage. Each droplet was kept at 20μl to prevent the impact of different volumes. The distance between the two droplets was kept at 7 mm in the middle of the specimen. The electrode pattern was a pair of plate-plate aluminum electrodes with the parallel distance of 21 mm. One electrode was connected to a dc high voltage stress through a series of protective resistors, while the other electrode was connected to the ground. An acoustic sensor with a series of de-noise and amplifier circuit was employed to obtain the acoustic waveform at the same time. The sensor was fixed up against the specimen and the distance was kept at 15 cm. The frequency bandwidth of applied acoustic acquisition was 20 kHz to 45 kHz. After the experiment arrangement was ready, the power source was turned on and raised at the rate of 0.1 kV/s until the flashover occurred.

Figure 2b shows the experimental arrangement of DDT method. The dimensions of specimens are 50 mm × 35 mm × 5 mm. After certain time

Figure 2. Schematic of experimental arrangements. (a) Static droplets test (b) Dynamic droplets test

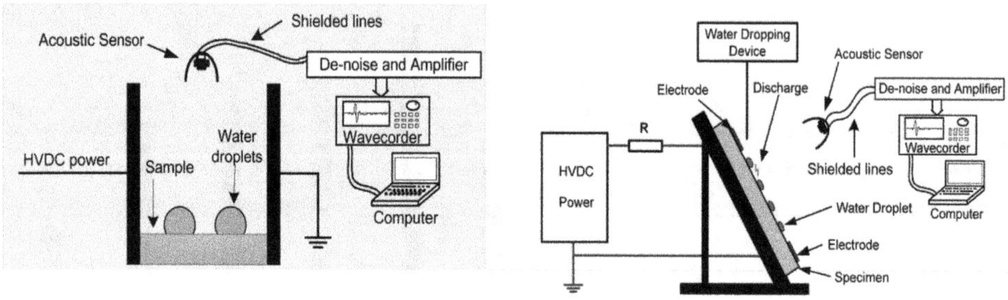

of corona-degradation, the aged specimen was put on an epoxy plane with the inclined angle of 60 degrees in a short time. The electrode pattern was a pair of plane-plane steel electrodes, with the parallel distance of 40 mm. The upper electrode was connected to the high voltage that was generated by a dc HV power source through a series of protective resistors, while the other electrode was connected to the ground. An acoustic sensor was fixed against the specimen, and the distance between them was kept at 15 cm. The sound of surface discharge was measured by the data detection system, whose sampling frequency was 200 kHz and the recording time was 20s. The water droplet was made of deionized water with certain NaCl, of which the conductivity was 2.0 mS/cm at 23 °C, and each droplet was kept at 40μl.

All the experiments were carried out in a shielding room to reduce the background noise for the surface discharge measurement. When the specimen was fixed between the electrodes, the contaminated droplet was dropped at the side of high-voltage electrode with the interval time of 2.0 s. After the droplet flowed down to the underside electrode for 30 s, dc voltage of 5.0 kV was turned on as the test voltage. Then, the droplet was still dropped to maintain the continuous effect on the surface discharges. At the same time, the acoustic signal was obtained by the system. The frequency bandwidth of the applied acoustic acquisition was 20 kHz to 45 kHz. Finally, the acoustic signal was analyzed to obtain the characteristic parameters for the hydrophobicity evaluation.

Acoustic Signal Treatment

Figure 3 shows the characteristic parameters of acoustic signal generated by surface discharge. To represent the strength of acoustic signal emitted from the

Figure 3. Characteristic parameters of acoustic signal

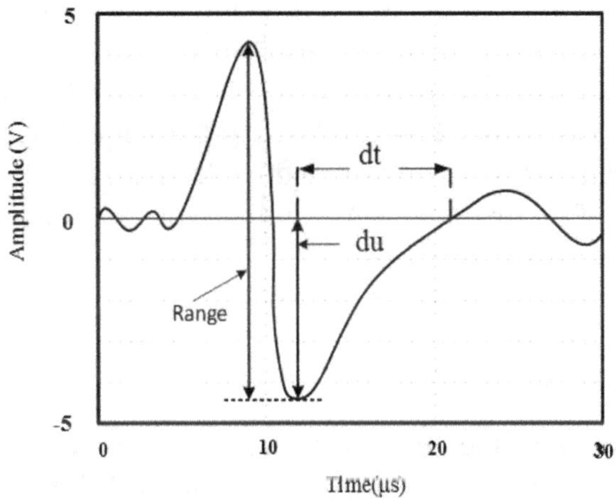

discharge, the voltage range (*Range*) of the signal was taken as a parameter. It was the difference between the maximum value and the minimum value in the signal data. As shown in figure 3, the maximum slope of the signal (*du/dt*) was calculated by dividing the changing value (*du*) with the corresponding time (*dt*). It was in relation with the time of the acoustic energy emitting.

To eliminate the parameter randomicity of acoustic signal, statistical parameters were employed to describe the signal in a period of time. 14 specimens were placed in the corona-degradation experiments for different time, to obtain the specimens with different degradation conditions. According to the different corona degradation time, the specimens were put into different groups. Because hydrophobicity would be losing owing to discharge and the changes are caused by the continuous surface discharge, the data of 5 seconds at the beginning of every signal recording were picked as the statistical object. Skewness (Sk), Kurtosis (Ku) and Weibull distribution parameter (β) were calculated. The quantities defining Sk and Ku are determined as follows:

$$Sk = \left(\sum_{i=1}^{n} (x_i - \mu)^3 p_i \right) \Big/ \sigma^3 \tag{1}$$

$$Ku = \left(\sum_{i=1}^{n} (x_i - \mu)^4 p_i \right) \Big/ \sigma^4 - 3 \tag{2}$$

where n is the number of different data, x_i is the value of data, p_i is the probability of x_i, μ is the mean value of data, and σ is the variance of data. Sk+ is defined as the Skewness of the positive data, while Sk- is for the negative. Ku+ is defined as the Kurtosis of the positive data, while Ku- is for the negative. The value of Sk is related to the value and number of signal peaks, and the value of Ku is determined by the distribution of signal peaks.

The function of two-parameter Weibull distribution is as follow:

$$F(x_i) = 1 - \exp[-(\frac{x_i}{\eta})^{\beta}] \tag{3}$$

where x_i is the value of the peaks, β is the shape parameter of distribution, and η is the dimension parameters of distribution. When the value of β is smaller than 1, the specimen is in the situation of premature failure. When β is equal to 1, it means the random failure. When the value of β is bigger than 1, it means the ageing failure.

SDT Hydrophobicity Evaluation

Figure 4 shows the relationship between the characteristic parameters and the corona degradation time. To obtain the mean value of the parameters, the experiments were carried out on three different places for each aged specimen.

The flashover voltage varying with the corona degradation time is shown in figure 4a. It is observed that the flashover voltage decreases with the increase of corona degradation time under both positive and negative voltage stress. Under a good condition of hydrophobicity, the flashover voltage under positive voltage stress is higher than that under negative voltage stress. While the corona degradation time is long, the flashover voltages are similar under both positive and negative voltage stress. Such difference between the dc flashover voltages is caused by a basic dc characteristic called the polarity effect, which is not consistent in the experimental models. In the condition of positive dc voltage, electrons emitted from the cathode are necessary to form the positive partial arcs on the specimen surface. While in the negative dc voltage, the anode can supply more electrons to maintain the high conductivity of negative partial arcs. As the negative partial arc has higher floating properties than the positive partial arc, when there are droplets on the surface, the negative partial arcs can easily bridge the droplets and result in the lower flashover voltage.

Figure 4. Relationship between the characteristic parameters and the corona degradation time.(a) flashover voltage (b) pulse number (c) energy of the acoustic signal

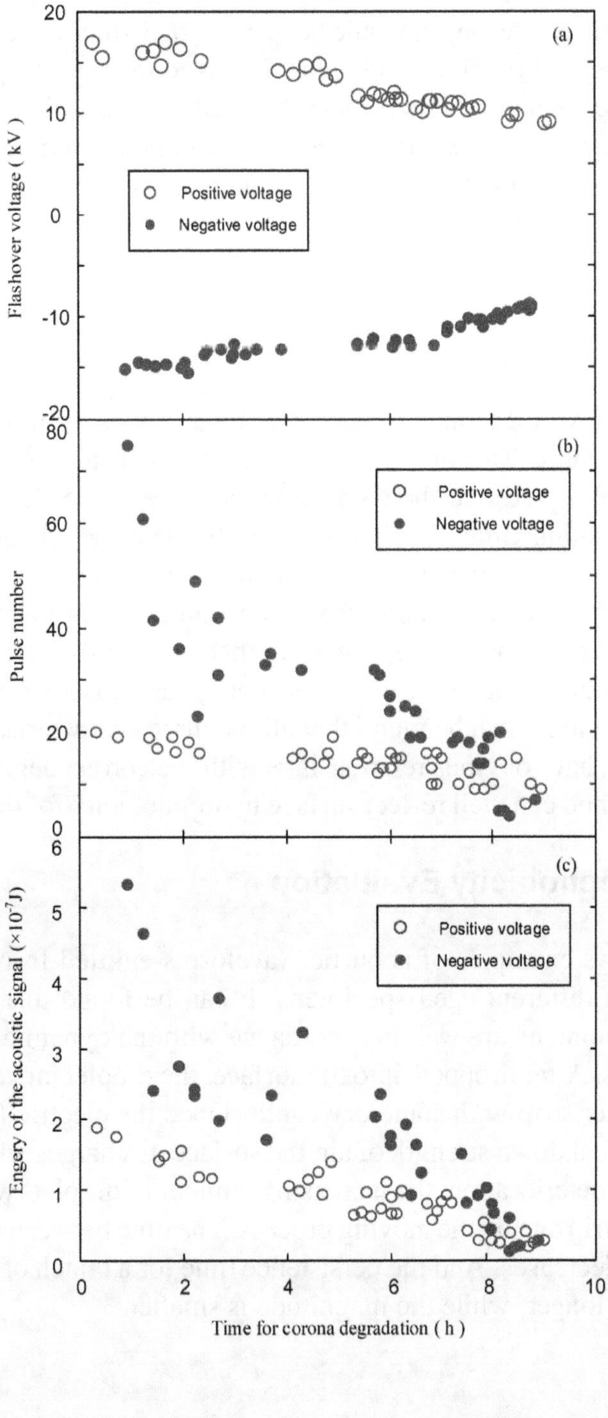

As shown in figure 4b, the pulse number of the signal decreases with the process of surface deterioration or hydrophobicity loss. When the corona degradation time was longer than 6h, it was observed in the experiment that the acoustic emitted from the discharge was weak. And once the flashover discharge occurred, the power would be turned off. So there were fewer pulses which were beyond the threshold value. It was found that the pulse number under negative voltage stress was more than that under positive voltage stress.

When surface discharge occurs, there is an instantaneous release of energy, some of which is emitted as the acoustic energy. As shown in figure 4c, the acoustic signal energy (*E*) was calculated by the following expression:

$$E = \frac{1}{R}\sum u_n^{\,2}t \tag{4}$$

where u_n is the voltage value of acoustic signal, *t* is the sampling time and *R* is the essential resistor of the sensor. *E* is in direct ratio with the emitted energy. It is observed that the signal energy decreases with increasing the corona degradation time. The reason may be deduced as following. The flashover voltage of a specimen under good hydrophobicity condition was higher, so there would be more charge accumulated on plates. When the discharge occurred, more energy was emitted. While the signal amplitude and the pulse number decrease, the signal energy decreases with the decrease of hydrophobicity. It can be found that all the three characteristics above for the evaluation have obvious relationships with the corona degradation time. The SDT method can well reflect surface hydrophobicity for the evaluation.

DDT Hydrophobicity Evaluation

Figure 5 shows examples of acoustic waveforms emitted from the surface discharges on different aged specimens. It can be found that the acoustic emission phenomena are well in accordance with the dynamic behaviors of water droplets. After dropped onto the surface, the droplet moved downward and left a water strip with some dry bands. Once the electric field strength reaches the breakdown strength of air, the surface discharges will occur. Due to the surface deterioration, the increasing number of droplets will lay on the specimen in strip during the moving process. The time between the discharge occurrences decreases. And the persistence time for a bunch of pulses in the waveforms is longer, while the magnitude is smaller.

Figure 5. Examples of acoustic waveforms of surface discharge on different aged specimens

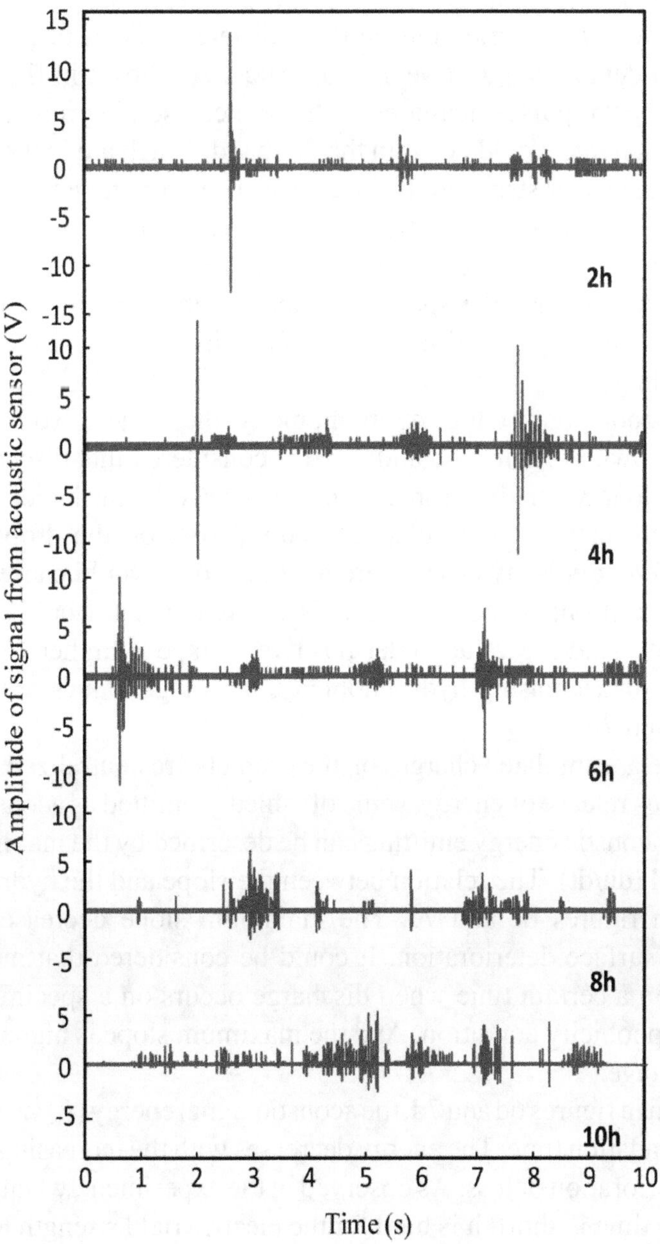

Figures 6 and 7 show the relationship between the statistical parameters and the corona degradation time under both positive and negative voltage. The study took every 10s signal recorder as a specimen to have the statistic results. The experiment was carried out on three different places of one specimen after corona depredated to have a mean value. As shown in figures 6a and 7a, the number of pulses increases with the increase of corona degradation time. The discharge would occur in the dry band, which are in the water strip or between the water strip and the electrode. The growing number indicated that the water move to the electrode more quickly and form the water strip with dry band easily. Compared with the negative voltage in figure 7a, the pulse number of the similar specimen under positive voltage in figure 6a is fewer. It can be conjectured that the surface discharge easily occur under negative voltage.

Under a good condition of hydrophobicity, the value of voltage range is higher, as shown in figures 6b and 7b. It is considered that discharge would emit sound loudly for the droplets on the surface being larger and laying further. There may be more charge accumulating on the droplets. When the two different polarity charges are neutralized, it would cause a massive shock. Then a stronger acoustic signal is emitted, which would be obtained by the sensor. And the value under positive voltage is higher. While under a condition of decreasing hydrophobicity, the range values are similar in figures 6b and 7b.

When the accumulated charges on the droplets are neutralized, there is an instantaneous release of energy, some of which is emitted as acoustic energy. The rate of acoustic energy emitting can be described by the maximum slope of the signal (du/dt). The relation between the slope and the hydrophobicity is shown in figures 6c and 7c. The maximum slope decreases with the increase of surface deterioration. It could be considered that more energy is released in a certain time when discharge occurs on a specimen under a good hydrophobicity condition. And the maximum slope is higher under the positive voltage.

As shown in figures 6d and 7d, the acoustic signal energy (E) varies with the corona degradation time. The energy decreases with the increasing process of surface deterioration or loss. As observed in the experiment, when the corona degradation time is short, it is hard for the electric field strength to reach the flashover strength. There would be more charge accumulated on droplets. Once the discharge occurs between the droplets, it would emit more energy. But when the corona degradation time is shorter than 4h, it is similar under both positive and negative voltage stress. Under a condition of decreasing

Figure 6. Relationship between the statistical parameters and the corona degradation time under positive voltage.(a) pulse number (b) range of acoustic signal (c) du/dt_{max} (d) energy of signal

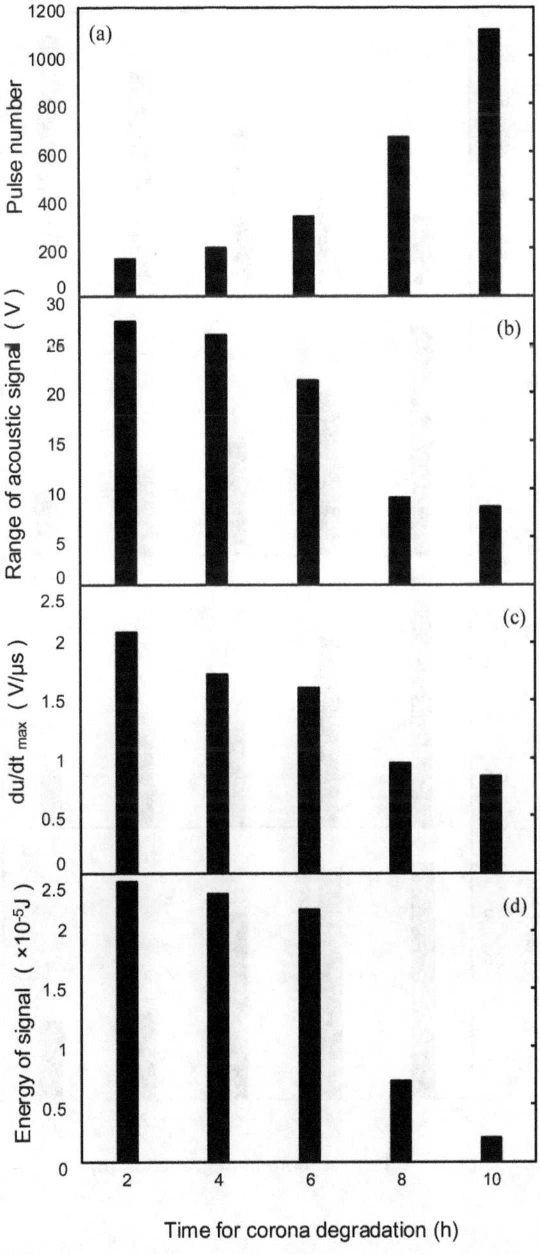

Figure 7. Relationship between the statistical parameters and the corona degradation time under negative voltage .(a) pulse number (b) range of acoustic signal (c)du/dt$_{max}$ (d) energy of signal

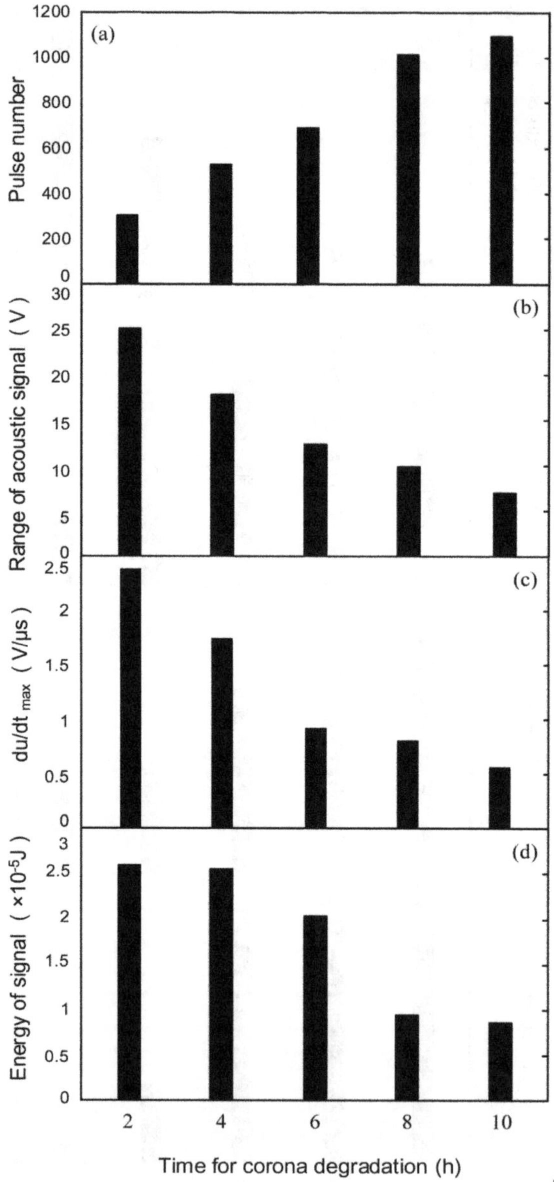

hydrophobicity, it is easy for discharge occurring. E is proportional to the emitted energy, so the obtained acoustic energy becomes smaller. Compared with the negative voltage in figure 7d, the value of signal energy in figure 6d is lower. The reason may be that the energy is influenced by the pulse number, the amplitude distribution and some other parameters. So there may be not a certain relationship between the emitted energy and the voltage polarity.

Figures 8 and 9 show the relationship between the statistical parameters and the corona degradation time under positive and negative voltage. Figures 8a and 9a show the Skewness of acoustic signal. Because the absolute value of Sk is related to the value and number of signal peaks, the larger absolute value of Sk means that the signal peak is fewer than a smaller Sk one. It can be clearly observed in figure 5 that the peak number of signal is larger, the corona degradation time is longer than 6h. Both positive and negative part of one signal are asymmetrical. So the Sk+ values from specimens at different ageing degrees are in a different range, compared with the Sk- values. As shown in figures 8a and 9a, the specimens at different ageing degrees can be separated.

The results of Kurtosis from the acoustic signal are shown in figures 8b and 9b. It shows the regular pattern of the value distribution of signal peaks. When the absolute value of Ku is low, the amplitudes of signal peaks in the signal are separated and the numbers of different amplitudes are similar, such as the signal from the specimen which has been aged for 2h in figure 5. As shown in figure 8b, some points of specimens at different ageing conditions are placed in an approximate area, but most of the points are separated according to the surface hydrophobicity. So Ku can be a parameter which is an inexact reference value for hydrophobicity evaluation.

The results of parameter (β) for the two - parameter Weibull distribution are shown in figures 8c and 9c. It can be observed that β value is larger than 1, which means the situation of ageing failure. With the ageing failure becomes more serious, the β value increases. Although the signal is asymmetrical, the parameters of both positive and negative sides increase with increase of surface deterioration. So in figures 8 and 9, the points present an ascending trend, which can well reflect the degradation degree of the acoustic signal in DDT method, which can be employed to evaluate the hydrophobicity.

Figure 8. Relationship between the statistical corona degradation time under positive voltage

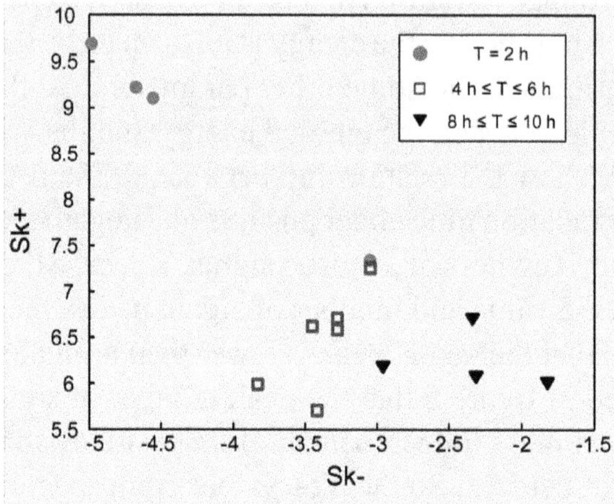

(a) Skewness

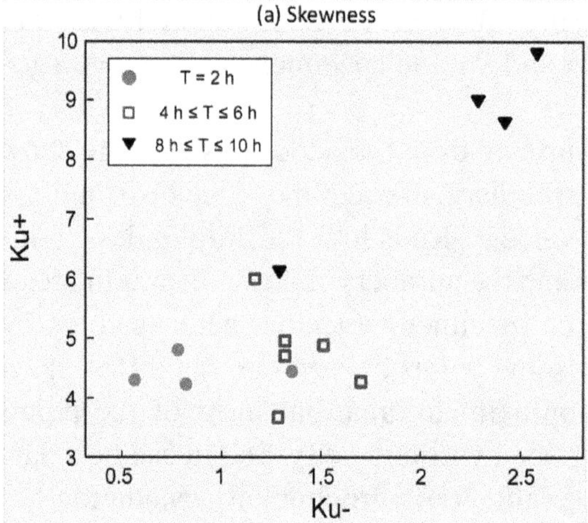

(b) Kurtosis

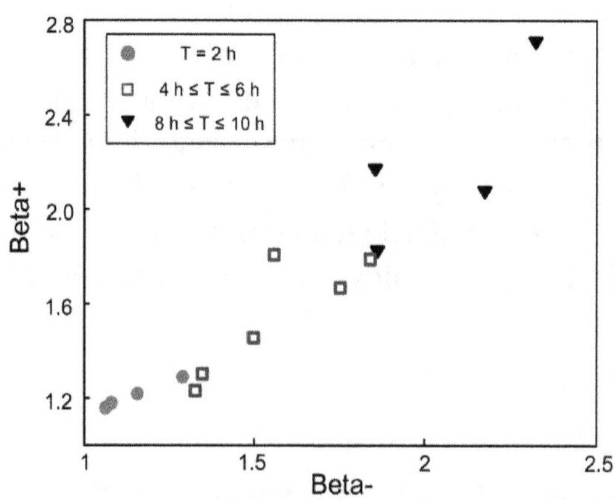

Figure 9. Relationship between the statistical parameters and the parameters and the corona degradation time undernegative voltage

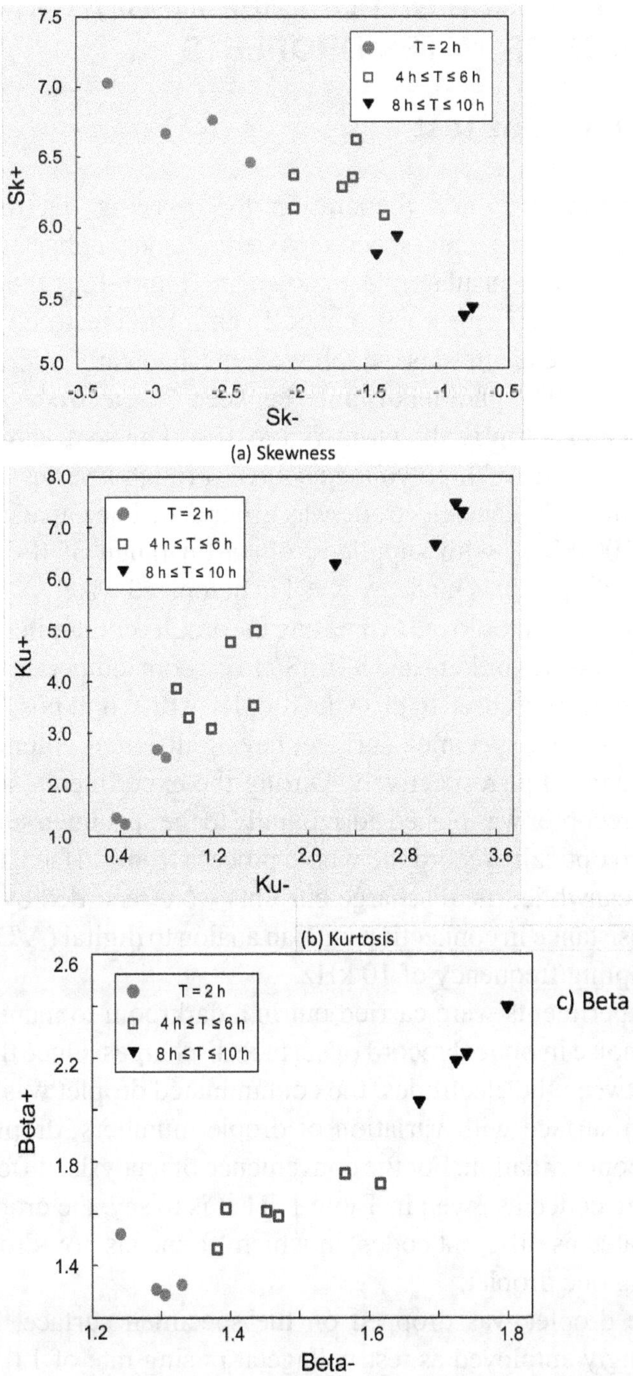

ENERGY EIGENVECTOR ANALYSIS OF SURFACE DISCHARGES FOR EVALUATING THE PERFORMANCE OF POLYMER INSULATOR IN PRESENCE OF WATER DROPLETS

The Preparation of Test

Figure 10 shows specimen preparation for carrying out the flashover experiments. The experimental specimens were silicone rubber slices removed from a new polymer insulator. As shown in figure 10a, the dimensions of specimens were 70 mm × 30 mm × 5 mm. The electrode pattern was semi-circle to plane electrodes, which were mechanically impressed on the specimen surface. The parallel distance between the electrodes was 40 mm. The experimental setup is shown in figure 10b. The semi-circle electrode was connected to an AC high voltage source through a series of protective resistance, while the plane electrode was grounded. The capacity of voltage source was 100 kV·A with supplied voltage in range of 0~100 kV. The discrete water droplet was made by NaCl in deionized water. A conductivity measuring meter was employed to measure the droplet contamination, making the conductivity of 1.0 mS/cm and 3.0 mS/cm at room temperature. And then, a micro-liter meter was used to drop the droplet at different position between the electrodes on the specimen surface, having different volumes of 10 μl, 20 μl, 30 μl and 40 μl respectively. During the experiments, a high-speed digital video recorder was placed horizontally to the specimen surface, which was applied to optically record the whole process from surface discharges to flashover. Meanwhile, the discharge currents were recorded by using a 1Ω measuring resistance in connection with an analog to digital (A/D) converter, with the sampling frequency of 10 kHz.

All the experiments were carried out in a darkroom to manually reduce background noise in optical record of surface discharges. Once the specimen was fixed between the electrodes, the contaminated droplet was dropped on the specimen surface with variation of droplet numbers, droplet location, volume and contamination. For the convenience of analysis, different droplet locations were coded as given in Table 1. That is to say, the droplet location can be separated as different codes, in which '0' means non-droplet and '1' means putting one droplet.

When the droplet was dropped on the specimen surface, AC voltage was increasingly employed as test voltage at raising rate of 1.0 kV/s. Once flashover occurred on the specimen surface, the applied voltage was recorded

Figure 10. Experimental arrangement. (a) Preparation of specimen (b) Experimental setup

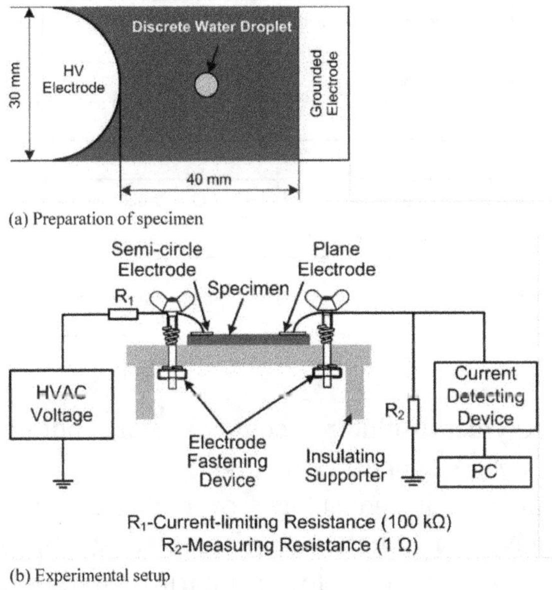

as flashover voltage. Both dynamic behaviors of water droplet and the induced discharge phenomena were captured by using the high-speed recorder. The phenomena and characteristics of surface discharges and surface flashover can be reflected by the optical images and discharge currents, which are the basis for further evaluation of insulator performance. For the purpose of determining reliability of experimental results, each test was repeated 5 times under the same experimental conditions. So the obtained results including flashover voltage and calculated characteristics of discharge current are the average value of these measurements.

Data Processing Method

It has been proved that time-frequency characteristics of leakage current on outdoor insulators can detail useful information of surface discharges (Chandrasekar et al., 2009; Liu et al., 2010; Du et al., 2009). The distortion of leakage current is associated with the activities (droplet deformation, corona discharge, dry-band discharge and flashover) on the insulator, which are in terms of different frequency properties. The extraction of these frequency components and analysis of their characteristics are useful for evaluating the

Table 1. Position codes of discrete water droplets on the specimen

Position Codes	Dropping Position of Discrete Water Droplets		
	Grounded Electrode Side	Middle between the Electrodes	HV Electrode Side
001	Non	Non	One droplet
010	Non	One droplet	Non
100	One droplet	Non	Non
011	Non	One droplet	One droplet
101	One droplet	Non	One droplet
110	One droplet	One droplet	Non
111	One droplet	One droplet	One droplet

performance of polymer insulator in contaminated conditions (Douar et al., 2010; Du et al., 2009; Meghnefi et al., 2007).

Wavelet transform, with advantages over Fourier transform for signal analysis with transient features, can provide substantial information for processing signals with property of localization in time and frequency domains (Du et al., 2005; Moula et al., 2013; Noskoski et al., 2013). It is considered as one of best solutions for extracting frequency characteristics and has been utilized as a practical scale for various signal processing. Wavelet analysis decomposes original signal into approximation and detail signals. The approximation signals correspond to low frequency components and detail signals correspond to high frequency components. Choosing different mother wavelet and number of decomposition level, different results will be obtained.

In this study, wavelet transform is applied to extract frequency components of discharge currents for further characteristic analysis. The chosen mother wavelet is Daubechies series mother of complex conjugate function with the formulation as follows:

$$W(a,b) = \frac{1}{\sqrt{a}} \int_{-\infty}^{+\infty} x(t) \cdot \Psi_g^* \left(\frac{t-b}{a} \right) dt \tag{5}$$

$$\Psi_g(t) = \exp\left(-\frac{t^2}{2} \right) \left\{ \exp\left[j\omega t \right] - \exp\left[-\frac{\omega^2}{2} \right] \right\} \tag{6}$$

where $W(a,b)$ is complex wavelet transform coefficient; $x(t)$ is time-domain leakage current; $\Psi^*_g(t)$ denotes complex conjugate of $\Psi_g(t)$, which is the chosen mother wavelet; ω is constant at 2π; a and b are respectively scale

Figure 11. Decomposition structure of wavelet transform

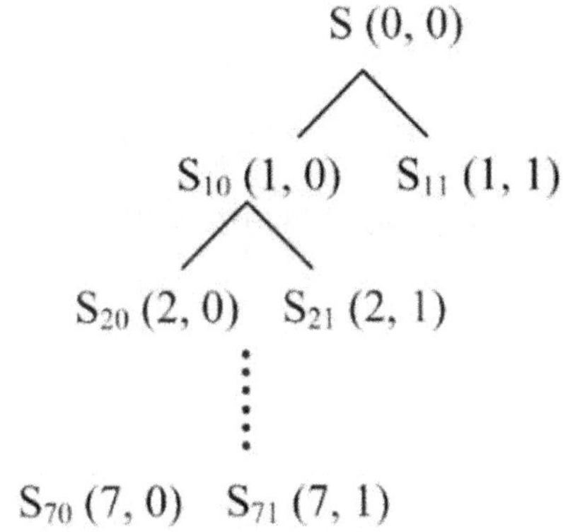

and translate. The modulus and phase of complex *W(a,b)* correspond to time-frequency information and localized amplitude regarding the current in cycle *a* vicinity in time *b*.

For extraction of discharge currents in this study, “Daubechies (db8)” was chosen as the mother wavelet and the decomposition was done up to 7 levels, with the structure shown in figure 11. According to sampling theorem and the applied sampling frequency, seven frequency components extracted from the original currents were selected for the analysis, where S_{70} shows 0-78 Hz components, S_{71} 78-156 Hz, S_{61} 156-312 Hz, S_{51} 312-625 Hz, S_{41} 625-1250 Hz, S_{31} 1.25-2.5 kHz and S_{21} 2.5-5 kHz.

Based on observation in the experiments, it is found that surface discharges are associated with variation of droplet parameters. The variation of droplet numbers, droplet location, volume and contamination causes different tokens in frequency components of discharge currents, such as the discharge energy. Therefore, energy eigenvector analysis was conducted on different frequency components of discharge currents to reveal effects of water droplet on the performance of polymer insulator. Taking energy calculation of S_{70} for example, by defining E_{70} is energy of S_{70}, then it can be obtained as follows.

$$E_{70} = \int | S_{70}(t) |^2 \, dt = \sum_{m=1}^{n} | X_m |^2 \tag{7}$$

where X_m (m=1, 2, …, n) is discrete digital current values of S_{70}.

Then, the energy eigenvector *T* can be obtained:

$$T = [E_{70}, E_{71}, E_{61}, E_{51}, E_{41}, E_{31}, E_{21}] \tag{8}$$

For the comparison of different discharge components, normalization processing of T was carried out to calculate the base value E.

$$E = (\sum_{j=0}^{7} | E_j |^2)^{1/2} \tag{9}$$

Finally, normalized energy eigenvector C of surface discharges can be obtained.

$$\begin{aligned} C &= [E_{70}, E_{71}, E_{61}, E_{51}, E_{41}, E_{31}, E_{21}, E_{37}] / E \\ &= [C_{70}, C_{71}, C_{61}, C_{51}, C_{41}, C_{31}, C_{21}] \end{aligned} \tag{10}$$

Process and Characteristics of Surface Discharges to Flashover

By raising the test voltage on the specimen, electric field around HV electrode and between the electrodes can be enhanced. Once the electric field strength is higher enough, both field emission and air ionization will occur to cause surface discharges. With the increasing strength and propagation of surface discharges, the discharges will pass across the droplet to connect the electrodes, which stands for the occurrence of surface flashover.

It has been investigated that strength and distribution characteristics of surface discharges are proportional to intensity and statistics of discharge luminance. Figure 12 shows one typical discharge phenomena during the flashover, which was continuously captured by the high-speed camera in the dark room. Different stages in the flashover process induced by a single discrete water droplet under the condition of 100, 3.0 mS/cm, 20 μL is taken as an example for analyzing the process of surface discharges to flashover.

As shown in figure 12a, at the initial stage, it can be observed that corona discharges firstly occurred at the tip of semi-circle electrode, which shows most brightness light emission around HV electrode in the image. Along the direction of HV electrode to grounded electrode, the filament brightness emitted by corona discharges can also be observed except the surrounded droplet. The filaments are conductive regions to reduce distance between the electrodes, which leads to the increase in electric field strength. With the increase in applied voltage and intensification of electric field, propagation of corona discharges can be enhanced, which emit brighter luminance between the electrodes at the middle stage, as shown in figure 12b. As resistance of water droplet is much lower than that of insulator surface, corona discharges

Figure 12. Process of surface discharges induced by single discrete droplet

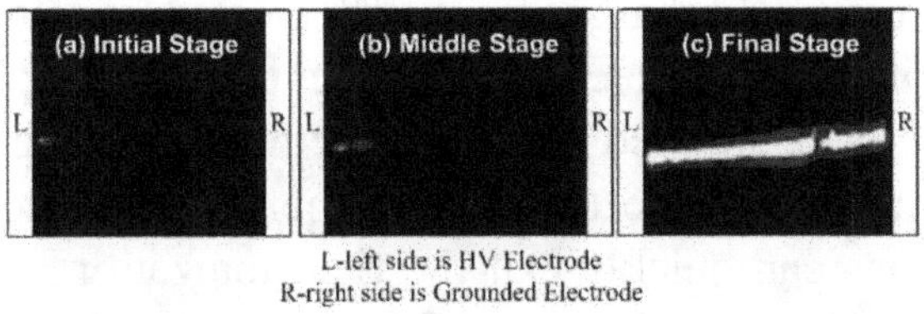

occurred around the tips of droplet in the direction of applied electric field. At this time, both the increasing corona discharge and the conductive water droplet existed on the specimen. The conductive droplet surface can provide a path for arcing discharges. Finally, when the applied voltage was higher than the arcing ignition value, arcing discharges can be observed to travel across the droplet to connect the electrodes, which forms a brightest discharge channel emitted by surface flashover, as shown in figure 12c.

Effects of Droplet on Surface Discharges

Single Droplet

Figures 13 to 15 give effects of location, volume and contamination of a single droplet on characteristics of discharge current. Due to the increasing application of test voltage at a constant speed, it is found that there is almost no deformation on the droplet based on observation of discharge phenomena during the flashover. Therefore, variation and characteristics of discharge current are related to droplet parameters.

As shown in figure 13, different dropping position between the electrodes causes different characteristics of surface discharges, in which relationship among maximum value of discharge current, cumulative charge and droplet location is shown in figure 13a, and figure 13b shows energy eigenvector of different frequency components of discharge current in relation with droplet location. The cumulative charge is obtained by integration of discharge current with lapse time, which can also be considered as accumulation of discharge current to reveal the quantity of charge contained in the discharges. With location of a single droplet from 100, 010 to 001, both maximum value of discharge current and cumulative charge show the decreasing tendency. It indicates that the droplet near HV electrode can make easy occurrence of surface flashover, while the droplet at grounded location needs more energy to induce the flashover.

Such variation is due to application of AC electric field, which makes propagation of surface discharges from both electrodes to droplet position. It is known that mobility of electrons is higher than that of positive charges. When the droplet was dropped near HV electrode, positive charges can be rapidly accumulated to make the enhancement of electric field, and the generated electrons can quickly exit the field ionization zone towards the anode, which will lead to development of discharge channel across the droplet to cause surface flashover. Therefore, different energy of discharge current can be found with variation of droplet location, as shown in figure 13b. From energy eigenvector of different frequency components, it is found that the low frequency component (C70) dominates discharge current under all the investigated droplet location. However, discharge energy under the condition of 100 shows the highest value at the components from C71 to C21 than those under the other locations. It means that although surface discharge shows higher value in maximum discharge current and cumulative charge, surface discharge are relatively weak and the main frequency component is 0-78 Hz, which is the conductive properties induced by movement of charges along the specimen surface.

Figure 13. Location effect of a single droplet on discharge current the condition of 1.0 mS/cm, 40 μL. (a) maximum value of discharge current and cumulative charge and (b) energy eigenvector under

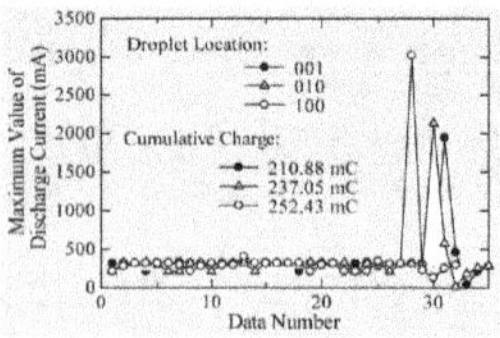

(a) Maximum value of discharge current and cumulative charge (top)

(b) Energy eigenvector (right)

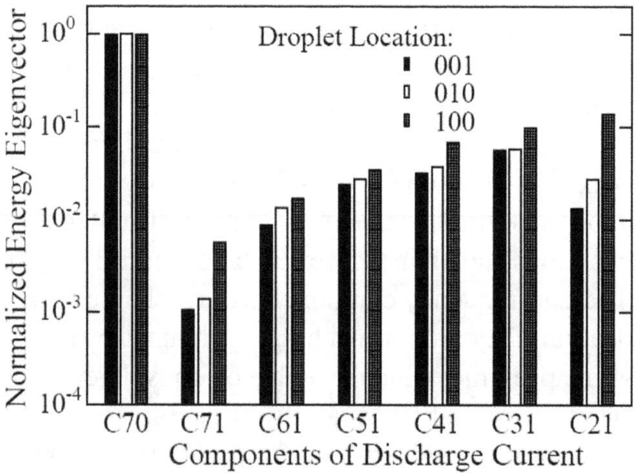

Figure 14 shows the volume effect of a single droplet at position 001 on discharge current, in which relationship among maximum value of discharge current, cumulative charge and droplet volume is shown in figure 14a, and figure 14b shows energy eigenvector of different frequency components in relation with droplet volume. With the droplet volume increasing from 10 μl to 30 μl, maximum value of discharge current and cumulative charge show a little increasing tendency, but show an obvious decrease from 30 μl to 40

Figure 14. Volume effect of a single droplet on discharge current under the condition of 001, 3.0 mS/cm. (a) maximum value of discharge current and cumulative charge and (b) energy eigenvector

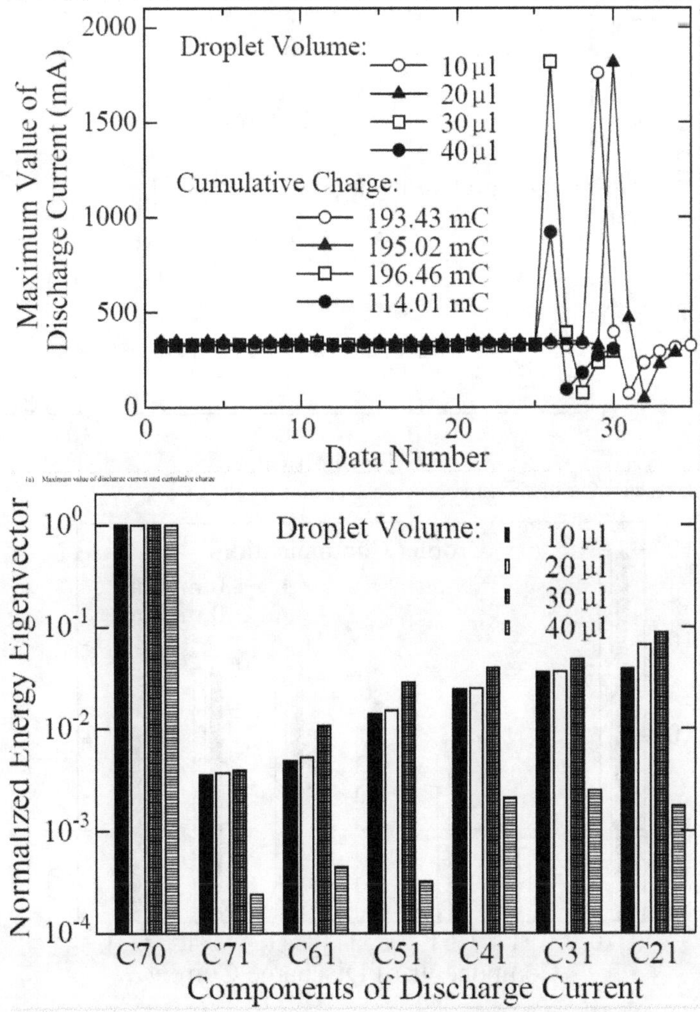

μl. It indicates that the increase in droplet volume can inhibit formation of discharge channel on the specimen surface, but the inhibition effect is to a little extent and also to the droplet volume. From energy eigenvector of different frequency components, it is found that discharge energy at C70 component is also almost the same under the investigated droplet volumes, but the other components (C71 to C21) under the condition of 40 μl show the

Figure 15. Contamination effect of single droplet on discharge current under the condition of 001, 40 μl. (a) maximum value of discharge current and cumulative charge and (b) energy eigenvector

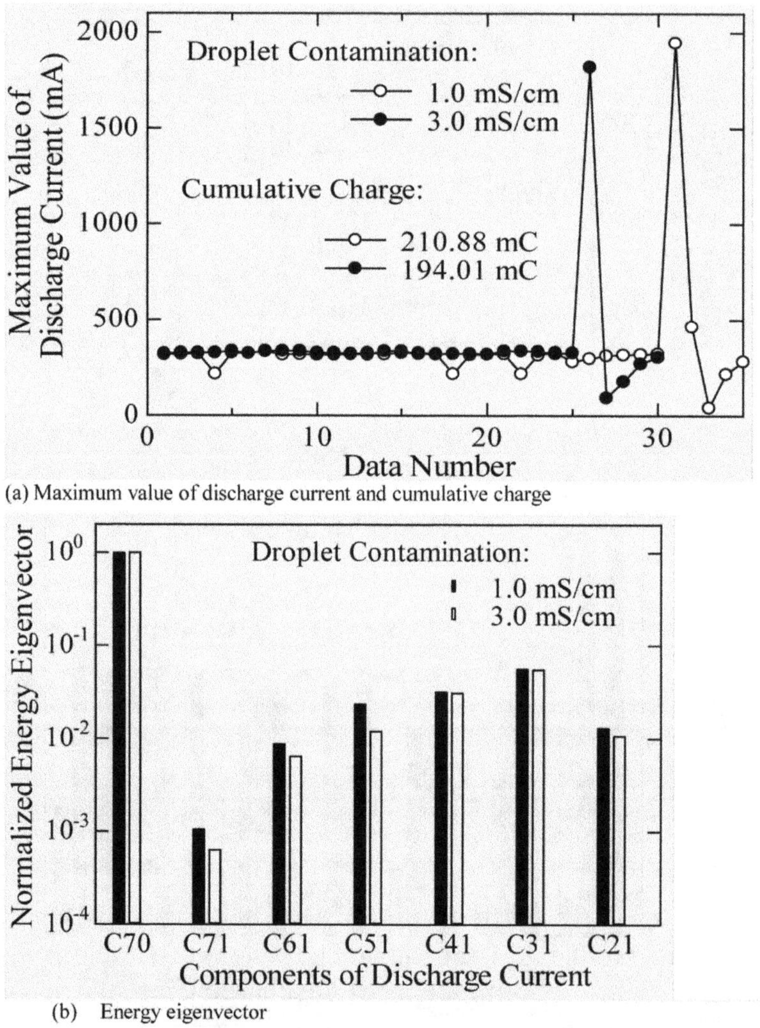

lowest discharge energy. With increasing the droplet volume from 10 μl to 30 μl, the discharge energy shows a little increasing tendency at the frequency components of C71 to C21.

Figure 15 shows the contamination effect of a single droplet at position 001 on discharge current, in which relationship among maximum value of discharge current, cumulative charge and droplet contamination is shown in figure 15a, and figure 15b shows energy eigenvector of different frequency components in relation with droplet contamination. It is found that both maximum value of discharge current and cumulative charge under the condition of 1.0 mS/cm show a little higher value than those under 3.0 mS/cm. It indicates that surface resistance between the droplet and the electrode is the same with the investigated droplet contamination, while only the resistivity of droplet has a little difference to restrict discharges across the droplet. Therefore, strength and distribution of surface discharges show a little decrease with increasing the droplet contamination.

From energy eigenvector of different frequency components, the discharge energy is almost the same at the low frequency component (C70) under the investigated droplet contamination. The discharge energy at other frequency components (C71 to C21) under 1.0 mS/cm is a little higher than that under 3.0 mS/cm.

Associated with characteristics of discharge currents, table 2 gives relationship between flashover voltage and parameters of a single droplet. It is found that with the droplet location from 001, 010 to 100, the flashover voltage show the increasing tendency, which has the apparently higher value at position 100. For variation of droplet volume, the flashover voltage increases

Table 2. Effects of a single droplet on flashover voltage

Droplet Contamination	Droplet Volume	Flashover Voltage (kV)		
		001	010	100
1.0 mS/cm	10 μL	22.1	22.9	24.5
	20 μL	22.7	23.5	25.1
	30 μL	23.5	23.8	25.7
	40 μL	19.4	21.6	22.7
3.0 mS/cm	10 μL	20.2	21.5	23.8
	20 μL	20.9	22.1	24.7
	30 μL	21.4	22.6	25.2
	40 μL	18.9	19.7	21.5

from 10 μL to 30 μL, but has the lowest value at 40 μL. With increasing the droplet contamination, the flashover voltage shows the decreasing tendency.

Double Droplets

Figure 16 shows maximum value of discharge current, cumulative charge and energy eigenvector of different frequency components in relation with location of double droplets under the condition of 1.0 mS/cm, 40 μL. It is found that maximum value of discharge current and cumulative charge show the decreasing tendency with the droplet location from 110, 011 to 101. It is considered that double droplets at position 110 can enhance inhibition of positive charges to pass across the specimen surface and more energy is needed to make occurrence of surface discharges across the droplets. From analysis of energy eigenvector, discharge energy at the low frequency component is almost the same under the investigated droplet location. And the discharge energy decreases with the droplet location from 110, 011 to 101.

Figure 17 shows maximum value of discharge current, cumulative charge and energy eigenvector of different frequency components in relation with volume of double droplets at position 110. Maximum value of discharge current and cumulative charge show a little increasing tendency with volume of double droplets from 10 to 30 μl, but show a little decreasing tendency from 30 to 40 μl. From energy eigenvector of different frequency components, it is found that the discharge energy at C70 component is also almost the same under the investigated volumes. With increasing the volume from 10 to 30 μl, the discharge energy at other components (C71 to C21) shows a little increasing tendency, but shows a little decreasing tendency with the volume from 30 to 40 μl.

Figure 18 shows the contamination effect of double droplets at position 101 on discharge current, in which relationship among maximum value of discharge current, cumulative charge and droplet contamination is shown in figure 18a, and figure 18b shows energy eigenvector of different frequency components in relation with the droplet contamination. It is found that maximum value of discharge current and cumulative charge under the condition of 1.0 mS/cm show a little higher value than those under the condition of 3.0 mS/cm. From energy eigenvector of different frequency components, the discharge energy is almost the same at C70 component under the investigated contamination. The discharge energy at other frequency components (C71 to C21) under the condition of 1.0 mS/cm is a little higher than that under 3.0 mS/cm.

Figure 16. Location effect of double droplets on discharge current under the condition of 1.0 mS/cm, 40 μL. (a) maximum value of discharge current and cumulative charge and (b) energy eigenvector

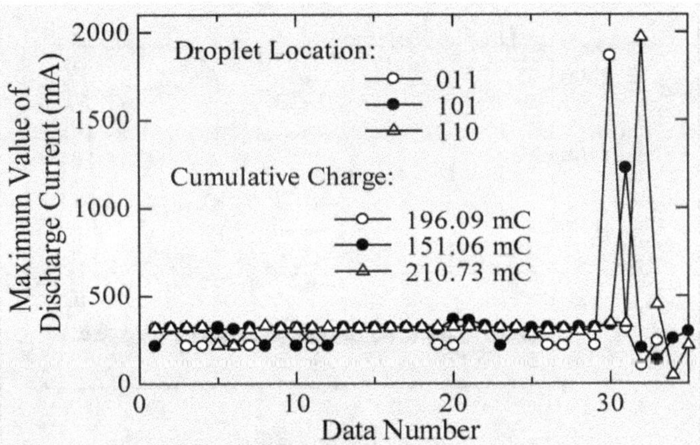

(a) Maximum value of discharge current and cumulative charge

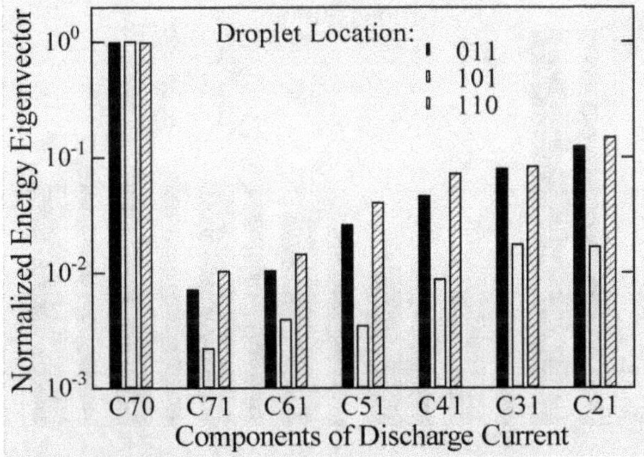

(b) Energy eigenvector

Table 3 gives the flashover voltages in relation with parameters of double droplets. The flashover voltage shows the lowest value at position 101, while it is a little higher at 110 than that at 011. With variation of droplet volume, the flashover voltage shows the increasing tendency with increasing the volume from 10 to 30 μL, but shows the lowest value under the condition of 40 μL. Meanwhile, the increase of droplet contamination from 1.0 to 3.0 mS/cm causes a little decrease in the flashover voltage.

Figure 17. Volume effect of double droplets on discharge current under the condition of 110, 1.0 mS/cm. (a) maximum value of discharge current and cumulative charge and (b) energy eigenvector

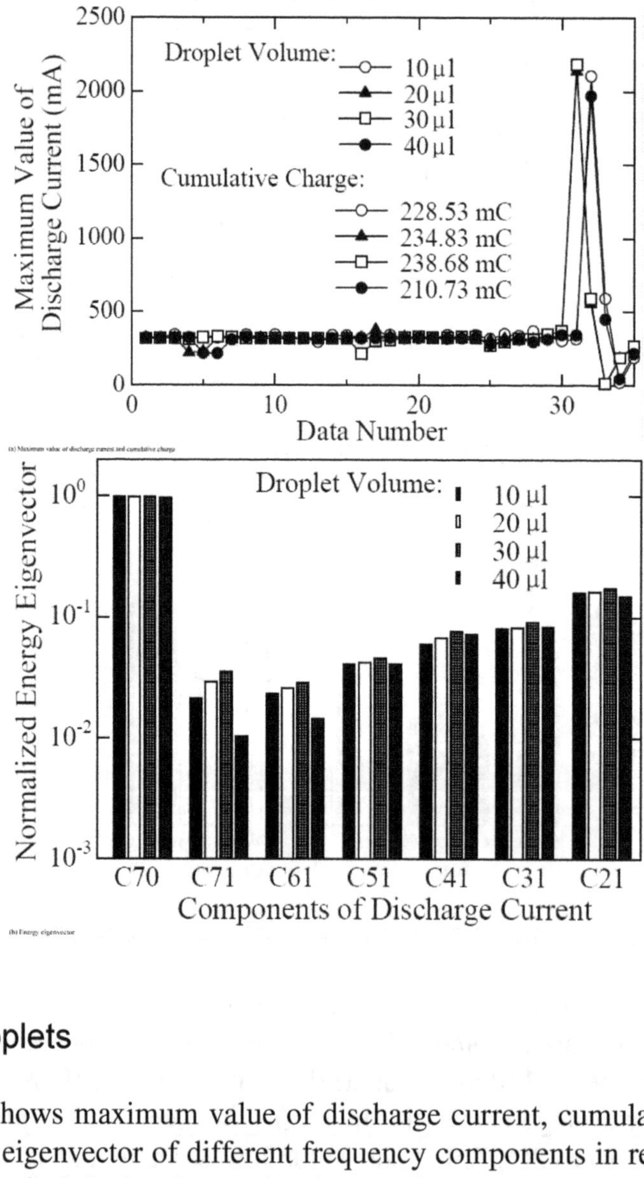

Triple Droplets

Figure 19 shows maximum value of discharge current, cumulative charge and energy eigenvector of different frequency components in relation with the volume of triple droplets under the condition of 111, 1.0 mS/cm. It is found that maximum value of discharge current and cumulative charge show the decreasing tendency with increasing the droplet volume, as shown in figure 19a. From analysis of energy eigenvector, the discharge energy at the low frequency component is almost the same under the investigated droplet

Figure 18. Contamination effect of double droplets on discharge current under the condition of 101, 40 μl. (a) maximum value of discharge current and cumulative charge and (b) energy eigenvector

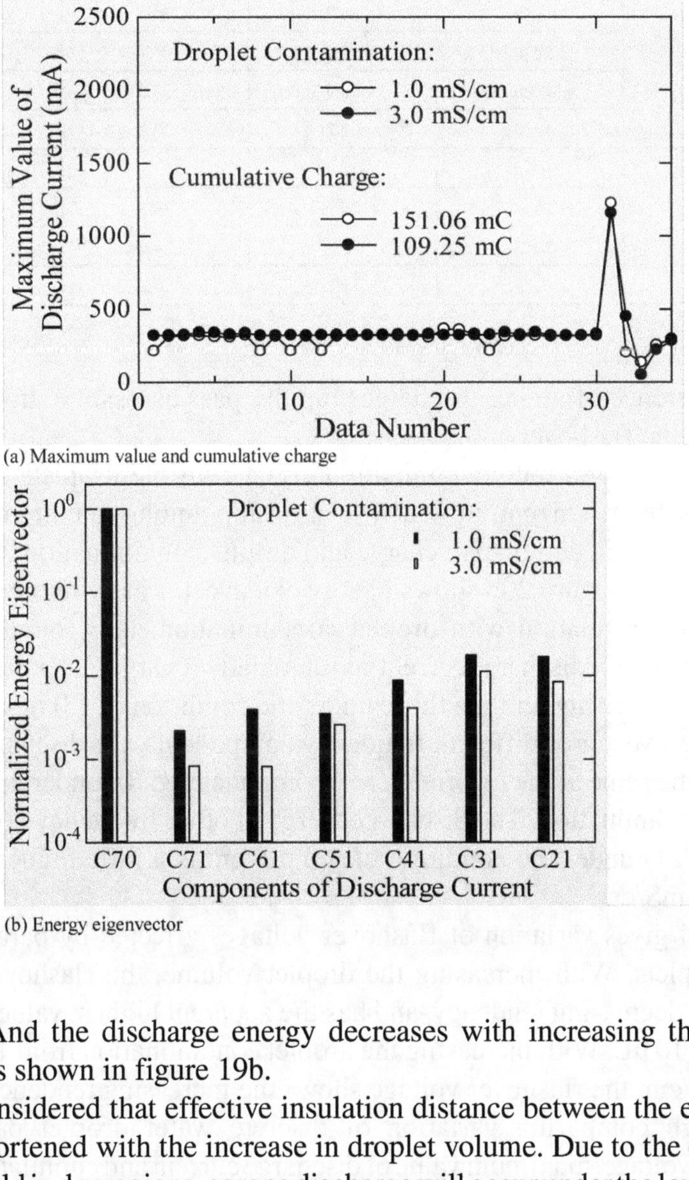

volume. And the discharge energy decreases with increasing the droplet volume, as shown in figure 19b.

It is considered that effective insulation distance between the electrodes will be shortened with the increase in droplet volume. Due to the enhanced electric field in dry regions, corona discharge will occur under the low voltage. However, when the droplet volume is big enough, effects of increasing droplet volume on the decrease in surface tension will be prominent. Once the electrostatic forces caused by applied electric field is higher than the surface tension, deformation of droplet will occur, which may be sufficient to induce

Table 3. Effects of double droplets on flashover voltage

Droplet Contamination	Droplet Volume	Flashover Voltage (kV)		
		011	101	110
1.0 mS/cm	10 μL	26.2	24.5	26.9
	20 μL	26.6	24.7	27.5
	30 μL	26.9	25.2	27.8
	40 μL	22.3	20.7	23.9
3.0 mS/cm	10 μL	25.8	23.7	26.4
	20 μL	26.1	24.2	27.3
	30 μL	26.7	24.6	27.4
	40 μL	21.2	20.2	23.6

easy occurrence of arcing discharges rapidly pass across the droplets, thus triggering the flashover.

Figure 20 shows the contamination effect of triple droplets at position 111 on discharge current, in which relationship among maximum value of discharge current, cumulative charge and droplet contamination is shown in figure 20a, and figure 20b shows energy eigenvector of different frequency components in relation with droplet contamination. It is found that both maximum value of discharge current and cumulative charge under the condition of 3.0 mS/cm are higher than those under the condition of 1.0 mS/cm. From energy eigenvector of different frequency components, the discharge energy is almost the same at the low frequency component (C70) under investigated droplet contamination. The dischargeenergy at other frequency components (C71 to C21) under the condition of 3.0 mS/cm is a little higher than that under 1.0 mS/cm.

Table 4 gives variation of flashover voltages affected by parameters of triple droplets. With increasing the droplet volume, the flashover voltage shows the decreasing tendency, and has the apparent highest value when the volume is 10 μL. With increasing the droplet contamination from 1.0 mS/cm to 3.0 mS/cm, the flashover voltage shows the increasing tendency.

Through comparing variation of discrete water droplet parameters, flashover voltage, maximum value of discharge current and cumulative charge show the highest value when triple droplets are dropped on the specimen at position 111. This is due to that the presence of three droplets can cause the most triple connection of droplet, specimen surface and air, the uniformity of electric field between the electrodes. The distortion of electric field can

Figure 19. Effect of droplet volume on discharge current under the condition of 111, 1.0 mS/cm. (a) maximum value of discharge current and cumulative charge and (b) energy eigenvector

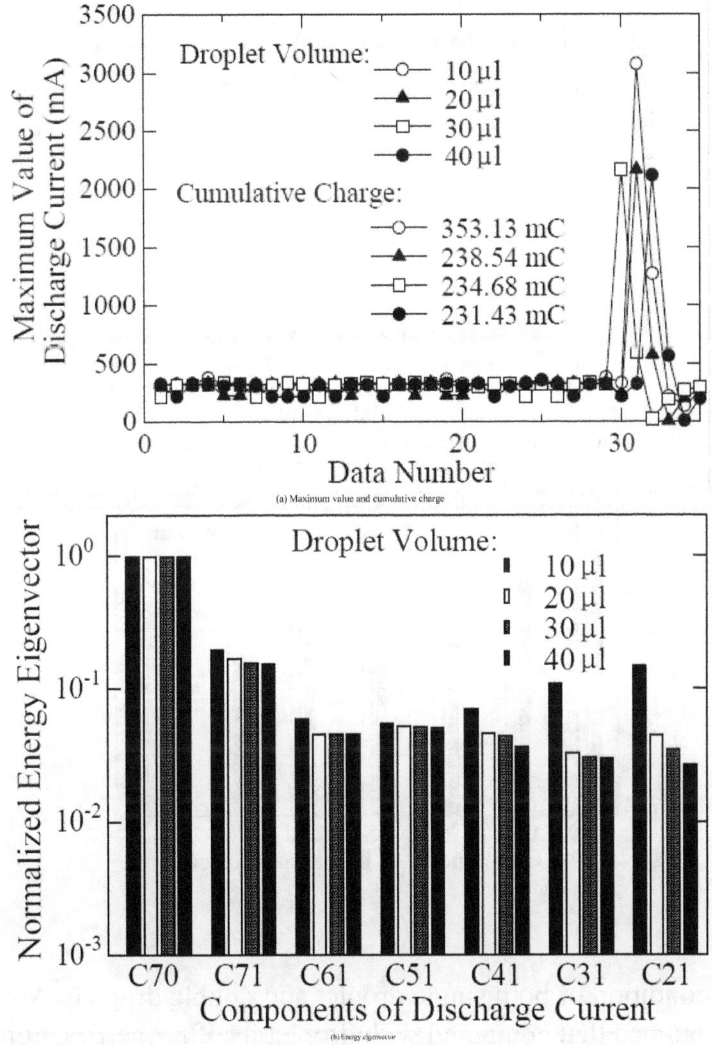

be enhanced to generate more free charges from the ionization of surrounded air. Once the applied voltage reaches a critical value to make the charges pass across the droplets, the ends of each arcing discharge will connect together to form surface flashover. Therefore, effects of droplet volume and droplet contamination on discharge characteristics are not the same with those effects

Figure 20. Effect of droplet contamination on discharge current under the condition of 111, 30 μl. (a) maximum value of discharge current and cumulative charge and (b) energy eigenvector

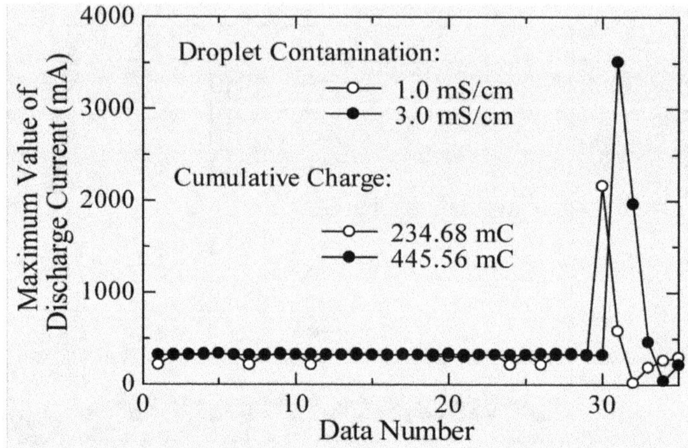

(a) Maximum value and cumulative charge

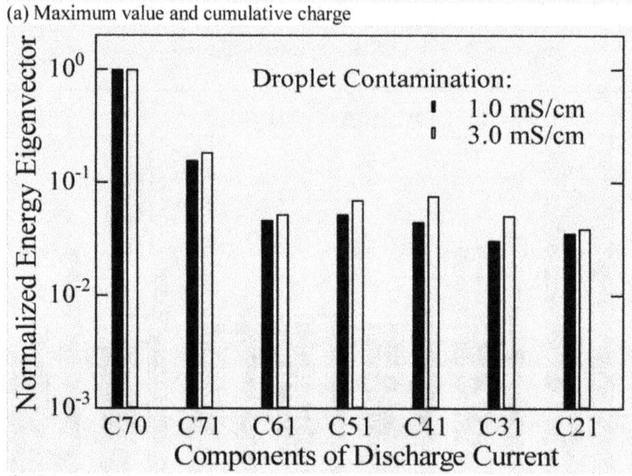

(b) Energy eigenvector

under the condition of both single droplet and double droplets. Meanwhile, it can be obtained that compared with droplet itself properties, number and location of droplets have most significant influence on surface discharges, even the flashover.

Table 4. Effects of triple droplets on flashover voltage

Droplet Contamination	Droplet Volume	Flashover Voltage (kV)
		111
1.0 mS/cm	10 μL	28.1
	20 μL	26.8
	30 μL	26.5
	40 μL	26.1
3.0 mS/cm	10 μL	29.8
	20 μL	27.4
	30 μL	27.1
	40 μL	26.7

EFFECTS OF WIND CONDITION ON HYDROPHOBICITY BEHAVIOR OF SILICONE RUBBER IN CORONA DISCHARGE ENVIRONMENT

The Preparation of Test

The test setup is shown in figure 21. The specimen was treated with corona in a wind tunnel and the wind was generated by an air compressor which was carefully adjusted to ensure the uniform distribution of air velocity in the wind tunnel. The wind speed V was controlled from 20 to 80 m/s in this test. A pair of needle-plate electrode was employed. The needle electrode, with 1 mm in diameter and a tip curvature radius of 0.1 mm, was connected to the high voltage positive dc power through a series of protective resistors, while the back electrode made of copper was grounded. The specimen was placed on the copper plate, and the distance between the grounded specimen surface and the tip of needle electrode was fixed to 5 mm. The corresponding discharge currents of corona discharge were monitored through a sampling resistance of 7 kΩ in connection with an analog-to-digital (A/D) converter. The sampling frequency of the converter was 10 MHz. An anemometer was employed to measure the wind speed. DC positive voltages of 7, 8.5, 10 kV were applied to the needle electrode for different times of 5, 15 and 30 min, respectively. The specimen and electrodes were rotated to make sure that the wind direction was parallel or vertical to the specimen during different tests. When the wind blew in the direction vertical to the specimen, air turbulence formed at the needle tip in a small zone as shown in figure 22 (Abdel-Salam et

Figure 21. Schematic diagram of test system

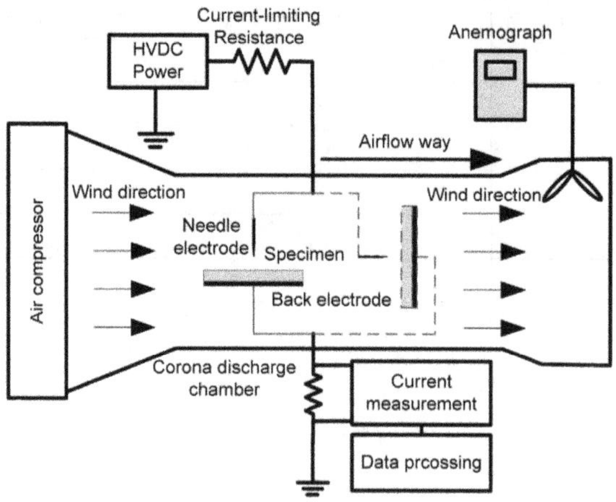

Figure 22. Stream lines of wind around the tip of the needle electrode

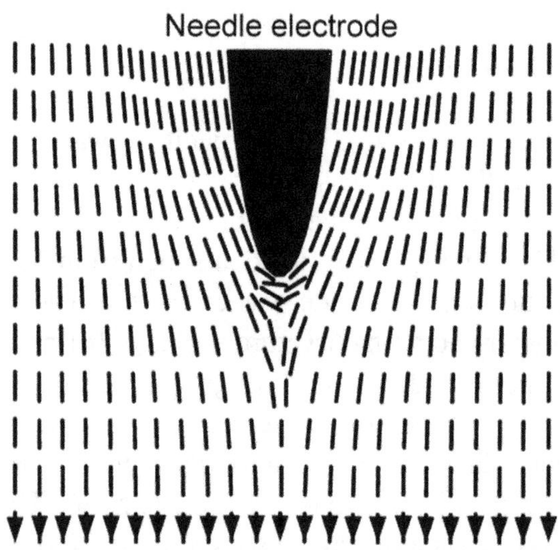

al., 1987). The air turbulence becomes weak in the space away from the needle tip. Considering the large gap between the needle electrode and specimen surface and the small needle tip curvature radius, the effect of air turbulence on the corona discharge appearance is negligible. The corona exposure test in the absence of wind was also conducted for comparison. A high-speed camera was fixed rightly in front of the specimen to record discharge during the test. The corona exposure test was conducted in a darkroom to get rid of the interference of background light.

In this section, according to the direction of wind parallel or vertical to the specimen during the corona treatment, specimens are called specimen under parallel wind (SPW), specimen under vertical wind (SVW) and specimen without wind (SWW), respectively.

RP Approach of Discharge Current

Recurrent plot (RP) is a useful graphical tool which can extract information from discharge current waveforms (Du et al., 2009; Liu et al., 2010). The non-linear behavior of dynamical systems can be visualized on the RP map, which composes of rectangular block structures with higher density of points. The RP method can be mathematically computed as shown in (11).

$$R_{i,j} = H(\varepsilon - \|x_i - x\|), i, j = (1, \ldots, N) \quad (11)$$

where ε represents the predefined threshold; $H(x)$ is the Heaviside function and x is the vector reconstructed from the studied discharge current; N is the total number of x_i. According to equation (11), a dot is placed at (i,j) in the RP map if the distance between x_i and x_j is less than ε, else a blank is placed.

Figure 23 shows an example of the RP maps and RP indicators of typical waveforms including signals of Sine, Rossler and Random. The RP maps of the typical waveforms are shown in figure 23(a). It is found that the recurrent points for the Sine signal is the most regular, followed by the Rossler signal, while the Random signal coming in the last. The decreasing tendency of regularity in RP maps indicates that the waveforms are becoming more complex and more unstable.

In order to characterize the RP maps quantitatively, two fundamental indicators of entropy (ENTR) and determinism (DET) are obtained according to the following equation:

1. DET is the ratio of recurrence points om the diagonal structures to all recurrence points. The value of DET can be obtained accorfing to equation (12), which reflects the predictability of the studied system.

$$DET = \frac{\sum_{l=l_{min}}^{N} lP(l)}{\sum_{l=1}^{N} P(l)} \quad (12)$$

where $P(l)$ is the frequency distribution of the lines with the length of l which are parallel to the diagonal line in the RP map.

2. ENTR denotes the Shannon entropy of the probability $p(l)$ to find a line with a length of l which is parallel to the diagonal in the RP map. ENTR can be calculated according to equations (13) and (14), which is related to the complexity of the RP concernig the diagonal lines.

$$p(l) = P(l) / \sum_{l=l_{min}}^{N} P(l) \tag{13}$$

$$ENTR = -\sum_{l=l_{min}}^{N} p(l) \ln p(l) \tag{14}$$

Figure 23 (b) shows RP indicators of the typical waveforms. Both ENTR and DET show an decreasing tendency. The varition of ENTR indicates the waveforms become non-periodical. The DET value of Sine is high suggesting the waveform is determinable and periodic. The decreasing DET value indicates that the waveforms become unstably. In this section, the ENTR and DET of discharge current are used to reveal the non-linear characteristics under different wind conditions.

Effects of Wind Direction and Speed on Corona Discharge

During the test, corona discharge with blue light was observed around the tip of the needle and the specimen. Figure 24 shows typical corona discharge light under different wind conditions with the applied voltage U of 10 kV. The wind speed V is fixed to 80 m/s. Compared to SWW and SVW, corona discharge of SPW is the weakest and the discharge on the left edge of specimen is almost distinguished as shown in figure 24(a). However, corona discharge of SVW is the strongest and a discharge passage can be observed in the gap between the needle electrode and specimen surface as shown in figure 24(c). It can be seen that discharge characteristics in high-velocity airflow are greatly different from that in stationary air.

The discharge with blue light that can be observed is a bunch of plasma, which is easy to be bent or distorted by external force. In the case of SPW, a part of ions are blown away which results in the extinguishment of corona discharge, so the discharge intensity is relatively weak. In the case of SVW, because the direction of wind and the electric field are nearly the same, the mobility of ions moving forward to the surface of specimen is accelerated by airflow. Besides, the density of positive charges near the tip of needle is

Figure 23. (a)The RP maps and (b)RP indicators of typical waveforms

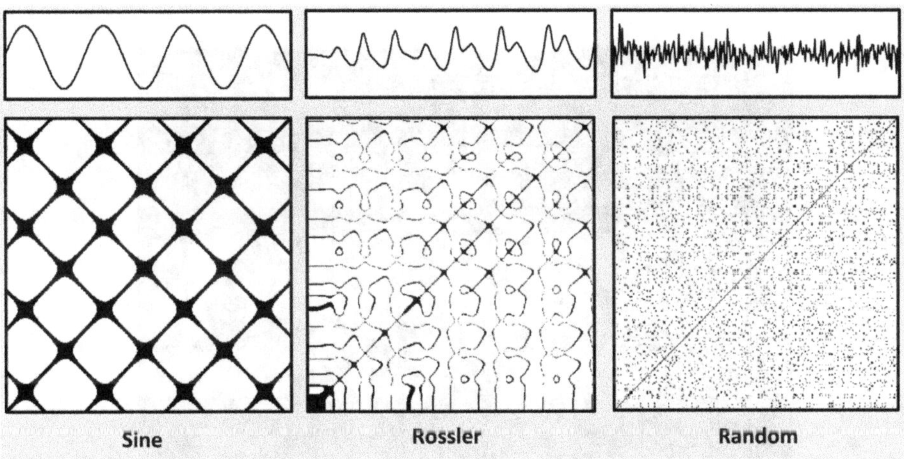

(a) RP maps of typical waveforms

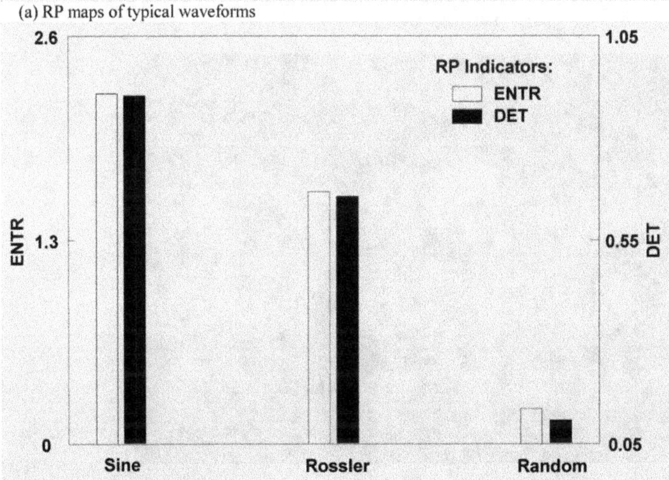

(b) RP indicators of typical waveforms

reduced due to the remove of space charges by wind, therefore, the field around the needle tip is enhanced and the inhibitory effect on corona discharge is eased up. So the discharge intensity of SVW is the highest.

Corona current can be observed when the positive dc voltage is applied to the needle electrode. This is due to the convection of ions from the needle electrode to ground electrode (Waters et al., 1973). In the absence of wind, the axial ion drift velocity can be calculated (Sigmond, 1986).

$$v = \mu V / d \quad (15)$$

Figure 24. An example of corona discharge images of different specimens. (a) SPW (b) SWW (c) SVW

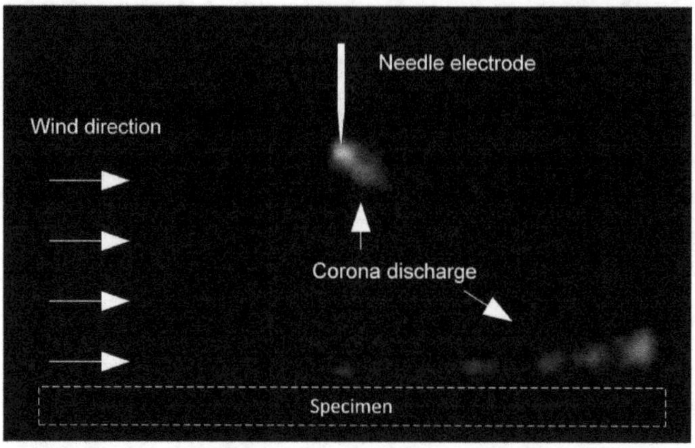

(a) SPW, U=10 kV, V=80 m/s

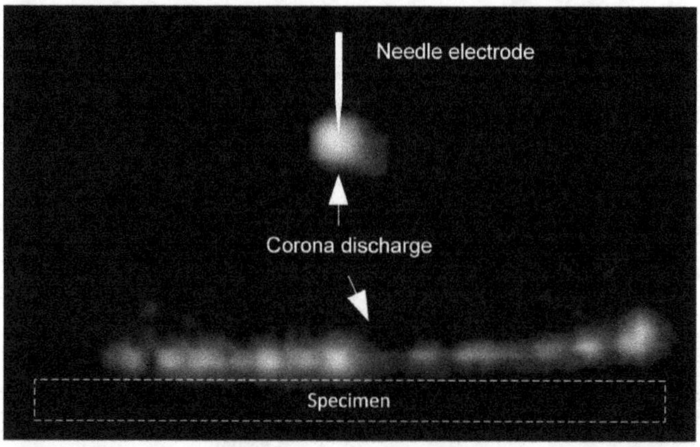

(b) SWW, U=10 kV, V=0 m/s

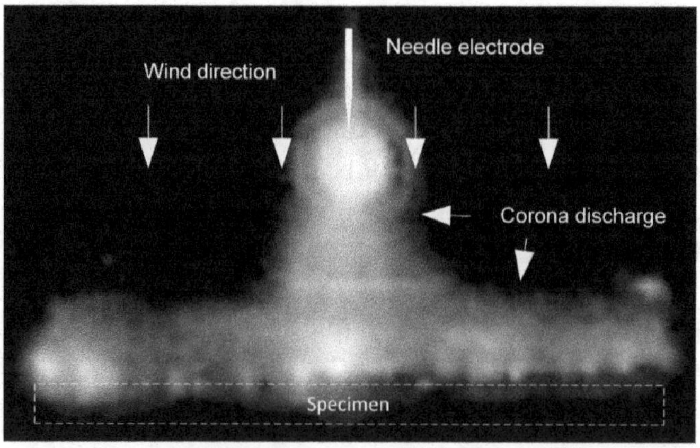

(c) SVW, U=10 kV, V=80 m/s

where μ is the mobility of atmospheric positive ions, which is about 1.5×10^{-4} m^2/Vs; V is the applied voltage; d is the gap distance. For example, in figure 24(b), where V is 10 kV and d is 5 mm, the axial ion drift velocity v is about 300 m/s.. When high-velocity airflow in the range of 20~80 m/s is introduced, ions filling the gap between the tip of needle electrode and the specimen surface will be redistributed, as a result, the corona current characteristics may change.

Figure 25 shows typical leakage current waveforms under different conditions and the applied voltage is 10 kV. Current pulses can be observed when corona occurs. Compared to those of SWW, the magnitude and pulse repetition of SVW increases a lot, while the fewest pulse peaks can be observed in the case of SPW. The introduction of wind can influence the drift of ions, therefore, the properties of corona current also change.

The corona current under different wind conditions were analyzed. Figure 26(a) shows the relationship between the pulse number and wind speed during the corona test. It can be seen obviously that the number of pulses is in this order: SPW< SWW< SVW. SPW with higher wind speed shows lower pulse number, while SVW shows an opposite tendency. The time-averaged value of corona current is shown in figure 26(b). With the applied voltage of 10 kV, the average corona current of SWW is 518.4 μA. However, the value of average corona current decreases with increasing the wind speed for SPW, and increases with increasing the wind speed for SVW. The total cumulate charge during the corona exposure test for 30 min was also calculated. Figure 26(c) shows the relationship between the total cumulate charge and the wind speed. With the applied voltage of 10 kV, the total cumulate charge of SWW is about 940.1 mC. It is observed that with the increase in wind speed, the total cumulate charge decreases in the case of SPW, while it increases in the case of SVW.

The reason for such a behavior can be explained in the viewpoint of the directional relation between airflow and the gap axis. In stationary air, the trajectory of positive ions is influenced by two forces: the electric fields caused by the applied voltage and the space charge filling the gap between the tip of needle and the specimen surface. $\vec{F}_1$ is considered as the resultant force of applied voltage and space charge. In the presence of wind, the neutral particles can impart momentum to the ions and influence their drift path. The force caused by wind is defined as $\vec{F}_2$, which is related to the wind speed. Figure 27 shows a simple model for ions' drift path line under different wind conditions. In the absence of wind, ions drift to the specimen under the force $\vec{F}_1$, as shown in figure 27(a). When the wind is blowing perpendicular to the

Figure 25. The typical leakage current waveforms of different specimens. (a) SPW (b) SWW (c) SVW

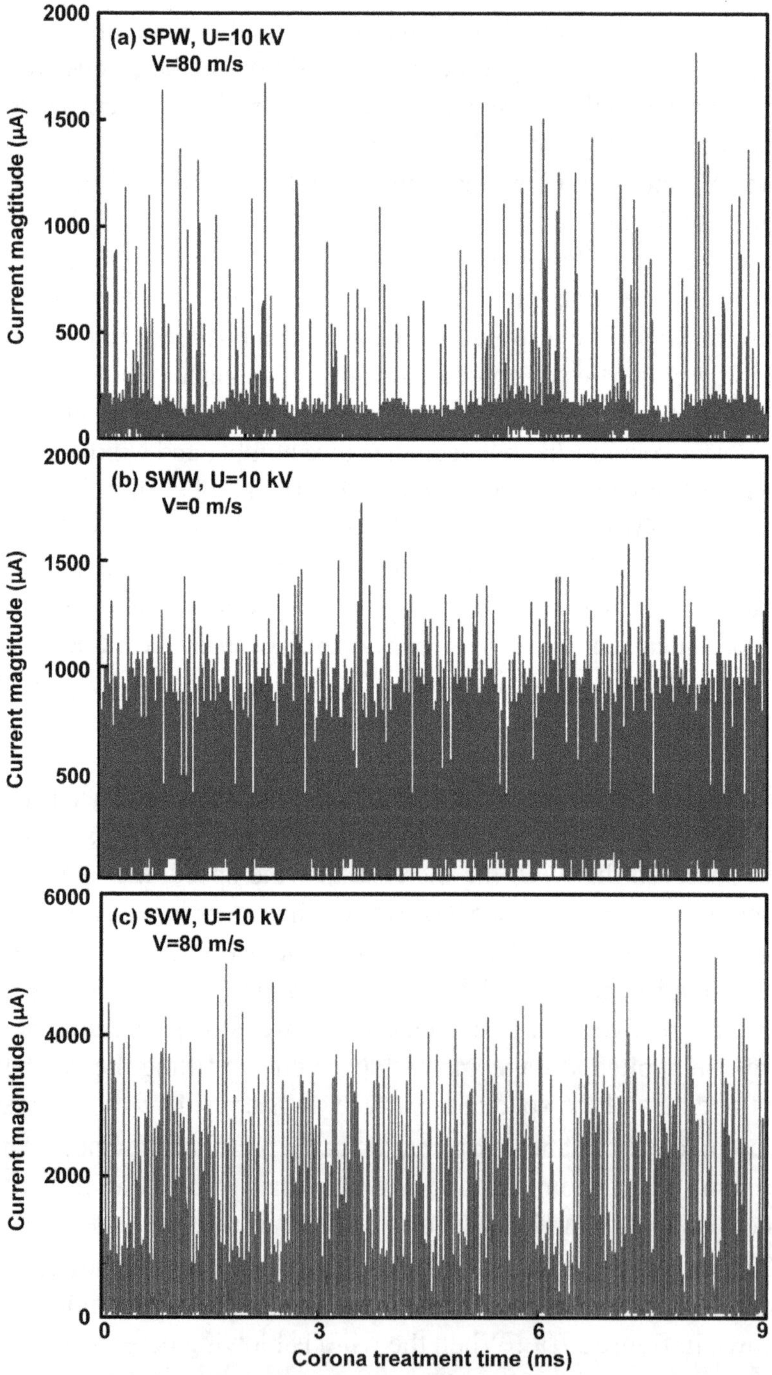

Figure 26. Relationship between the discharge characteristics and the wind speed. (a) number of corona current pulses (b)average corona current (c) cumulative charge

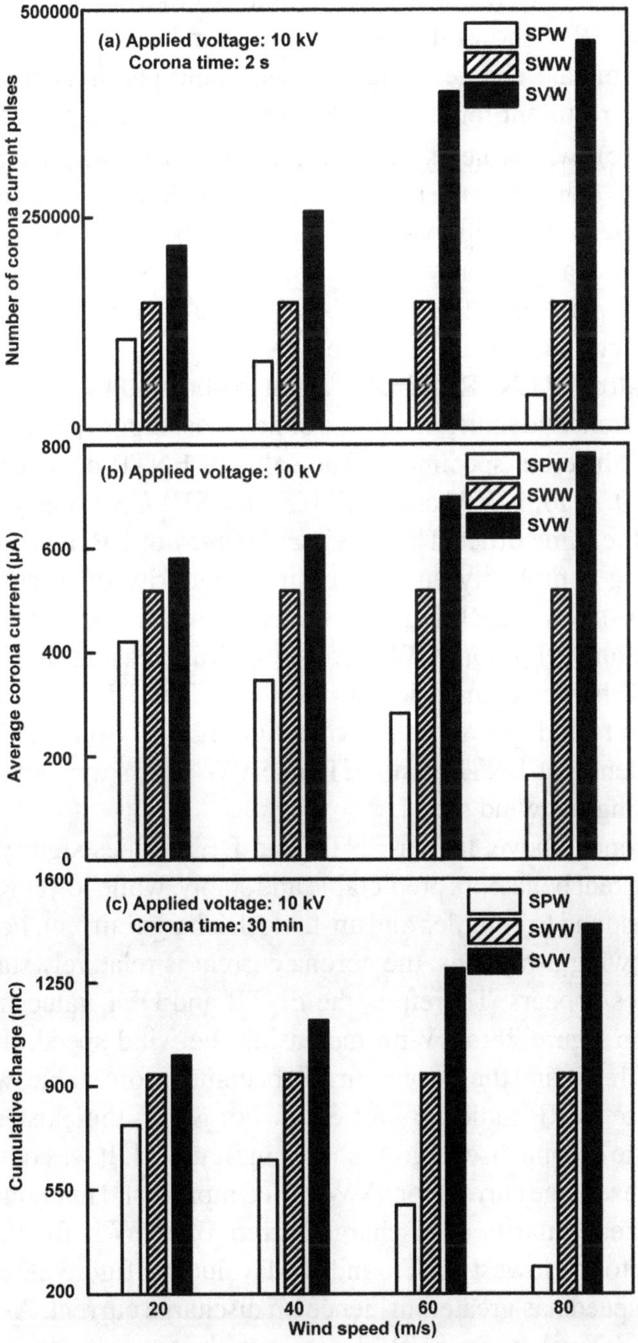

gap axis, the resultant force $\vec{F}_1+\vec{F}_2$ can be deflected in the direction of wind, as shown in figure 27(b), as a result, ions are blown away and fewer ions can reach the surface of specimen and migrate to the ground electrode. This explains why SPW shows the lowest pulse number, time-averaged current value and cumulate charge. When the wind and gap axis are in the same direction, the resultant force $\vec{F}_1+\vec{F}_2$ is contracted closer to gap axis as shown in figure 27(c), which helps positive ions to move away from the needle electrode, thus enhancing the field around the needle electrode and increasing the corona stress. The high value of pulse number, average corona current and cumulate charge clearly reveals the effect of ionization enhancement when wind is parallel to the gap axis. With increasing the wind speed, the influence on corona current is greater.

RP indicators of ENTR and DET are also obtained to analyze the corona discharge current under different wind conditions. Figure 28(a) shows the RP indicators of different specimens. The value of ENTR decreases from 0.95 for SPW, to 0.74 for SWW and until 0.55 for SVW. Similarly, DET values decrease in the same order. The varying tendency of RP indicators signifies the increasing complexity and decreasing regularity of discharge current. Figure 28(b) shows the RP indicators of SPW under different wind speeds. Both ENTR and DET for SPW increase with increasing the wind speeds, which are all higher than indicators of SWW. The RP indicators of SVW under different wind speeds are shown in figure 28(c). Compared to SPW, the varying tendency of ENTR and DET for SVW shows opposite phenomena with increasing the wind speed.

As mentioned above, higher ENTR and DET values signify the studied discharge current behaviors predictable and stably, while lower RP indicators always correspond to complex and unstable discharge current. For SPW, with the wind blowing away ions, the corona current is relatively stable and few current pulses appears. Therefore, the ENTR and DET values for SPW are the highest in figure 28(a). With increasing the wind speed, the effects of wind also increase and the corona current becomes more stable, which results in an increase in RP indicators for SPW. For SVW, the existence of wind intensified the corona discharge as shown in figure 24. It was observed in the test that the discharge current for SVW varied rapidly and many current pulses appeared. The regularity of discharge current of SVW is the lowest, which corresponds to the lowest ENTR and DET values in figure 28(a). The wind with higher speed has greater influence on discharge current. As a result, the ENTR and DET values for SVW decrease with increasing the wind speed.

Figure 27. A model of the ions' drift path line of different specimens. (a) SWW (b) SPW (c) SVW

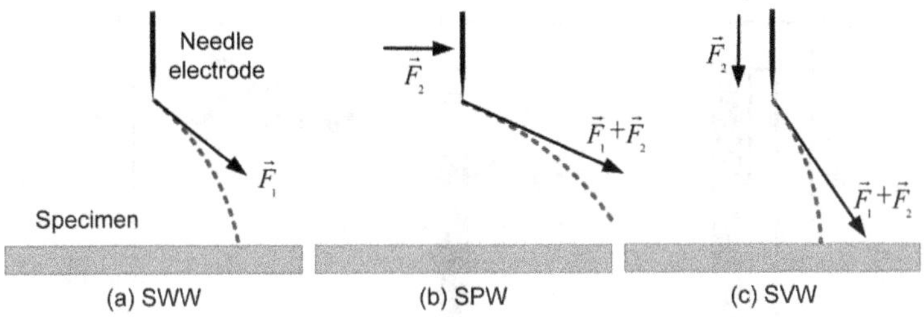

Effects of Wind Direction and Speed on Contact Angles

The static contact angle is an effective parameter for characterizing the hydrophobicity of the SiR specimen. Before corona treatment, the contact angle of specimen is about 118 °. Figure 29(a) shows the variation in the contact angles of specimens after corona treatment with different applied voltage. Compared to SPW and SWW, more reduction in contact angles can be found in the case of SVW. Under the same wind condition, the contact angles decrease with the increase in applied voltage. The higher applied voltage means that the corona stress is stronger. It appears that the stronger corona stress results in more hydrophobicity loss.

Figure 29(b) shows the variation in the contacts angles with respect to corona treatment time. The decreasing tendency indicates that the specimens changes from a hydrophobic state to a hydrophilic state with the lapse of corona treatment time. Compared to the contact angles in the absence of wind, it is found that the vertical wind accelerates the reduction in contact angles, while the parallel wind inhibits the reduction.

The contact angles of specimens after corona treatment under different wind conditions are shown in figure 29(c). In the absence of wind, the static contact angles are reduced to 62.6 °. In the presence of wind, the reduction in contact angles is inhibited or accelerated to varying degrees depending on the wind direction and wind speed. For SPW, the hydrophobicity loss caused by corona is inhibited and higher wind speed corresponds to less reduction in contact angles. For SVW, the hydrophobicity loss becomes more severe with increasing the wind speed.

In order to furtherly study the effects of wind on hydrophobicity loss, another test was conducted. SiR specimens were placed in parallel or vertical wind for several hours without corona discharge. The obtained results revealed that

Figure 28. Relationship between the RP indicators and the corona discharge parameters with applied voltage of 10 kV. (a) SWW (b) SPW (c) SVW

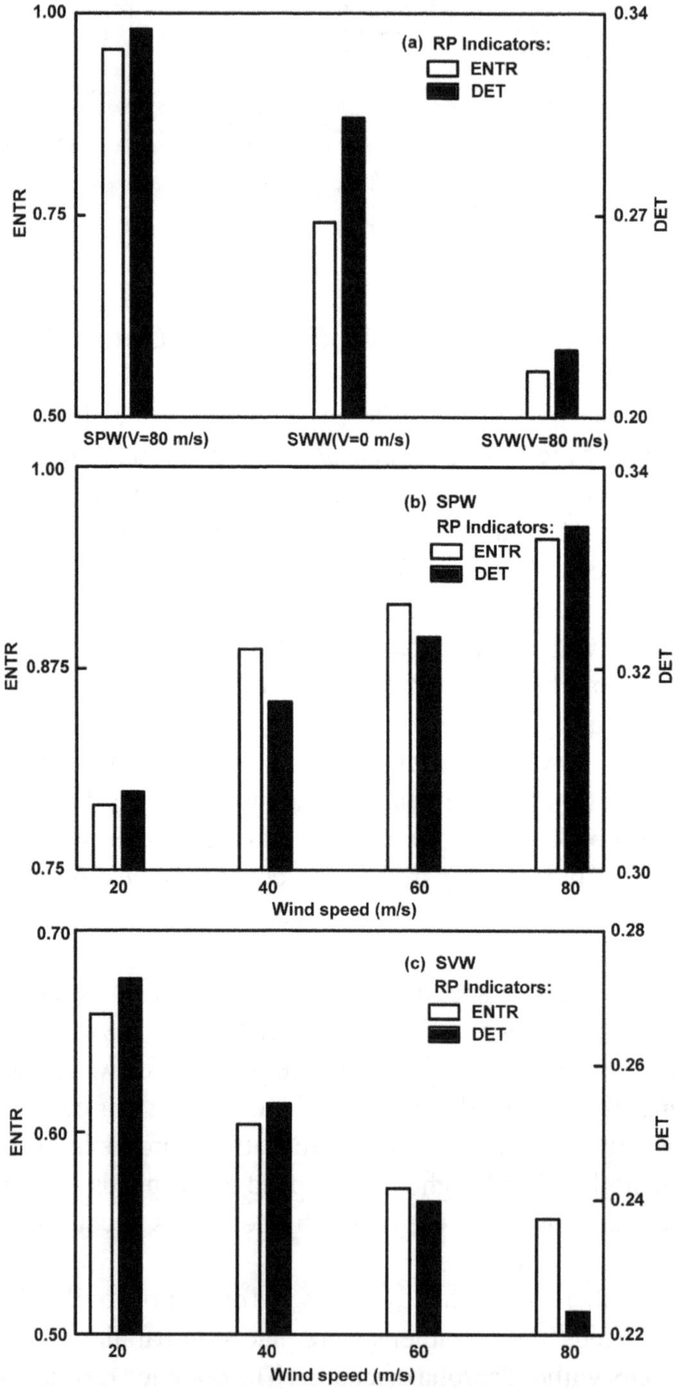

Figure 29. Relationship between the contact angles and (a)applied voltage, (b) corona time and (c) wind speed

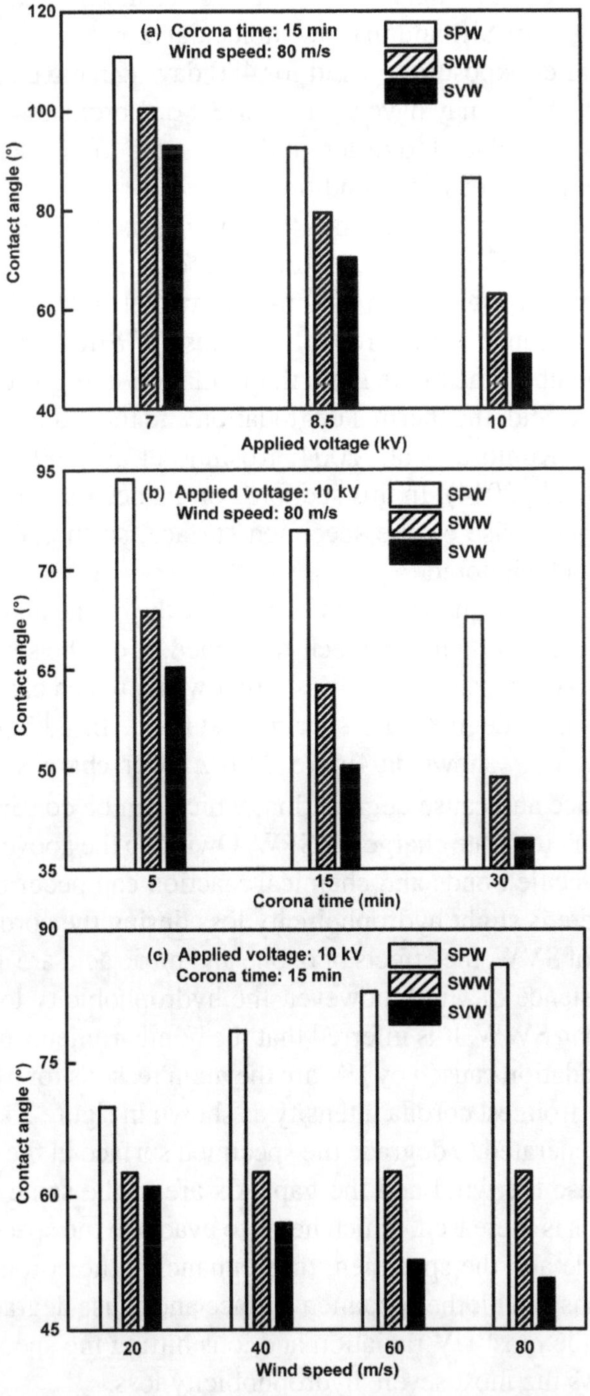

static contact angles on the surface of specimens didn't change a lot after wind treatment. So it can be concluded that there is no direct relationship between the wind and the contact angles of SiR. This is consistent with the results that the contact angles of SiR and the low molecular weight fluid content didn't change much after exposure to wind for 480 days (Zhang et al., 2000). It is inferred that the wind may have a direct effect on corona discharge, and the contact angles are reduced to different degrees due to the change of corona intensity under different wind conditions.

All the results above reveal that the specimens perform in this order: SPW> SWW> SVW. During the corona process, a mixture of species is simultaneously generated, such as ions, ultraviolet light (UV) and ozone. The species can cause the hydrophobicity loss of SiR from the following aspects: the bombardments of ions, the oxidation and corrosion effect of ozone and nitric acid, the thermal degradation and the radiation degradation caused by UV (Kumosa et al., 2004; Koshino et al., 1998; Hernandez et al., 1999; Du et al., 2014). In this section, corona occurs in the air resulting in low temperature rise on the specimen surface, so the effect of thermal degradation can be ignored.

In the case of SPW, the wind can blow away the ozone and nitric acid, as a result, the effects of them are much weakened. It can be seen in figure 24 that the discharge intensity for SPW is rather weak, which can generate only a little radiation to degrade the specimen surface. In addition, under the resultant force $\vec{F}_1+\vec{F}_2$ shown in figure 27(b), fewer charges can reach the specimen surface and cause degradation, which can be confirmed from the above results of cumulate charge for SPW. Owing to the above reasons, little rupture of molecule bonds and chemical reaction can occur on the surface of SPW, so there is slight hydrophobicity loss during the corona process.

In the case of SVW, the effects of ozone and nitric acid are also weakened due to the existence of wind, however, the hydrophobicity loss of SVW is more severe than SWW. It is inferred that the bombardments of the ions and radiation degradation caused by UV are the main reasons for hydrophobicity loss. With the strongest corona intensity as shown in figure 24, the most UV radiation are generated to degrade the specimen surface in the case of SVW. Besides, because the wind and the gap axis are in the same direction, the mobility of ions is increased, which helps to evacuate the space between the needle electrode and the specimen, thus enhancing the corona stress. As a result, more ions can hit the specimen surface and cause degradation on SiR surface. With the most UV radiation and ions hitting the specimen surface, the SVW shows the most severe hydrophobicity loss.

Effects of Wind Direction on Hydrophobicity Recovery

Before the hydrophobicity recovery process, specimens were corona treated for 30 min in the absence of wind, and the contact angles were reduced to about 62°. Then the specimens were put in different wind conditions. The contact angles with the lapse of recovery time are shown in figure 30(a). All contact angles increase with recovery time. It can be seen that the hydrophobicity recovery rate decrease in this order: SVW> SWW> SPW. The transfer of low molecular weight (LMW) components from the bulk to the surface is greatly related to hydrophobicity recovery (Díaz et al., 2014; Takara, 2001;

Figure 30. Relationship between the contact angles and the recovery time under (a) wind condition and (b) test condition

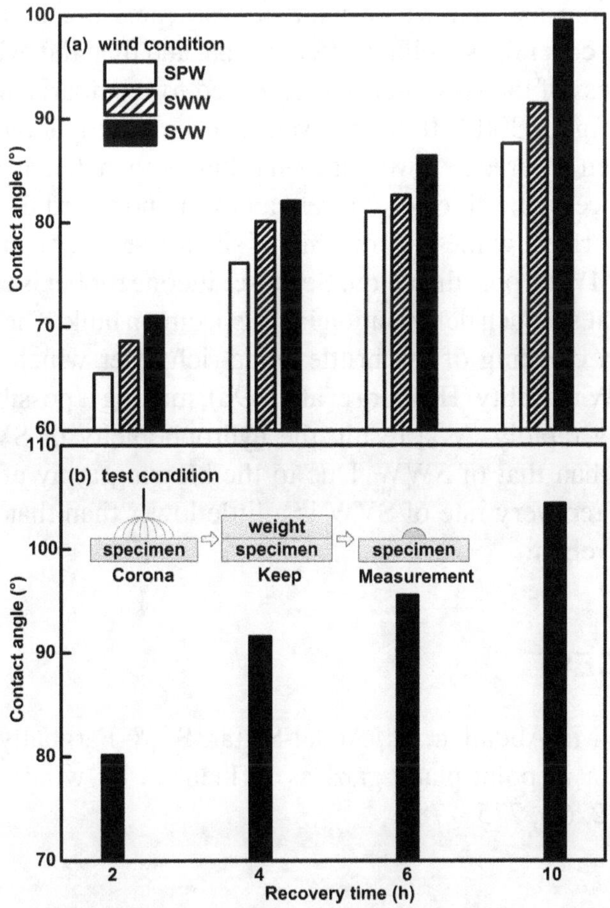

Ali et al., 2009; Ali et al., 2008; Zhang et al., 2016). Dai et al. have proved the volatilization of the LMW, which can diffuse from bulk across the SiR surface to the surrounding air. In the process of volatilization of the LMW, a part of LMW remains on the surface and makes it hydrophobic, while the rest diffuses to the surrounding air. It is inferred that the wind can blow away the LMW, so there is a high concentration gradient between the specimen surface and the surrounding air. The continuous volatilization of LMW tends to decrease the concentration gradient. In case of SPW, the higher concentration gradient of LMW makes more LMW diffuse to the surrounding air rather than remain on the surface, so the hydrophobicity recovery rate of SPW is lower than that of SWW.

Though the effects of wind on LMW also exist in the case of SVW, the recovery rate of SVW is higher than that of SWW. This is contributed to the pressure imposed on the SiR surface by vertical wind. With a wind speed of 80 m/s, the wind pressure intensity is about 4 kN/m^2. In order to verify the effects of pressure, a weight which could impose a pressure intensity of 4 kN/m^2 was used to cover the specimen after corona, and then the hydrophobicity recovery process of the specimen was recorded for comparison. The results are shown in figure 30(b). It is observed that the hydrophobicity recovery rate of specimen covered by a weight even is higher than that of SPW. During the corona process, the silica-like layer can form and retard the diffusion of LMW to the surface, while the cracking of silica-like layer can promote the transport of LMW by pore diffusion. Because silicone rubber is elastomer, the pressure can cause much deformation in the specimen bulk. The deformation can result in a cracking of the brittle silica-rich layer which significantly increases LMW mobility (Hillborg et al., 1998), making it possible to recover hydrophobicity rapidly. As a result, the hydrophobicity of SVW recovers more rapidly than that of SWW. Due to the blowing away effect of wind on LMW, the recovery rate of SVW is a little lower than that of specimen covered by a weight.

REFERENCES

Abdel-Salam, M., Abdallah, H., Abdel-Sattar, S., & Farghally, M. (1987). Positive corona in point-plane gaps as influenced by wind. *IEEE Trans. Electr. Insul., 22*(6), 775 – 786.

Ali, M., & Hackam, R. (2008). Effects of Saline Water and Temperature on Surface Properties of HTV Silicone Rubber. *IEEE Transactions on Dielectrics and Electrical Insulation*, *15*(5), 1368–1378. doi:10.1109/TDEI.2008.4656246

Ali, M., & Hackam, R. (2009). Recovery of Hydrophobicity of HTV Silicone Rubber after Accelerated Aging in Saline Solutions. *IEEE Transactions on Dielectrics and Electrical Insulation*, *16*(3), 842–852. doi:10.1109/TDEI.2009.5128525

Avella, F., Díaz, J., Ariza, D., Escobar, O., & Román, F. (2012). Influence of Thunderstorm-like Wind Velocities on Point Discharge Corona Currents. *IEEE Int. Conf. Light. Prot.*, 4. doi:10.1109/ICLP.2012.6344399

Baker, A. C., Farzaneh, M., Gorur, R. S., Gubanski, S. M., Hill, R. J., Karady, G. G., & Schneider, H. M. (2009). Insulator Selection for AC Overhead Lines with Respect to Contamination. *IEEE Transactions on Power Delivery*, *24*(3), 1633–1641. doi:10.1109/TPWRD.2009.2024666

Barsch, R., Jahn, H., Lambrecht, J., & Schmuck, F. (1999). Test Methods for Polymeric Insulating Materials for Outdoor HV Insulation. *IEEE Transactions on Dielectrics and Electrical Insulation*, *6*(5), 668–675. doi:10.1109/94.798123

Boczar, T., & Zmarzły, D. (2004). Application of Wavelet Analysis to Acoustic Emission Pulses Generated by Partial Discharges. *IEEE Transactions on Dielectrics and Electrical Insulation*, *11*(3), 433–449. doi:10.1109/TDEI.2004.1306722

Chandrasekar, S., Kalaivanan, C., Cavallini, A., & Montanari, C. G. (2009). Investigations on Leakage Current and Phase Angle Characteristics of Porcelain and Polymeric Insulator under Contaminated Conditions. *IEEE Trans. Dielectr. Electrl. Insul.*, *16*(2), 574-583.

Chandrasekar, S., Kalaivanan, C., Cavallini, A., & Montanari, G. (2009). Investigations on Leakage Current and Phase Angle Characteristics of Porcelain and Polymeric Insulator under Contaminated Conditions. *IEEE Transactions on Dielectrics and Electrical Insulation*, *16*(2), 574–583. doi:10.1109/TDEI.2009.4815193

Díaz, J., Ocampo, G., Chaves, S., & Román, F. (2014). Positive and Negative Point Discharge Corona Currents: Influence of Thunderstorm-like Wind Velocities. *IEEE Int. Conf. Light. Prot.*, 1708-1711.

Douar, M. A., Mekhaldi, A., & Bouzidi, M. C. (2010). Flashover Process and Frequency Analysis of the Leakage Current on Insulator Model under Non-Uniform Pollution Conditions. *IEEE Transactions on Dielectrics and Electrical Insulation*, *17*(4), 1284–1297. doi:10.1109/TDEI.2010.5539701

Du, B. X., Liu, Y., Liu, H. J., & Yang, Y. J. (2009). Recurrent Plot Analysis of Leakage Current for Monitoring Outdoor Insulator Performance. *IEEE Trans. Dielec. Electr. Insul*, *16*(1), 139-146.

Du, B. X., Liu, Y., Liu, H. J., & Yang, Y. J. (2009). Recurrent Plot Analysis of Leakage Current for Monitoring Outdoor Insulator Performance. *IEEE Trans. Dielectr. Electrl. Insul.*, *16*(1), 139-146.

Du, B. X., & Hang Xu, . (2014). Effects of Thermal Conductivity on Dc Resistance to Erosion of Silicone Rubber/BN Nanocomposites. *IEEE Transactions on Dielectrics and Electrical Insulation*, *21*(2), 511–518. doi:10.1109/TDEI.2013.004139

Du, B. X., Liu, Y., Liu, H. J., & Yang, Y. (2009). Recurrent Plot Analysis of Leakage Current for Monitoring Outdoor Insulator Performance. *IEEE Transactions on Dielectrics and Electrical Insulation*, *16*(1), 139–146. doi:10.1109/TDEI.2009.4784561

Du, B. X., & Yamano, Y. (2005). Effects of Atmospheric Pressure on dc Resistance to Tracking of Polymer Insulating Materials. *IEEE Transactions on Dielectrics and Electrical Insulation*, *12*, 1162–1171. doi:10.1109/TDEI.2005.1561796

Du, B. X., & Yong Liu, . (2009). Frequency Distribution of Leakage Current on Silicone Rubber Insulator in Salt-Fog Environments. *IEEE Transactions on Power Delivery*, *24*(3), 1458–1464. doi:10.1109/TPWRD.2009.2016626

Du, B. X., & Yong Liu, . (2009). Frequency Distribution of Leakage Current on Silicone Rubber Insulator in Salt-Fog Environments. *IEEE Transactions on Power Delivery*, *24*(3), 1458–1464. doi:10.1109/TPWRD.2009.2016626

Fernando, M. A. R. M., & Gubanski, S. M. (2010). Ageing of Silicone Rubber Insulators in Coastal and Inland Tropical Environment. *IEEE Transactions on Dielectrics and Electrical Insulation*, *17*(2), 326–333. doi:10.1109/TDEI.2010.5448085

Fritz, J. I., & Owen, M. J. (1995). Hydrophobicity Recovery of Plasma-Treated Polydimethylsiloxane. *The Journal of Adhesion*, *54*(1), 33–45. doi:10.1080/00218469508014379

Fu, G. J., Wang, L. M., & Guan, Z. C. (2011). Feasibility of Application of Composite Insulators in ±800 kV UHV DC Tension Strings[in Chinese]. *Proceedings of the CSEE*, *31*(22), 119–125.

Guide, S. T. R. I. (1992). *Hydrophobicity Classification Guide*. Swedish Transmission Research Institute.

Hernandez, R., Moreno, V. M., Ponce, M. A., Valle, E., & Jimenez, D. (1999). *A Study on The Effect of Environmental Stresses on The Hydrophobic Characteristics and Field Performance of RTV Silicone Rubber Coatings*. Austin, TX: Annu. Rep. Conf. Electr. Insul. Dielectr. Phenom. doi:10.1109/CEIDP.1999.807910

Hillborg, H., & Gedde, U. W. (1998). Hydrophobicity Recovery of Polydimethylsiloxane after Exposure to Corona Discharges. *Polymer*, *39*(10), 1991–1998. doi:10.1016/S0032-3861(97)00484-9

Hillborg, H., & Gedde, U. W. (1999). Hydrophobicity Changes in Silicone Rubber. *IEEE Transactions on Dielectrics and Electrical Insulation*, *6*(5), 703–717. doi:10.1109/94.798127

IEC/TS 62073-2003. (2003). *Guidance on the Measurement to Wettability of Insulator Surfaces*. International Electrotechnical Commission.

Khattak, A., & Amin, M. (2016). Accelerated aging investigation of high voltage EPDM/silica composite insulators. *Journal of Polymer Engineering*, *36*(2), 199–209. doi:10.1515/polyeng-2015-0004

Kim, S. H., & Cherney, E. A. (1994). Chemical Changes at the Surface of RTV Silicone Rubber Coatings on Insulators during Dry-Band Arcing. *IEEE Transactions on Dielectrics and Electrical Insulation*, *1*(1), 106–123. doi:10.1109/94.300238

Kim, S. H., Cherney, E. A., Hackam, R., & Rutherford, K. G. (1994). Chemical Changes at the Surface of RTV Silicone Rubber Coating on Insulators during Dry-band Arcing. *IEEE Transactions on Dielectrics and Electrical Insulation*, *1*(1), 106–123. doi:10.1109/94.300238

Kim, T. Y., Suh, K. S., Nam, J. H., & Takada, T. (2003). Acoustic Monitoring of HV Equipment with Optical Fiber Sensors. *IEEE Transactions on Dielectrics and Electrical Insulation*, *10*(2), 266–270. doi:10.1109/TDEI.2003.1194109

Koshino, Y., Umeda, I., & Ishiwari, M. (1998). *Deterioration of Silicone Rubber for Polymer Insulators by Corona Discharge and Effect of Fillers*. Atlanta, GA: Annu. Rep. Conf. Electr. Insul. Dielectr. Phenom. doi:10.1109/CEIDP.1998.733853

Kumosa, M., Kumosa, L., & Armentrout, D. (2004). Causes and Potential Remedies of Brittle Fracture Failure of Composite (Non-Ceramic). *Insulator. IEEE Trans. Dielec. Electr. Insul.*, *11*(6), 1037–1048. doi:10.1109/TDEI.2004.1387827

Li, C. R., Wang, X. X., Li, M., Zhan, H. M., Xu, J. B., Liang, Y., & Liang, X. D. (2008). Dielectric Barrier Discharge Using Corona-Modified Silicone Rubber. *Europhysics Letters*, *84*(2), 25002. doi:10.1209/0295-5075/84/25002

Li, Y. H., Wang, J., Wang, C., An, Z. Y., Hou, S. L., & Xing, F. (2010). Properties of Surface Arc Discharge in A Supersonic Airflow. *Plasma Sources Science & Technology*, *19*(2), 025016. doi:10.1088/0963-0252/19/2/025016

Liu, Y., & Du, B. X. (2010). Recurrent Plot Analysis of Leakage Current on Flashover Performance of Rime-Iced Composite Insulator. *IEEE Transactions on Dielectrics and Electrical Insulation*, *17*(2), 465–472. doi:10.1109/TDEI.2010.5448102

Liu, Y., Xia, B., Du, B., & Farzaneh, M. (2016). Influence of fine metal particles on surface discharge characteristics of outdoor insulators. *Energies*, *9*(2), 1–13. doi:10.3390/en9020087

Lopes, I. J. S., Jayaram, S. H., & Cherney, E. A. (2001). A Study of Partial Discharges from Water Droplets on a Silicone Rubber Insulating Surface. *IEEE Transactions on Dielectrics and Electrical Insulation*, *8*(2), 262–268. doi:10.1109/94.919951

Lopes, I. J. S., Jayaram, S. H., & Cherney, E. A. (2002). A Method for Detecting: The Transition from Corona from Water Droplets to Dry-Band Arcing on Silicone Rubber Insulators. *IEEE Transactions on Dielectrics and Electrical Insulation*, *9*(4), 964–971. doi:10.1109/TDEI.2002.1115491

Lundgaard, L. E. (1992). Partial discharge XIII: Acoustic partial discharge detection-fundamental considerations. *IEEE Electrical Insulation Magazine*, 25-31.

Lundgaard, L. E. (1992). Partial discharge XIV: Acoustic partial discharge detection-practical application. *IEEE Electrical Insulation Magazine*, 34-43.

Lundgaard, L. E., & Hansen, W. (1998). Acoustic Method for Quality Control and In-Service Periodic Monitoring of Medium Voltage Cable Terminations. *Conference Record of the 1998 IEEE International Symposium on Electrical Insulation*, 130-133.

Markalous, S. M., Tenbohlen, S., & Feser, K. (2008). Detection and Location of Partial Discharges in Power Transformers using Acoustic and Electromagnetic Signals. *IEEE Transactions on Dielectrics and Electrical Insulation, 15*(6), 1576–1583. doi:10.1109/TDEI.2008.4712660

Matsuoka, R., Naito, K., Irie, T., & Kondo, K. (2002). Evaluation Methods of Polymer Insulators under Contaminated Conditions. *IEEE/PES Transmission and Distribution Conference and Exhibition, Asia Pacific*, 2197-2202.

Meghnefi, F., Volat, C., & Farzaneh, M. (2007). Temporal and Frequency Analysis of the Leakage Current of a Station Post Insulator during Ice Accretion. *IEEE Transactions on Dielectrics and Electrical Insulation, 14*(6), 1381–1389. doi:10.1109/TDEI.2007.4401220

Moreno, V. M., & Gorur, R. S. (2001). Effect of Long-term Corona on Non-Ceramic Outdoor Insulator Housing Materials. *IEEE Transactions on Dielectrics and Electrical Insulation, 8*(1), 117–128. doi:10.1109/94.910434

Moula, B., Mekhaldi, A., Teguar, M., & Haddad, A. (2013). Characterization of Discharges on Non-Uniformly Polluted Glass Surfaces Using a Wavelet Transform Approach. *IEEE Transactions on Dielectrics and Electrical Insulation, 20*(4), 1457–1466. doi:10.1109/TDEI.2013.6571469

Nazir, M. T., Phung, B. T., & Hoffman, M. (2016). Performance of Silicone Rubber Composites with SiO2 Micro/Nano-filler under AC Corona Discharge. *IEEE Transactions on Dielectrics and Electrical Insulation, 23*(5), 2804–2815. doi:10.1109/TDEI.2016.7736840

Noskoski, O. A., Bermudez, J. C. M., & de Almeida, S. J. M. (2013). Region-Based Wavelet-Packet Adaptive Algorithm for Identification of Sparse Impulse Responses. *IEEE Transactions on Signal Processing*, *61*(13), 3321–3333. doi:10.1109/TSP.2013.2257763

Phillips, A. J., Childs, D. J., & Schneider, H. M. (1999). Aging of Non-Ceramic Insulators due to Corona from Water Drops. *IEEE Transactions on Power Delivery*, *14*(3), 1081–1089. doi:10.1109/61.772357

Rajini, V., & Udayakumar, K. (2009). Degradation of Silicone Rubber under AC or DC Voltages in Radiation Environment. *IEEE Transactions on Dielectrics and Electrical Insulation*, *16*(3), 834–841. doi:10.1109/TDEI.2009.5128524

Sabri, Y., Farzaneh, M., & Zhang, J. (2010). Application of Identification Methods for Predicting the Flashover Voltage of Contaminated Insulators Covered with Ice. *IEEE Transactions on Dielectrics and Electrical Insulation*, *17*(2), 451–457. doi:10.1109/TDEI.2010.5448100

Sakoda, T., Nakashima, T., Nainggolan, J. M., Otsubo, M., Kurihara, S., Nagasato, S., & Yarimitsu, T. (2010). Diagnostics of Insulation Deterioration of Ethylene Propylene Rubber Using an Acoustic Emission Technique. *IEEE Transactions on Dielectrics and Electrical Insulation*, *17*(4), 1242–1248. doi:10.1109/TDEI.2010.5539696

Shang, C. (2006). Development of Ultra-high Voltage Transmission Technology in China Southern Power Grid.[in Chinese]. *High Voltage Engineering*, *32*(1), 35–37.

Sharkawy, R. M., Abdel-Galil, T. K., Mangoubi, R. S., Salama, M. M., & Bartnikas, R. (2008). Particle Identification in Terms of Acoustic Partial Discharge Measurements in Transformer Oils. *IEEE Transactions on Dielectrics and Electrical Insulation*, *15*(6), 1649–1656. doi:10.1109/TDEI.2008.4712669

Shu, Y., Liu, Z., Gao, L., & Wang, S. (2006). A Preliminary Exploration for Design of ±800kV UHVDC Project with Transmission Capacity of 6400MW. [in Chinese]. *Power System Technology*, *30*(1), 1–8.

Sigmond, R. S. (1986). The unipolar corona space charge flow problem. *Journal of Electrostatics*, *18*(3), 249–272. doi:10.1016/0304-3886(86)90021-5

Takara, T. (2001). Loss and Recovery of Hydrophobicity and Surface Energy of HTV Silicone Rubber. *IEEE Transactions on Dielectrics and Electrical Insulation*, *8*(6), 1088–1097. doi:10.1109/94.971469

Wang, W., Yang, B., Li, C. R., Luo, B., & Li, X. (2008). The Effect of Wind on Corona Current Generated from Conductor in Corona Cage. *IEEE Int. Conf. High Voltage Eng. Appl.*, 152-153.

Waters, R. T., & Stark, W. B. (1973). Control of corona and sparkover of coaxial systems by irradiation. *Proceedings of the Institution of Electrical Engineers*, *120*(4), 519–522. doi:10.1049/piee.1973.0108

Liu, Y., & Du, B. X. (2010). Recurrent Plot Analysis of Leakage Current on Flashover Performance of Rime-Iced Composite Insulator. *IEEE Trans. Dielec. Electr. Insul.*, *17*(2), 465-472.

Liu, Y., & Du, B. X. (2010). Recurrent Plot Analysis of Leakage Current on Flashover Performance of Rime-Iced Composite Insulator. *IEEE Trans. Dielectr. Electrl. Insul.*, *17*(2), 465-472.

Zhang, H., & Hackam, R. (2000). The Effect of Low Molecular Weight Fluid on Aging Performance of HTV Silicone Rubber. *IEEE Int. Symp. Electr. Insul.*, 184-187. doi:10.1109/ELINSL.2000.845485

Zhang, R., Zhou, X., Xia, Y., Liu, H., & Shen, Q. (2015). Acceleration of hydrophobicity transfer rate of contaminated HTV silicone rubber by plasma jet.*Proceedings of the IEEE International Conference on Properties and Applications of Dielectric Materials*, 756-759. doi:10.1109/ICPADM.2015.7295382

Zhang, R., Zhou, X., Xia, Y., Su, S., Guan, Z., Chen, J., & Liu, Y. (2016). Hydrophobicity improvement of contaminated HTV silicone rubber by atmospheric plasma jet treatment. *IEEE Transactions on Dielectrics and Electrical Insulation*, *23*(1), 377–384. doi:10.1109/TDEI.2015.005117

Zhu, Y., Haji, K., Otsubo, M., & Honda, C. (2006). Surface Degradation of Silicone Rubber Exposed to Corona Discharge. *IEEE Transactions on Plasma Science*, *34*(4), 1094–1098. doi:10.1109/TPS.2006.876498

Chapter 4
Polypropylene Insulation for Cables

ABSTRACT

Cable insulation made of polypropylene is of great prospective since it is recyclable to some degree. There are many problem the cable insulation need to deal with in HV system. The tree properties of polypropylene with repetitive pulse voltage at low temperature is discussed firstly. At different temperature, the pattern and length of electrical tree are shown in our results. This research analyzed the influence of temperature on the grow speed of electrical tree. The effects of mechanical stretching on space charge behaviors of pp/poe blend is also included in this chapter. This part presents the influence of stretching rate on electric properties. Finally, this chapter also provided some research of PP/POE/ZnO nanocomposites for HVDC cables. This part includes the space charge behavior analysis considering nano fillers.

INTRODUCTION

HVDC transmission develops in a fast speed due to its advantages of large electric energy conveying with the minimum loss. Therefore it is urgent to research and develop HVDC plastic cables (Murata et al., 2006; Chen et al., 2015). The operating impulse voltage is a kind of overvoltage that often occurs in HVDC converter transformers because of the on and off states of thyristors and polarity reversal (Gao et al., 2013; He et al., 2013). The repeated impact of pulse voltage on HVDC cable insulation could not

DOI: 10.4018/978-1-5225-2309-3.ch004

be ignored. Considering the lowest temperatures can reach -52.3 and -80.6 °C in the northeast China and Antarctic, HVDC cable may work under low temperature (Du et al., 2014). Over years, many researchers have investigated the tree growth mechanism in dielectric polymers. It was found that many factors such as the voltage, temperature, frequency and so on affected the treeing process. Chen et al have examined the effect of frequency with the applied AC voltage on electrical tree characteristics in the XLPE and found that higher frequency can accelerate the insulating breakdown, while the fractal dimension is mainly affected by the magnitude of AC voltage (Chen et al., 2009). Stone et al investigated the epoxy resin to understand the aging phenomenon with positive and negative pulse voltage. The epoxy resin lifetime showed a decrease tendency with the increase of repetition surge voltage (Stone et al., 1992). Holto et al investigated the electrical tree characteristics in extruded syndiotactic polypropylene (s-PP) at room temperature and found two types of electrical trees. One type was the simple straight tree - once a single branch reached the ground electrode and the breakdown occurred. Another type was the multiple branches tree - breakdown occurred after several branches reached the ground electrode (Holto et al., 2010; Holto et al., 2012). In our previous researches, the influence of high temperature up to 90 °C and low temperature down to -196 °C on the tree structure in silicon rubber has been concluded (Du et al., 2011; Du et al., 2015; Du et al., 2014). However, few researches are done about the low temperature effect on the treeing process in the PP with repetitive pulse voltage. This chapter compares the treeing process in the PP with that in the XLPE under low temperature and investigates the treeing process of the PP with different pulse frequencies and amplitudes under low temperature. The test temperature was set to three gradient, which was -30, -90 and -196 °C. It indicates that the tree is more difficult to be generated from the needle tip in the PP than in the XLPE. With the same treeing time, the tree length and fractal dimension in the PP are smaller than these in the XLPE. Also, the temperature, pulse frequency and amplitude have great effects on the tree characteristics.

The recyclable blend of PP and POE has attracted much attention. It not only modifies the brittleness and stiffness of pure PP, but also shows many outstanding performances in electrical and thermal properties. The insulation material of extruded cable has to face the challenge of mechanical stretching in some cases (Mazzanti et al., 2011; Dissado et al., 2002; Tanaka et al., 1976). For example, during the cooling process of fabrication, mechanical strain can be generated by the different cooling rates of inner and outer insulation due to temperature gradients. Besides, the cables may be bent when installed. During the operation of cables, thermomechanical strain can be created by the different thermal expansion between conductor and the insulation material. In addition, the electric field can also induce mechanical stress in polymers. Such mechanical stretching can result in complex microstructure changes in the semi-crystalline polymers (Crine, 2005; Yahagi., 1980; Li et al., 2015). For example, the amorphous regions may be elongated and the crystalline lamella may be broken, leading to significant influence on the electrical performance of polymers. David et al investigated the effect of internal mechanical strain on electrical tree characteristics of polyethylene, and found that with increasing the value of residual stresses, the tree inception time decreased and the length of trees increased (David et al., 1996). Mita et al reported the dc and ac breakdown strength of LDPE decreased with increasing elongation ratio up to 30% and then increased until the elongation ratio reached to 160%, while both of them decreased monotonically in HDPE (Mita et al., 1975). However, little work has been done about the influence of mechanical stretching on space charges. A thorough understanding of space charge behaviors in stretched PP/POE blend is necessary for the safe operation of HVDC cables. This chapter investigates the correlation between space charge behavior and the mechanical stress. PP/POE blends with different elongation ratios were prepared. The results indicate a larger number of space charges accumulate in stretched specimens after polarization, and they decay faster than those in original specimens during depolarization process. In addition, the charge mobility and trap depth show a dependence on elongation ratio, which is related to the morphology change caused by mechanical stretching.

In the operation of DC cables, space charge accumulation is a great concern, which can cause electric field distortion and increase the local field within the insulating materials, leading to partial discharge, faster degradation and even premature breakdown (Tanaka et al., 2004; Vu et al., 2015; Fabiani et al., 2004; Mazzanti et al., 2011; Mazzanti et al., 2001; Dissado et al., 1997; Montanari et al., 2011; Fu et al., 2008). Many studies indicated that nanoparticles doping showed great performance in suppressing space charge accumulation in dielectric materials (Tanaka et al., 2011; Calebrese et al., 2011; Maezawa et al., 2007; Li et al., 2010). Dakka et al observed that the quantity of space charges was remarkably reduced by adding a small amount of nanoparticles to the polyethylene (Dakka et al., 2010). Montanari et al investigated the electrical properties of PP filled by nanosized layered silicates and achieved an improved space charge behavior (Montanari et al., 2004).

It was reported the extensive interface regions between nanoparticles and neighboring materials were closely related to the space charge suppression (Tanaka et al., 2008; Tanaka et al., 2008; Tanakai, 2005; Nelson et al., 2004; Roy et al., 2007). Generally, less accumulation of space charge makes it possible for insulating material to be applied in higher field with the same thickness, decreasing the cost and weight of power equipment. Therefore, it is critical to investigate the space charge behavior of PP/POE blend and its nanocomposites for developing new insulating materials for HVDC cables. In this chapter, pure PP, PP/POE blend and PP/POE/ZnO nanocomposites with different contents were prepared. Meanwhile, mechanical properties, space charge behaviors and trap level distribution were investigated. It is observed that the PP/POE blend and its nanocomposites show enhanced mechanical flexibility. In addition, the introduction of POE in PP leads to severe space charge accumulation while nanoparticles doping reduces the quantity of space charge remarkably, which is related to the increase of trap level density. The results indicate the PP/POE/ZnO nanocomposites have much potential for HVDC cable application, which show high mechanical flexibility as well as excellent space charge performance.

EFFECT OF LOW TEMPERATURE ON ELECTRICAL TREEING OF POLYPROPYLENE WITH REPETITIVE PULSE VOLTAGE

Tree Growth Characteristics

A typical measuring system for electrical tree can be referred to (Du et al., 2015). The pulse voltage waveform is shown in figure 1. The rise of the pulse is about 100 μs and the fall time is 120 μs. The pulse amplitudes were 12 and 15 kV with frequencies of 400, 800 and 1200 Hz.

Figure 2 shows the typical tree structures in the PP and the XLPE tested in different temperatures. The pulse amplitude and frequency is 12 kV and 400 Hz, respectively. It is obtained that the tree structure changes with the temperature in the XLPE but is invariant with the temperature in the PP with the pulse amplitude of 12 kV. In the XLPE, the bush tree structure presents at 30 °C and the branch tree structure presents at -30 and -196 °C. While it is at -90 °C, the stagnated and branch-pine tree are observed. The tree structure is accounting for 50% of all samples respectively. At 30 °C, bush tree obtained is

Figure 1. The waveform of the repetitive pulse voltage

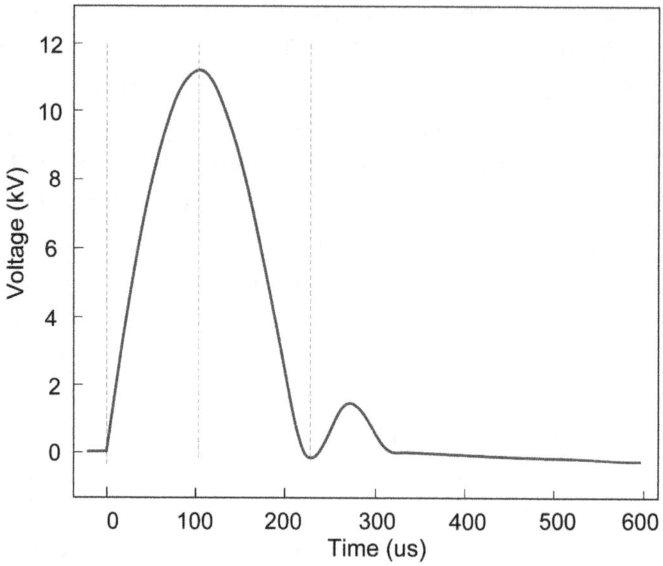

densely packed structures near the needle tip. At -30 and -196 °C, the branch trees are different from each other. At -30 °C, some channels are overlapped, while there are gaps among the main branches when the temperature is -196 °C. At -90 °C, the branch-pine tree is double structure tree that one part is branch structure and the other is pine structure. Another tree observed at -90 °C is the stagnated tree, and the length of the tree would stagnate when it reaches a certain extent. In the PP, the tree structure is branch tree at 30, -30, -90 and -196 °C. While the density of branch trees at different temperatures is different from each other. At -90 °C, the tree branches become denser and side branches generated from main branches are more and longer than these at -30 °C. At -196 °C, the number of main branches is smaller than these at -30 and -90 °C, and gaps among main branches are larger. At the same temperature, tree structure in the XLPE appears to be denser and a larger number of small branches are generated from main branches.

Figure 3 shows propagation characteristics of trees in the XLPE and the PP. The pulse voltage of figure 3 is the same as that of figure 2. The tree length is referring to the longest tree branch along the direction of the electric field. The fractal dimension can measure the branches complexity. Therefore, the box-counting method is employed, the value of fractal dimension varies between 1 and 2, which is consistent with the previous literatures (Lanca et al., 2001; Kudo, 1998).

At 30 °C, when the tree length reaches 2000 μm in the XLPE at 7 min, the tree length of the PP is only 1120 μm. When the treeing time reaches 16 min, the tree length of the PP is 1907 μm. Just before the breakdown occurs in the XLPE at 30 °C, the fractal dimension is 1.767 in the XLPE, and the fractal dimension is only 1.392 in the PP at that time, and it increases slowly afterwards. At -30 °C, when the tree breakdown occurs in the XLPE at 26 min, the tree length of the PP is only 198 μm, and it grows slowly afterwards. When the treeing time reaches 78 min, the tree length of the PP is 650 μm. Just before the breakdown occurs in the XLPE at -30 °C, the fractal dimension is 1.59 in the XLPE, and the fractal dimension is only 1.16 in the PP at that moment. At -90 °C, there are two types of trees growing in the XLPE. The branch-pine tree in the XLPE has been chosen as the typical tree to compare with the branch tree in the PP. The breakdown of branch-pine tree occurs in the XLPE at 75 min, and the branch tree length of PP at -90 °C is only 760 μm at 75 min. When the treeing time reaches 174 min, the tree length of the PP at -90 °C is only 1016 μm. Just before the breakdown of branch-pine tree occurs in the XLPE at -90 °C, the fractal dimension is 1.67 in the XLPE, and the fractal dimension is only 1.42 in the PP at that time. At -196 °C, when the tree length reaches 2000 μm in the XLPE at 220 min, the tree length of the PP is only 570 μm, and it grows slowly afterwards. When the treeing time reaches 334 min, the tree length of the PP is only 779 μm. Just before the breakdown occurs in the XLPE at -196 °C, the fractal dimension is 1.58 in the XLPE, and the fractal dimension is only 1.25 in the PP at that time. It can be seen from these figures that with the same treeing time, the tree length and fractal dimension in the PP are absolutely smaller than those in the XLPE and the electrical tree grows in a significant slower speed under the lower temperature. Due to the by-products produced during the crosslinking process, more defects would be produced in the XLPE. Charges would be trapped by these defects and accumulated in the bulk which leads to more partial discharges. Therefore, the molecular chains in XLPE is easier to be broken. That's why the tree length and fractal dimension in the XLPE are absolutely bigger than those in the PP with the same treeing time.

In order to investigate the difference between early tree growth in the PP and theXLPE under low temperature, the inception probability of electrical tree is calculated. It is the percentage of initial tree that gets 30 μm generated for 10 minutes taking the proportion of 20 samples. As shown in figure 4, with temperature varying from -30 to -196 °C, the tree inception probability decreases obviously. Taking the tree inception probability in the PP for an example, it is only 5% and 65% at -196 and -90 °C respectively, while the

Figure 2. Typical profiles of electrical trees in the XLPE and the PP under different temperatures with the pulse voltage of 12 kV and 400 Hz; (a) bush tree 30°C in XLPE; (b) branch tree 30°C in PP; (c) branch tree -30°C in XLPE; (d) branch tree -30°C in PP; (e) stagnated tree -90°C in XLPE; (f) branch tree -90°C in PP; (g) branch-pine tree -90°C in XLPE; (h) Branch tree -90°C in PP; (i) Branch tree -196°C in XLPE; (j) Branch tree -196°C in PP

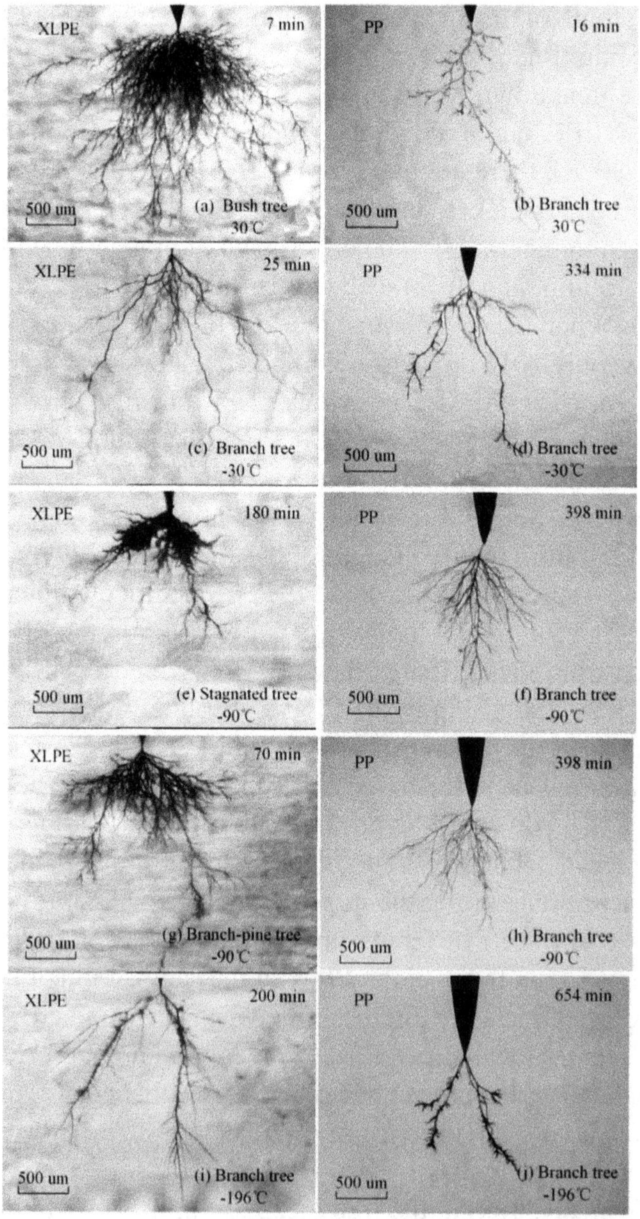

Figure 3. The propagation characteristics of electrical trees in the PP and the XLPE at 30, -30, -90 and -196 °C

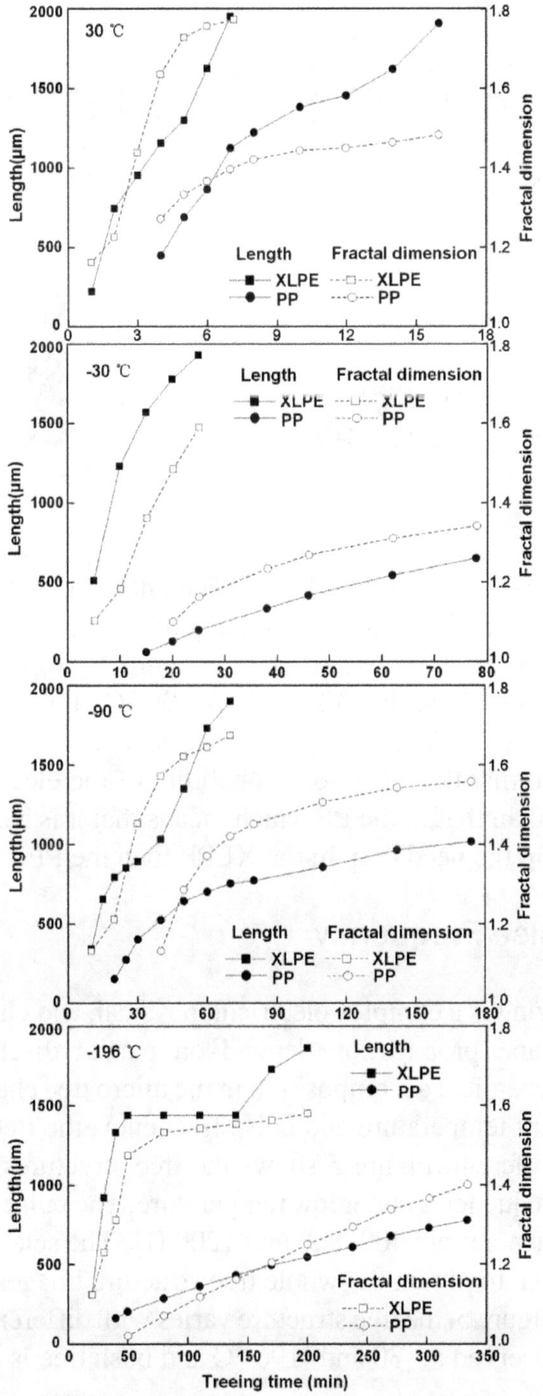

Figure 4. The tree inception probability for 10 minutes in the PP and the XLPE under low temperature

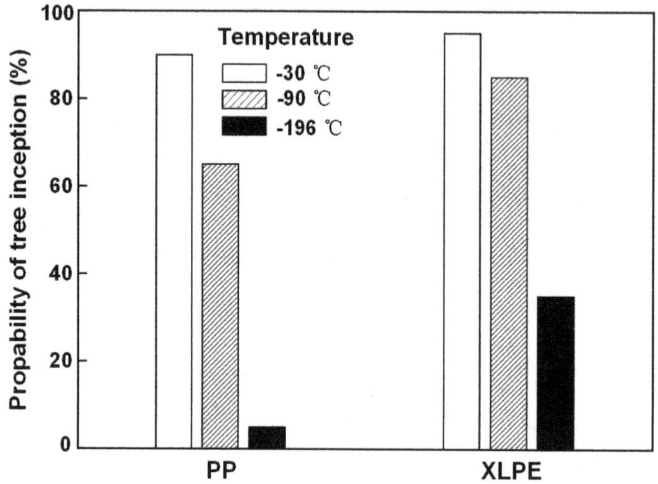

value can be up to 90% at -30 °C. It can been analyzed in figure 3 that it is more difficult for the electrical trees to be generated from the needle tip at the lower temperature. It can be seen from figure 4 the probability is 95% in the XLPE at - 30 °C, while 90% in PP. At -90 °C, it is 85% in the XLPE, while 65% in PP. At -196 °C, it is 35% in the XLPE, while 5% in the PP. It can be concluded that the inception probability of the electrical tree in the XLPE is higher than that in the PP which means that it is easier for a tree to be generated from the needle tip in the XLPE than the PP.

Effects of Pulse Frequency

Electrical tree aging is a complex electrical, physical, and chemical process. Tree initiation and propagation depend on partial discharges activity, oxidization and chemical decomposition in the micro tree channels. With the change of ambient temperature and pulse frequency, the tree structure may be completely different. Figure 5 shows the tree structures in the PP with different pulse frequencies under low temperature. The pulse amplitude is 15 kV and the frequencies are 400, 800 and 1200 Hz. The selected tree images are pre-breakdown to obtain the whole tree structure images.

As shown in figure 5, the tree structure varies with different temperatures. Branch tree is observed at -30 and -196 °C and bush tree is observed at -90

°C. At -196 °C, the branch tree is different from that at -30 °C. At -196 °C, the gaps between main branches are larger and more small branches are appeared among the main branches. Moreover, the tree channels become darker and thicker at the lower temperature. At -90 °C, bush tree is densely packed structure. At the same temperature, the tree structure is invariant with frequency, while the density of trees changes. With higher frequency, the number of main branches and side branches gets larger and more tree channels are overlapping together resulting in denser trees.

When the pulse voltage is stressed, charges are injected into the insulation material from needle tip and the energy releasing from the process of trapping and detraping of these charges would break molecules chains around the needle tip and the micro tree channels are formed. Besides, the partial discharges would release more charges into tree channels, releasing local heat and high pressures resulting in the further destruction to molecules chains. Higher frequency promotes the partial discharges activity and the amount of space charge produced by partial discharges becomes larger. Moreover, more charges are injected in unit time, colliding the molecular chains. As a

Figure 5. Typical profiles of electrical trees with different pulse frequencies under different temperatures with the pulse voltage of 15 kV; (a) branch, -30°C, 400Hz; (b) branch, -30°C, 800Hz; (c) branch, -30°C, 1200Hz; (d) bush, -90°C, 400Hz; (e) bush, -90°C, 800Hz; (f) branch, -90°C, 1200Hz; (g) branch, -196°C, 400Hz; (h) branch, -196°C, 800Hz; (i) branch, -196°C, 1200Hz

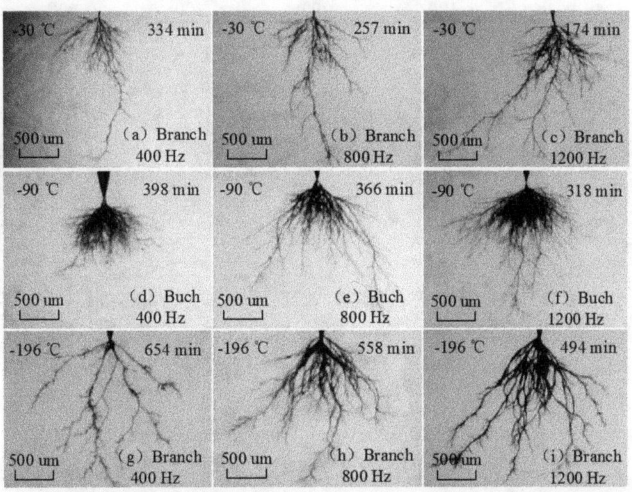

Figure 6. The propagation characteristics of electrical trees in the PP with the pulse frequency of 400, 800 and 1200 Hz at -30, -90 and -196 °C

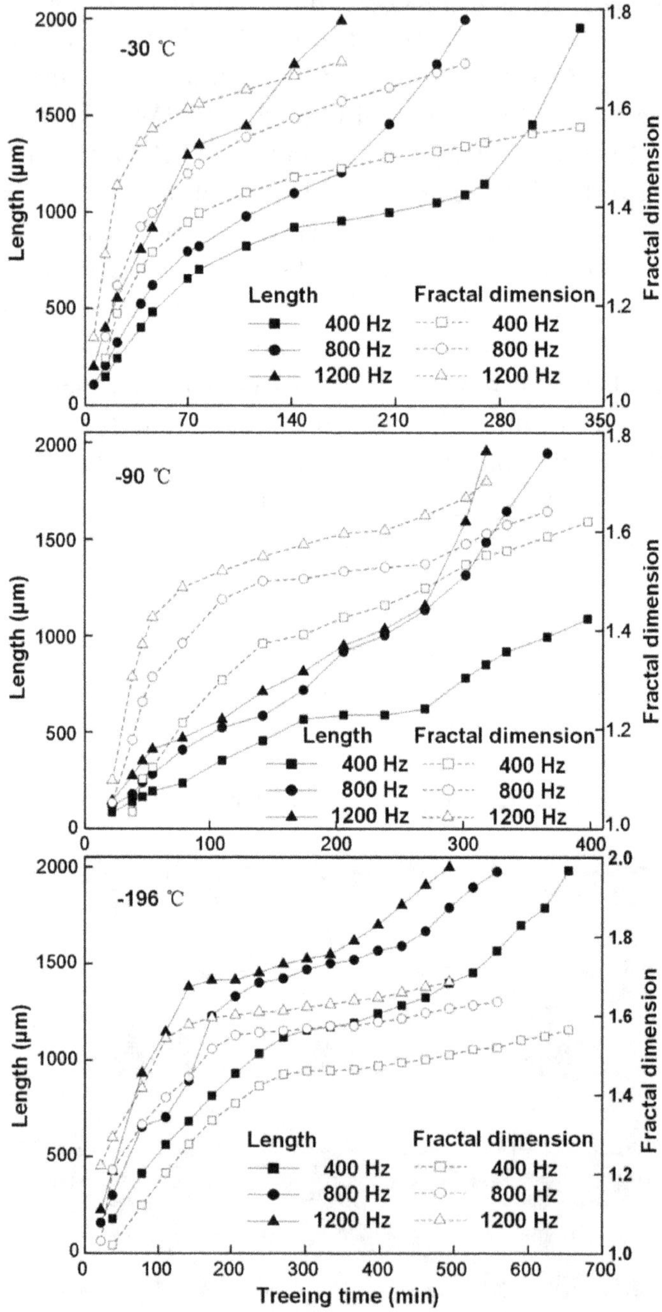

result, the number of main branches and side branches gets larger and more tree channels are overlapping together with higher frequency.

Figure 6 shows the propagation characteristics of trees in PP with different frequency. The pulse amplitude is 15 kV and the frequencies are 400, 800 and 1200 Hz. As shown in figure 6, the electrical tree grows in an obviously slower speed under the lower temperature. Under the low temperature, the partial discharges are greatly weakened. In addition, the tree growth patterns are different with the variation of temperature and frequency. At -30 °C, the growing trends of tree length are independent from frequency. At first, the tree length increases with a high speed. And then main branches grow slowly and new channels are generated from main branches and cross with each other. During the last period, the tree grows rapidly again. The tree conducting characteristics can be inferred from the tree channels color. The tree conductivity at -30, -90 and -196 °C is relatively high because of the black tree channels implying heavy carbon left in tree channels, which can act as an equivalent needle electrode. According to the maximum electric field strength equation at the needle tip in the reference (Noto et al., 1977), the electric field would be higher as the distance between tip and ground electrode becomes shorter. With 800 and 1200 Hz, the growing trends of -90 °C are dependent with that of -30 °C. But with frequency of 400 Hz at -90 °C, the tree length increases rapidly at first, and then nearly stops increasing, and at last increases again. The tree length nearly stops increasing because the bush area forms during this period. With higher frequency, the charges and partial discharges are enough to form the bush area and promote main branches to grow at the same time. That's why the tree length increases slowly during the second part with frequency of 800 and 1200 Hz, but nearly stops increasing with the frequency of 400 Hz. The growing trends of -196 °C with different frequency are dependent with that of -90 °C with 400 Hz. The partial discharges decrease with the decrease of temperature, and the heat and pressures during discharges are lower, and it is not adequate to promote the tree to grow. When the local heat and high pressures accumulated during the partial discharges, the electric field would be higher as the distance between tree tip and ground electrode becomes shorter. The two factors promote trees to grow rapidly again at -196 °C. The conclusion of the treeing tendency is that the tree length increases in a high speed at first. And then main branches grow slowly. During the last period, the tree grows rapidly again at -30, -90 and -196 °C.

As shown in figure 6, the length increases at a higher speed with the frequency increasing at the same temperature. As the frequency increases from 400 to 1200 Hz, charges injected from the needle tip become more in unit time, and activity of partial discharges increases which accelerates the degradation of PP molecules. In this case, the electrical tree length tends to be longer with higher frequency.

As shown in figure 6, the fractal dimension changing trends are similar at each condition. During the initial period, the fractal dimension increases rapidly and then continues to increase in a speed slower than the initial period in the later time. During the initial period, the strong partial discharges break molecules chains, and main branches are initialed to form a branch structure rapidly. During the propagation period, although the tree length increases slowly or nearly stops growing, more branches are generated leading to the slow increase of fractal dimension. It is noted that the fractal dimension at -196 °C increases much slower than that at -30 and -90 °C. Because the partial discharges decrease with temperature and the gap between main branches are larger at -196 °C, it is difficult for branches to be generated and cross with each other. During the last period, some slim channels grow rapidly to the ground electrode, which causes the fractal dimension increasing slowly. It can be seen from figure 6 that the fractal dimension becomes larger with frequency increasing. As the frequency increases, numerous charges may collide with polymers, breaking the molecular chain. Thus, the new degradation zone forms. There are more side branches generated from main branches and these small branches may be overlapping with each other during their extension processes. This makes the tree branches packed and the complexity of electrical tree increases accordingly.

Effects of Pulse Amplitude

Figure 7 shows typical profiles of trees with different pulse amplitudes in the PP under different temperatures. The pulse amplitudes are 12 and 15 kV with the frequency of 400 Hz. At -30 and -196 °C, the tree structure is branch tree, but the density of branches increases obviously with higher amplitude. In addition, at the same treeing time, the tree length with the amplitude of 15 kV is longer than that with12 kV at -30 and -196 °C. At -90 °C, the tree structure changes from branch to bush tree with higher amplitude. Branches develop mainly along the direction of electrical field with 12 kV. More side branches are extended in a wider arc angle direction around the needle tip, even perpendicular to the electric field

Figure 7. Typical profiles of electrical trees with different pulse amplitudes under different temperatures with the pulse frequency of 400 Hz; (a) branch, -30°C, 12kV; (b) branch, -30°C, 15kV; (c) branch, -90°C, 12kV; (d) branch, -90°C, 15kV; (e) branch, -196°C, 12kV; (f) branch, -196°C, 15kV

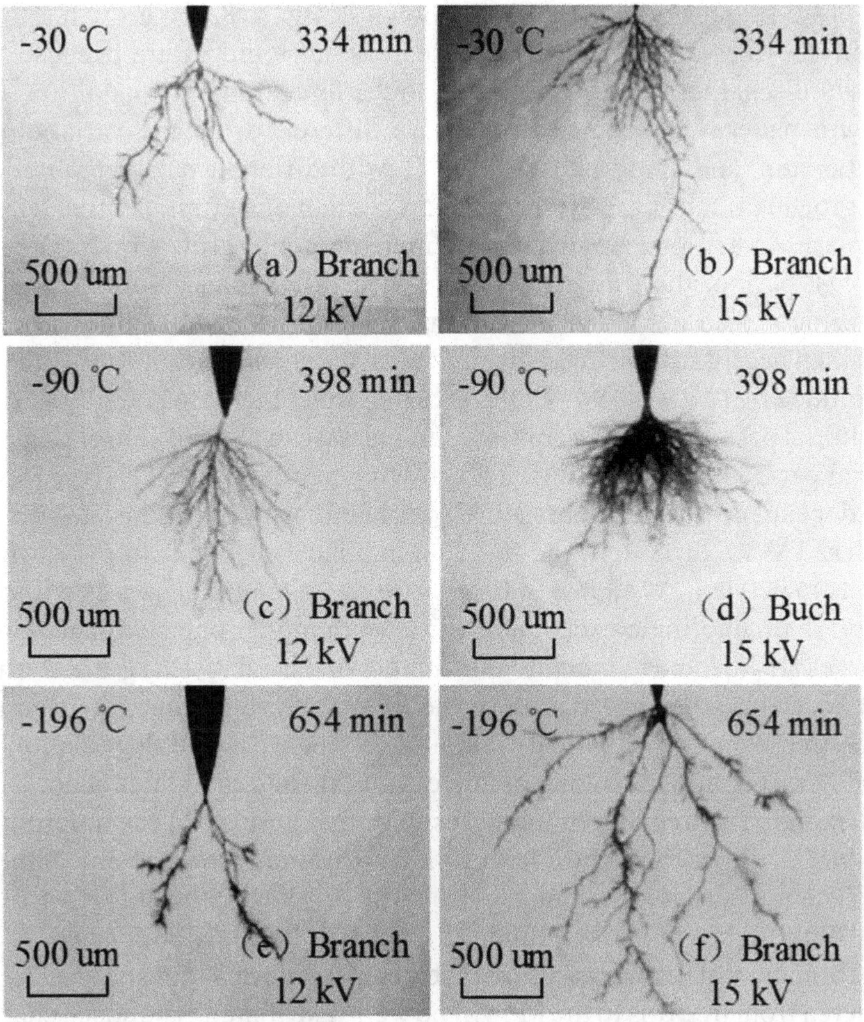

direction with the amplitude of 15 kV. With higher amplitude, charge actions are more intensive leading to more partial discharges and higher gas pressures and heat generated in tree channels would break molecules chains in random directions, so it is easier for tree channels to grow in a wider arc angle direction.

Figure 8 shows the propagation characteristics of trees in the PP with different pulse amplitudes. The pulse amplitudes are 12 and 15 kV, while the pulse frequency is 400 Hz. As shown in figure 8, the growth rate of electrical trees decreases with the decrease of temperature because the partial discharges decrease with the lower temperature. In addition, the growth patterns of electrical trees are different from the variation of temperature and amplitude. At -30 °C, with different pulse amplitudes, the growing trends of the tree length are the same. At first, the tree length increases in a high speed. And then main branches grow slowly. During the last period, the tree grows in a speed faster than before. With the amplitude of 15 kV at -90 °C, the tree length increases rapidly at first, and then nearly stops increasing, and at last increases again. But with the amplitude of 12 kV at -90 °C, the growing trend increases rapidly during the initial period, and then continues to increase in a speed slower than the initial period in the later time. The growing trends at -196 °C with 15 kV are dependent with those at -30 °C, while the growing trends at -196 °C with12 kV are different. The tree length almost suspends to grow when it gets to 920 μm. As shown in figure 8, the tree length increases with the increase of amplitude except at -90 °C. With higher pulse voltage, more charges are injected to increase the number of partial discharges activities and accelerate the degradation of the PP. At the same time, the charges also strengthen the electrical field around tree tips. The degradation of the PP will continue as long as the electrical field is high enough. This can explain the tree length increases with the increase of the amplitude. At -90 °C, the tree structure with 15 kV is different from the one with 12 kV. The tree structure is branch tree with 12 kV. It is bush tree with 15 kV. With higher pulse amplitude, the growth rate is slower because of the formation of bush area. When the bush area forms, the space charges injected from tree tips to the PP will be well-distributed. The space charge layer will form an electric field shield, resulting in the decrease of the electric field at tree tips surrounded by the space charge layer. Therefore, the growth rate of the bush tree in 15 kV is slower than that of the branch in 12 kV. It indicates that the growth rate is not only connected with the amplitude but also with the tree structure.

At different temperatures, the fractal dimension changing trends are similar. It increases rapidly during the initial period, and then continues to increase in a speed slower than the initial period in the later time.

Figure 8. The propagation characteristics of electrical trees in the PP with the pulse voltage of 12 and 15 kV at -30, -90 and -196 °C

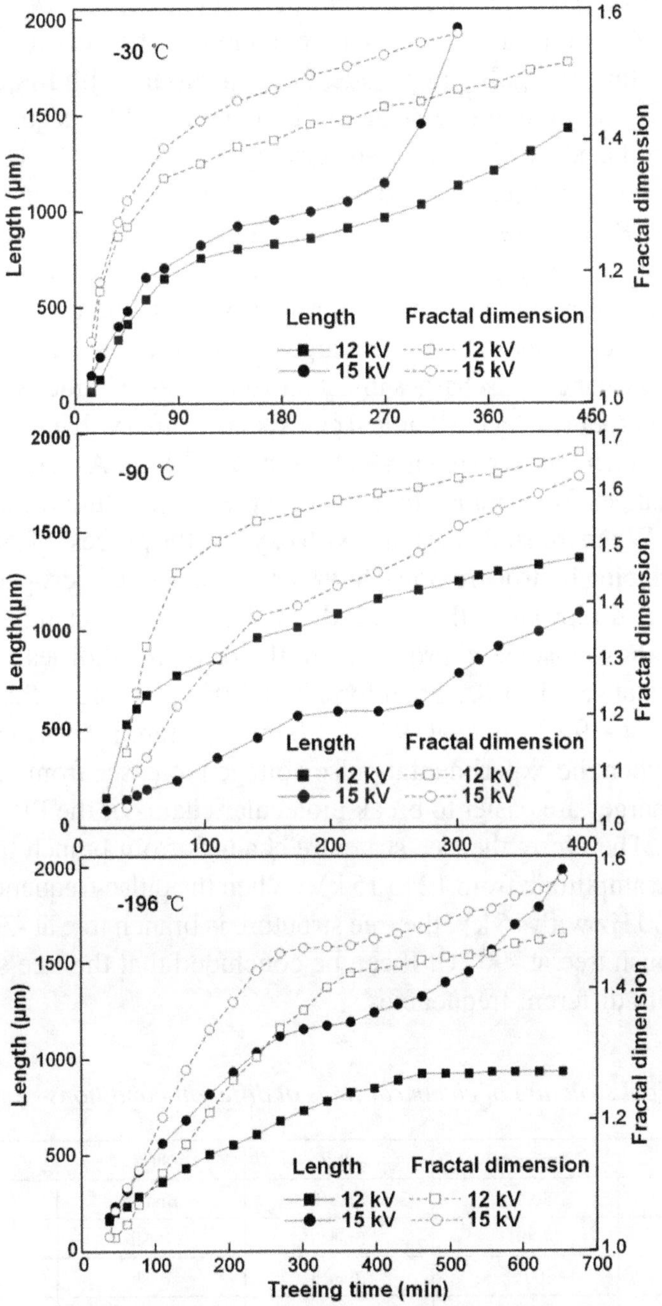

Tree Structure Distribution

Table 1 shows typical tree structure at different conditions, which occupies more than 80% of all samples. As shown in table 1, the tree structure varies with temperature and pulse amplitude, but is invariant with frequency.

At -30 °C, the branch tree structure is related to the adequate heat and pressures produced by discharges stressed by 15 kV with 400 Hz. With the decrease of temperature, the hardness and elastic modulus of PP increase which means PP is more fragile and cracks more easily at -90 °C than at -30 °C. Due to the charge injection or ultraviolet degradation, molecules chains are broken, resulting in the formation of micro voids, and then there will be electron avalanches and partial discharges in these micro voids. If charges produced by avalanche discharges inject into the existing branches or the local micro cracks produced by mechanical crack, denser area will form. Therefore, it is easier to form denser area at -90 °C than at -30 °C. As a result, the tree structure changes from branch to bush with the temperature from -30 to -90 °C in 15 kV. When free radicals react with oxygen, the process of autoxidation occurs, producing hydroperoxide. However, the unstable hydroperoxide may decompose and reproduce the free radical. The polymer chains are broken during this above discussed process. But the oxygen solidifies in polymers in the glass state at -196 °C. Therefore, it is difficult to break the molecular chains of PP at -196 °C than at -90 °C. So the structure is branch at -196 °C. At -90 °C, when the experimental pulse voltage increases from 12 to 15 kV, partial discharges are easier to break molecular chains of the PP to form the denser area. Therefore, the tree structure changes from branch tree to bush tree with the amplitude from 12 to 15 kV. When the pulse frequency are 400, 800 and 1200 Hz with 15 kV, the tree structure is branch tree at -30 and -196 °C, and is bush tree at -90 °C. It can be concluded that the tree structure is invariant with different frequencies.

Table 1. Typical structure of electrical trees at different conditions

Amplitude	Frequency	-30 °C	-90 °C	-196 °C
12 kV	400Hz	Branch	Branch	Branch
15 kV	400Hz	Branch	Bush	Branch
15 kV	800 Hz	Branch	Bush	Branch
15 kV	1200 Hz	Branch	Bush	Branch

EFFECTS OF MECHANICAL STRETCHING ON SPACE CHARGE BEHAVIORS OF PP/POE BLEND FOR HVDC CABLES

PP/POE Blend Characteristics

The uniform dispersion of POE is important to modify the mechanical flexibility of PP. The original PP/POE blend was fractured in liquid nitrogen and then etched in heptane under ultrasonic condition for 20 min to observe the phase structure by Scanning Electron Microscope (SEM) test. In order to investigate the effect of mechanical stretching, the PP/POE blend with elongation ratio λ of 1.4 was also fractured and etched, with the fracture surface parallel to the stretching direction. Figure 9a shows the dispersion of POE in original PP/POE blend. The dark holes represent POE particles, which were removed by heptane etching. It can be observed that POE particles disperse uniformly in the PP matrix with an average diameter of 1 μm. Figure 9b shows the dispersion of POE in stretched blend. Due to the external mechanical stretching, the POE particles are elongated along the stretching direction.

Space Charge Characteristic

The dynamic space charge distributions for different specimens during polarization process are shown in figure 10. In order to represent the results

Figure 9. SEM pictures of the fracture surface with elongation ratio of (a) 1 and (b) 1.4

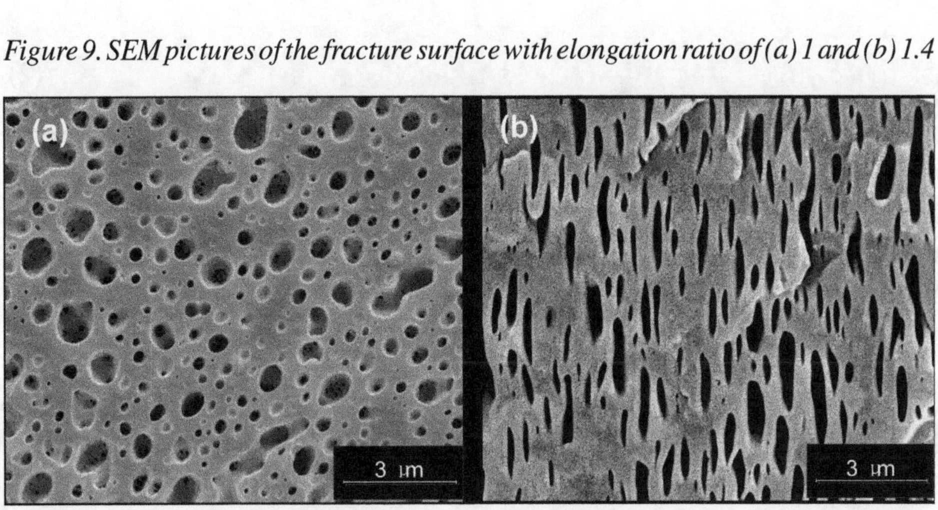

clearly, space charge distributions adjacent to cathode are also shown in the partly-enlarged graphs in each figure.

Figure 10 a shows the space charge behavior of original PP/POE blend. Homocharges are injected from the cathode, and the number of injected charges increases with increasing the polarization time. Few charges can be observed in the vicinity of anode. In the PP/POE blend, not only the chemical defects (impurities, residual catalysts, etc) can capture charges, but also the physical defects play an important role as trap sites, which include chain folding, microcavities, defects in amorphous/crystalline boundary and the interfaces between POE and PP.

Figures 10b, 10c and 10d show the space charge distributions in stretched PP/POE blend. Like the original specimen, negative charges are injected from the cathode, while the depth of charges injected and the number of space charges accumulated vary with the elongation ratios. As shown in figures 10b and 10c, negative charges can be observed adjacent to the anode within ~ 600 s, suggesting the migration of space charges is much faster than that in the original specimen. With increasing the elongation ratio to 1.4, negative charges stay near the cathode after polarization for 3600 s, indicating the migration of space charges gets slower.

For further analysis, the total number of space charges in the bulk was calculated to obtain the relationship between space charge accumulation and the elongation ratio. The following equations were used:

$$Q_{total}(t) = \int_0^L |\rho(x,t)| S dx \tag{1}$$

$$Q'_{total}(t) = Q_{total}(t)\frac{L_0}{L} \tag{2}$$

where $\rho(x,t)$ is the measured charge density in depth x (distance from the cathode) at polarization time t, S is the area of electrode and L is the thickness of specimen. Considering the difference in specimen thickness, the total number of space charges obtained were normalized to value of the same specimen thickness of L_0=180 μm.

Figure 11 shows the relationship between the normalized charge amount of different specimens and the polarization time. It is observed that the number of charges in stretched specimens is evidently higher than that in original specimens, which means the mechanical stretching can significantly promote

Figure 10. Effect of the polarization time on space charge distributions at 70 kV/mm DC field with elongation ratio of (a) λ=1; (b) λ=1.1; (c) λ=1.2; (d) λ=1.4

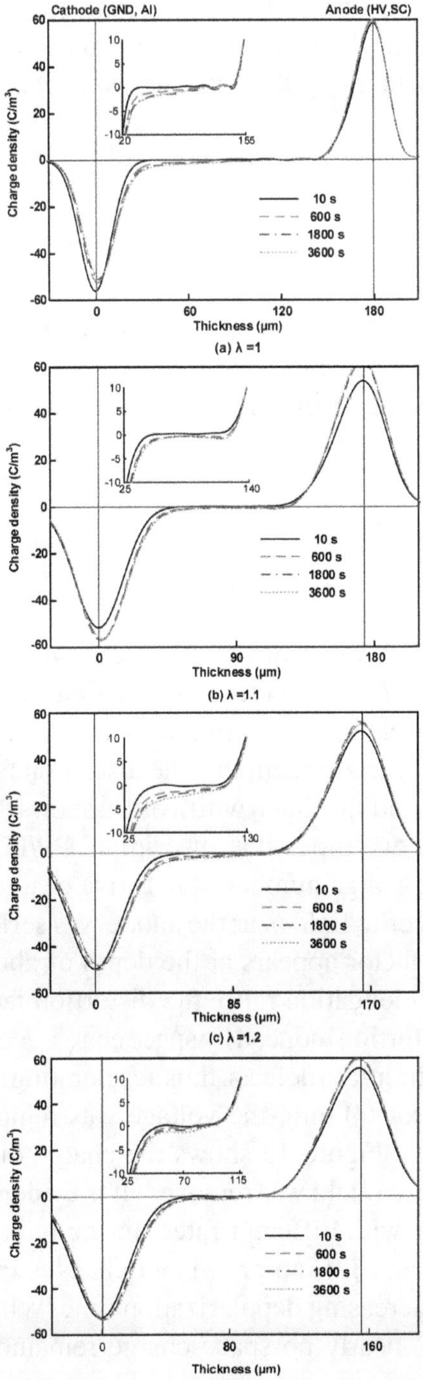

the space charge accumulation. The specimen with elongation ratio of 1.2 shows the highest charge amount during the whole polarization period.

The electric field distortion is closely related to the accumulation of space charge in specimen bulk, which is a great concern in the operation of HVDC cables. The electric field along the specimen was obtained from the space charge distribution:

$$E(x) = \frac{1}{\varepsilon_0 \varepsilon_r} \int_0^x \rho(x) dx \tag{3}$$

where $E(x)$ is the electric field in depth x, $\rho(x)$ is the measured charge density, ε_0 is the vacuum permittivity and ε_r is the relative permittivity of the specimen. Thus the distortion factor of the electric field ΔE can be obtained by the following equation:

$$\Delta E = \frac{E_{\max} - E_{av}}{E_{av}} \tag{4}$$

where E_{max} is the highest value of the electric field in specimen after DC polarization for 60 min, E_{av} is the ideal electric field without space charges.

Figure 12 shows the relationship between the electric field distortion factor and the elongation ratio. It is evident that the distortion factor is enhanced by mechanical stretching, and specimen with λ of 1.2 shows the highest distortion factor, which is about two times that of original PP/POE blend. It can be seen in figure 10c that a large number of negative charges are injected from cathode, as a result, electric field near the anode was seriously distorted, and the highest distortion factor appears at the depth of about 106.8 μm. With further increasing the elongation ratio, the distortion factor decreases. The severe electric field distortion induced by space charge accumulation may lead to partial discharge in the micro defects, thus accelerating the insulation aging.

After polarization for 60 min, the voltage was removed to observe the charge decay behaviors. Figure 13 shows the charge distributions at 3, 30 and 600 s after the applied field was removed. It is evident that space charges in specimens dissipate with different rates. In the bulk of specimens with elongation ratios of 1 and 1.4, no prominent changes in space charges can be observed with the increasing depolarization time. While for the specimen with λ of 1.2, there is nearly no space charge remained in the bulk after depolarization of 600 s.

Figure 11. Relationship between the total charge amount and the poling time

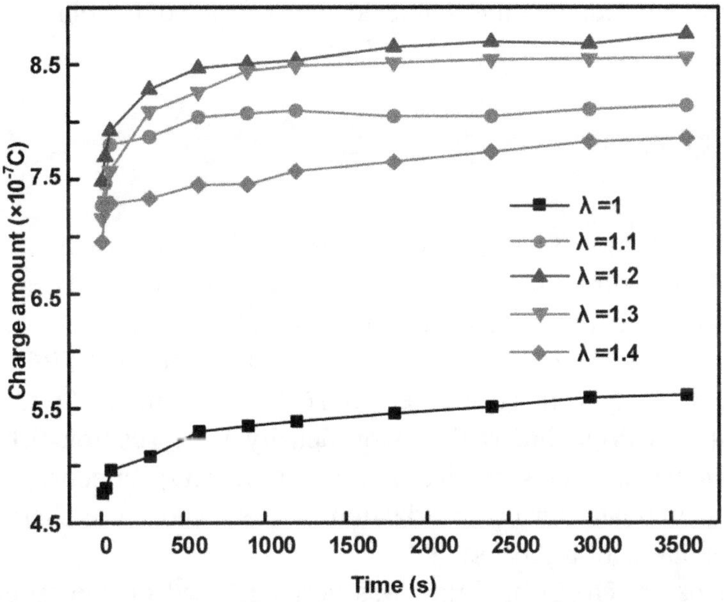

Figure 12. Distortion factor of the electric field after 1 h of DC polarization

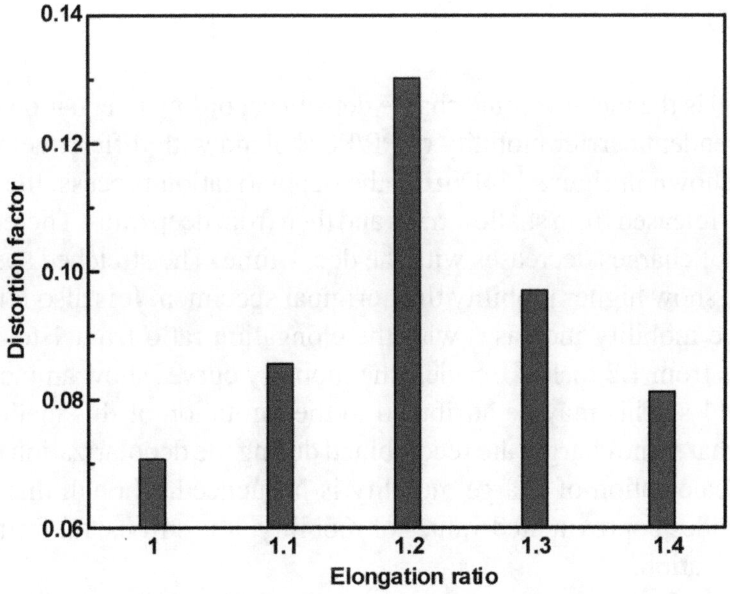

In order to quantify the space charge decay characteristics of different specimens, the mean volume charge density at time *t* of depolarization, *q(t)*, is calculated according to equation (5).

$$q(t) = \frac{1}{L}\int_0^L \left|\rho(x,t)\right| dx \tag{5}$$

The relationship between the mean volume charge density and the decay time is shown in figure 14. It is obvious that the charge density decays quickly at first and then becomes relatively stable for all the specimens. Compared with the stretched specimen, the charge density in original PP/POE blend decays very slowly, which decreases from 6.1 to 1.9 C/m^3 in 600 s. However, for 20% stretched specimens, the charge density decreases from 9.1 to 0.9 C/m^3. It obviously indicates that in the case of stretched specimens, more space charges accumulate during polarization process, and more charges decay during depolarization process.

The apparent charge mobility was derived based on the space charge decay characteristics:

$$\mu(t) = \frac{\varepsilon}{q^2(t)} \cdot \frac{dq(t)}{dt} \tag{6}$$

where *q(t)* is the mean volume charge density according to equation (5). The time-dependent carrier mobility of PP/POE blend with different elongation ratios is shown in figure 15. During the depolarization process, the charges are firstly released from shallow traps and then from deep traps. Therefore, the mobility of charges decreases with the decay time. The stretched specimens generally show higher mobility than original specimen. It is also found that the charge mobility increases with the elongation ratio from 1 to 1.2, and decreases from 1.2 to 1.4. Besides, the mobility curves show an increase in the first 11 s. This may be attributed to the limitation of this method. It is inferred that some charges are recombined during the depolarization process, thus the calculation of charge mobility is influenced. Though this method only provides approximated values of mobility, it is still useful for material characterization.

The trap depth was approximately evaluated according to the discrete-trap model proposed in (Mazzanti et al., 2003). The decay of mean volume

Figure 13. Effect of the depolarization time on space charge distributions after 1 h of DC polarization with elongation ratio of (a) λ=1; (b) λ=1.1; (c) λ=1.2; (d) λ=1.4

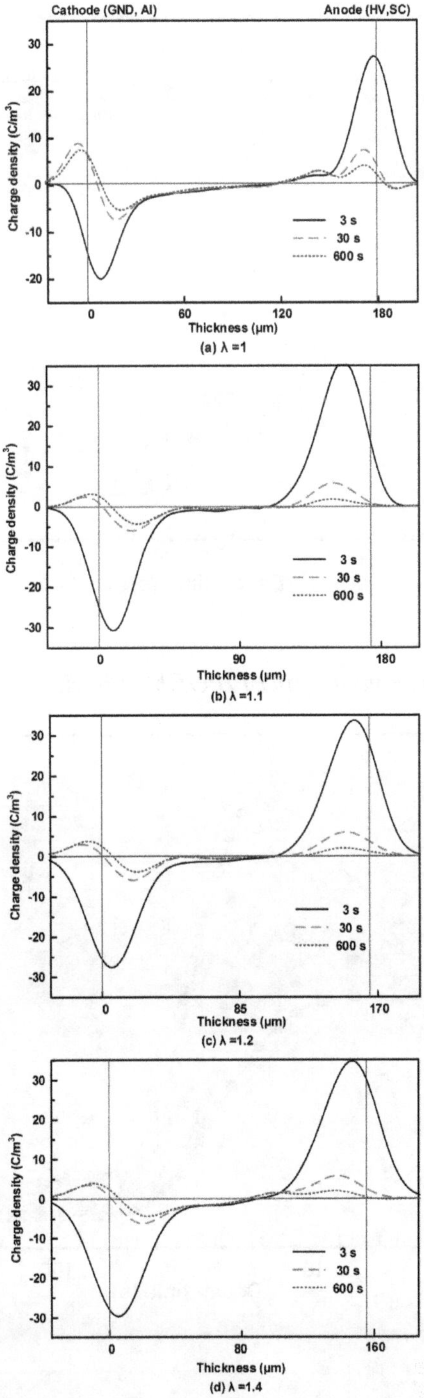

Figure 14. Relationship between the mean volume charge density and the depolarization time

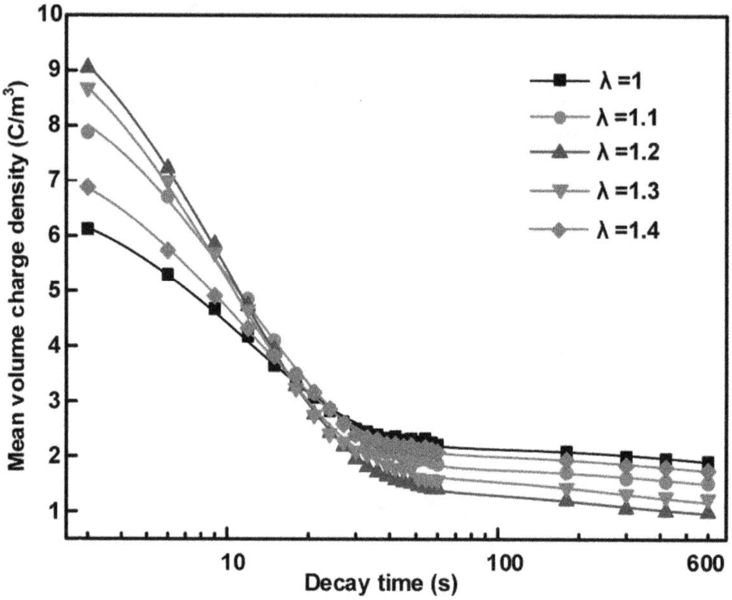

Figure 15. Apparent charge mobility in PP/POE blends with different elongation ratios

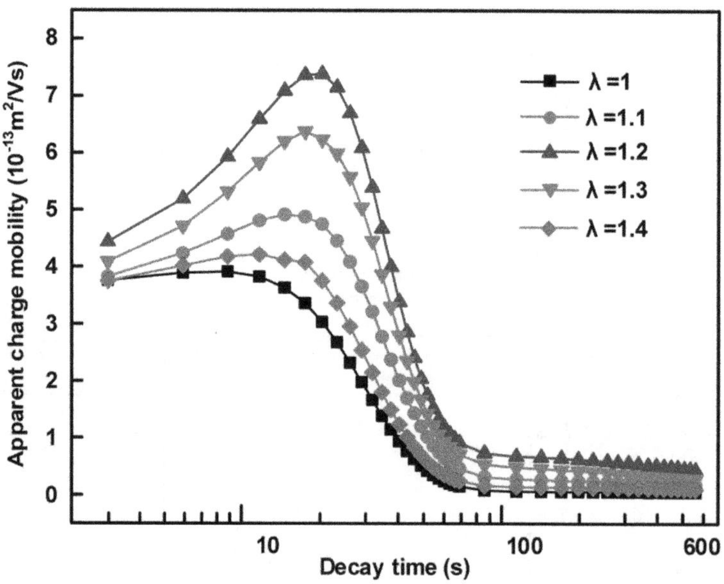

Figure 16. Apparent trap depths in PP/POE blends with different elongation ratios

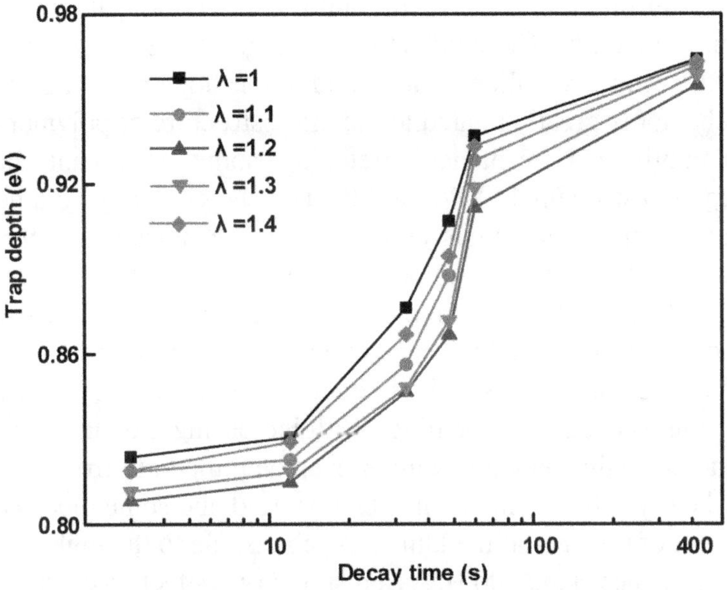

charge density can be divided into several segments, and then trap level can be estimated from each of them according to the following equation:

$$n_i(t) = n_i(0)\,e^{-vt\exp(\frac{\Delta U_i}{kT})} \quad (7)$$

where n_i is the trapped charge number density in the i th trap, t is decay time, v is the attempt frequency, k is the Boltzman constant, T is the temperature andΔU_i is the trap depth. From figure 16, it is observed that the trap depth decreases with increasing the elongation ratio from 1 to 1.2. When the elongation ratio exceeds 1.2, the trap depth increases.

Discussion

All the stretched specimens reveal a completely different behavior from the original specimen, with more space charge accumulation and faster charge decay. The change in microstructure may be responsible for the results. As semicrystalline polymers, PP usually contains crystalline lamella and amorphous region as illustrated in figure 17. The crystalline lamellar ribbons are usually in spherulitic arrays, and amorphous regions exist between the lamellar ribbons. In the crystalline lamella, molecular chains are arrayed compactly and orderly, while in the amorphous region, they are arrayed loosely and randomly. The conduction mechanisms of mobile charges are rather different in crystalline lamella and amorphous region, so the space charge behavior is greatly related to the microstructure of polymers.

Under continuous mechanical stretching, some significant changes in the aggregation structure of PP/POE blend can be gradually obtained, and a simplified schematic diagram is shown in figure 18. Influenced by relatively weak Van der Waals bonds, amorphous regions would firstly endure the external mechanical stress. Tie molecules, which transverse the amorphous region and connect two neighboring lamellas, would be extended, resulting in extended amorphous regions in the stretched polymer. The crystallites would tilt due to the mechanical stretching. With increasing the mechanical stress further, the tie chains may break up or even be pulled out from the lamella. Besides, the crystalline lamella may be distorted and some blocks can even be pulled out of the crystalline lamellar ribbons due to the highly extension of tie chains (Jones et al., 2005). As a result, more defective crystallites with smaller size can be generated during the drawing process, which provides more interfaces of crystalline/amorphous phase. The final consequence of successive mechanical stretching is the formation of cracks and fracture.

Figure 17. Idealized elements of PP morphology

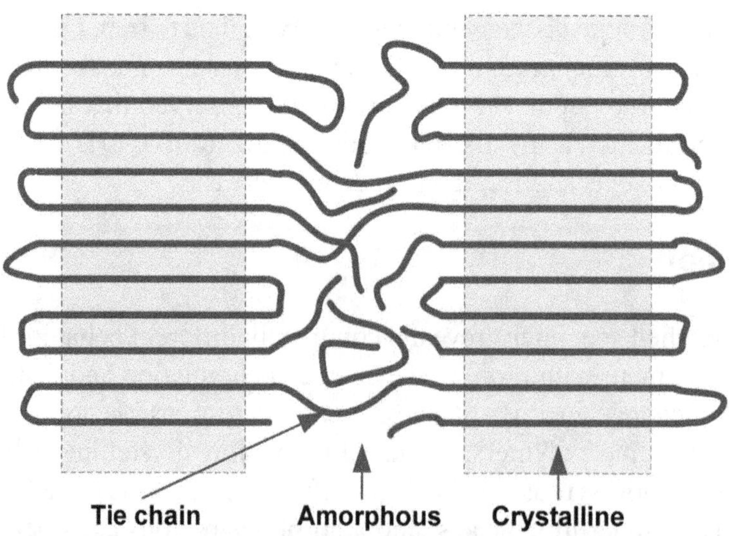

Figure 18. Morphology response of crystalline and amorphous regions to mechanical stretching

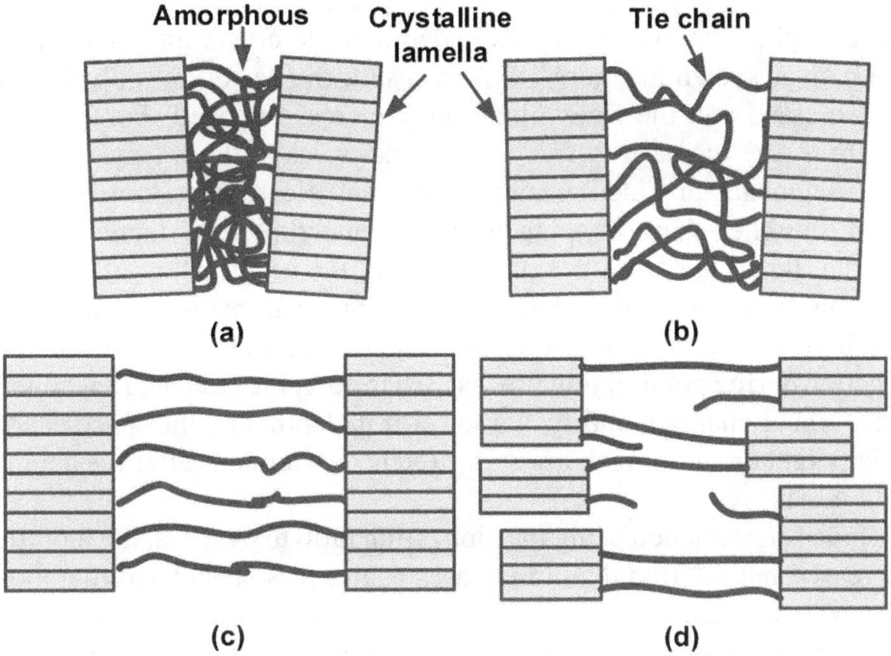

In the POE phase, most of the molecular chains array loosely and irregularly. During the mechanical stretching process, the curly molecular chains can be stretched or even broken up, and the POE particles can deform easily as shown in figure 9b. These deformed POE particles act as stress concentrators, resulting in shear yielding and/or formation of crazes in the surrounding PP matrix (Fu et al., 2011). As a result, the crack propagation resistance and toughness of PP/POE blend can be greatly enhanced.

The increased charge mobility and decreased trap depth in stretched specimens can be interpreted according to the nature of charge transport. In the specimen bulk, mobile charges consist of three distinct classes, massive molecular ions, electrons and holes, and they show very different transport characteristics. Due to the steric factors, diffusion of massive ions generally take place in free volume paths to avoid the space occupied by the polymer. As saturated hydrocarbon, both PP and POE have the negative charge affinity, making electrons transport in the intermolecular spaces between the polymer chains. In contrast to ions and electron transport, hole transport is confined to polymer chains, and long-range transport would require inter-chain hopping. All these charge transports are more likely to take place in the amorphous regions. After stretching, the amorphous region can be extended and the crystalline lamella can break up as shown in figure 18. As a result, the density of polymer can be decreased and the free volume may increase as well. For ions and electrons, the increase in free volume can weaken the interaction with chain segments, making it easier for ions and electrons to transport. For holes, the inter-chain hopping would be interrupted by increased free volume. It can therefore be concluded that the acceleration of ions and electrons transport increases the charge mobility more than the inhibition of holes transport decreases it, which may attribute to few positive charges injection during polarization process. Therefore, the stretched specimens show higher charge mobility and lower trap depth, and the space charge density remaining is much lower in stretched specimens after the applied field is removed.

However, with increasing the elongation ratio to over 1.2, the mobility decreases and the trap depth increases again. It is speculated that more microvoids, defective crystallites, interfaces of crystalline/amorphous phase are generated in highly stretched specimens, which can introduce much more trap sites and furtherly decreases the charge mobility. Besides, the binding force between the PP and POE also decreases due to stretching, and more trap sites may be introduced to inhibit the charge transport and increase the trap depth. Under continuous mechanical stretching, the POE particles play an important role in preventing crack formation, and the extensive fractures of molecular chains are resisted, which may help to resist the formation of deep trap sites.

During the polarization process, a large number of negative charges are injected from the cathode and accumulate close to the electrode in all the specimens. For the original specimen, the charge mobility is relatively lower. Therefore, fewer injected charges move into the bulk and most of them accumulate close to cathode, decreasing the effective electric field between the electrodes and the surface of the specimens and thus suppressing further space charge injection. However, in stretched specimens, higher charge mobility leads to more charges migrating into the bulk. Therefore, the inhibitory effect of firstly injected charge on charge injection is greatly migrated, as a consequence, more charges are injected into the stretched specimen bulk.

Table 2. Each type of the specimen studied

Designation	PP (wt%)	POE (wt%)	ZnO(wt%)
P100	100	0	0
P60E40	60	40	0
P60E40N0.5	60	40	0.5
P60E40N2	60	40	2

SPACE CHARGE BEHAVIORS OF PP/POE/ZnO NANOCOMPOSITES FOR HVDC CABLES

Analysis Methods

The designation, composition and weight ratio of each type of blend are given in table 2.

The trap level distributions in PP/POE blend and its nanocomposites were measured with the isothermal discharge current (MIDC) test, according to Lei's theory (Lei et al., 2010). Tian et al successfully measured the trap level distributions of LDPE and its nanocomposites with this method (Tian et al., 2015; Tian et al., 2012). In the MIDC test, charges were firstly injected into the specimen under -50 kV/mm for 60 min at 50 °C. Then at the same temperature, the specimen was short-circuited to record the discharge current caused by detraped electrons.

In the case of a continual trap level distribution with the function $N_t(E)$, which is spatially uniform deep into the sample with a distance l, the induced current in the external circuit during IDC measurement can be approximatively calculated as follows (Li et al., 1994):

$$I = \frac{el^2kT}{2dt} f_0(E)N_t(E) \tag{8}$$

where e is the electronic charge quantity; k is the Boltzmann constant; t is the measurement time; d is the thickness of the film; E is the trap level; $f_0(E)$ is the initial occupancy of a trap level and is approximately equal to 1 based on the assumption that all the traps are fulfilled in this experiment.

The energy gap can be defined as (Sjöstedt et al., 2009; Watson, 1990):

$$E_t = E_c - E_m = kT\ln(\nu t) \tag{9}$$

Then the trap level density distribution can be calculated based on equations (8) and (9):

$$\frac{N_t(E)}{E_t} = \frac{It}{\frac{el^2kT}{2d} f_0(E)kT\ln(\nu t)} \tag{10}$$

Sample Characteristic

In order to investigate the phase structure of PP/POE blend, the specimen fractured surface were etched by heptane in ultrasonic condition for 20 min and then washed with alcohol before Scanning Electron Microscope (SEM) test. The dispersion of POE in the blend can be observed in figure 19a. The dark holes represent POE particles, which were removed by heptane etching. POE particles disperse uniformly in the PP matrix with an average diameter of 1 μm, forming a typical sea-island structure. Figure 19b shows the distribution of nano-ZnO particles in the PP bulk. It can be seen that the distribution is satisfactory and the nano particles are uniformly scattered.

Figure 20 shows the relationship between the relative permittivity and the frequency. It can be seen that though nano-ZnO particles with higher

Figure 19. SEM pictures of the specimen fracture surface; (a) etched PP/POE blend; (b) nano-ZnO particles distribution

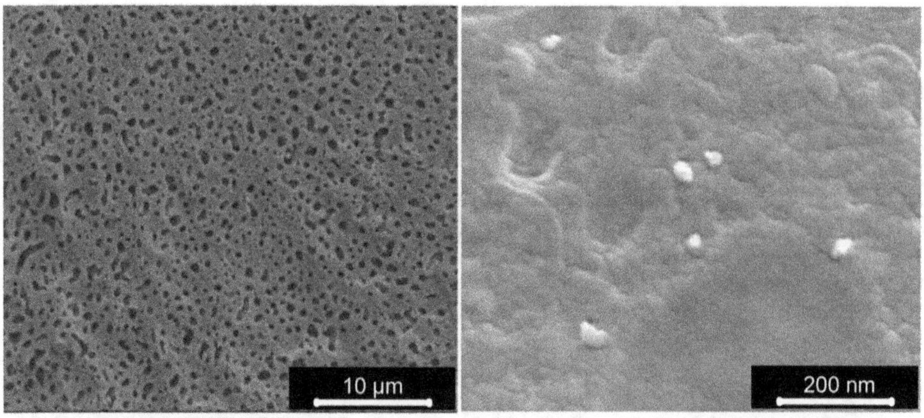

(a) etched PP/POE blend (b) nano-ZnO particles distribution

Figure 20. Relationship between the relative permittivity and the frequency

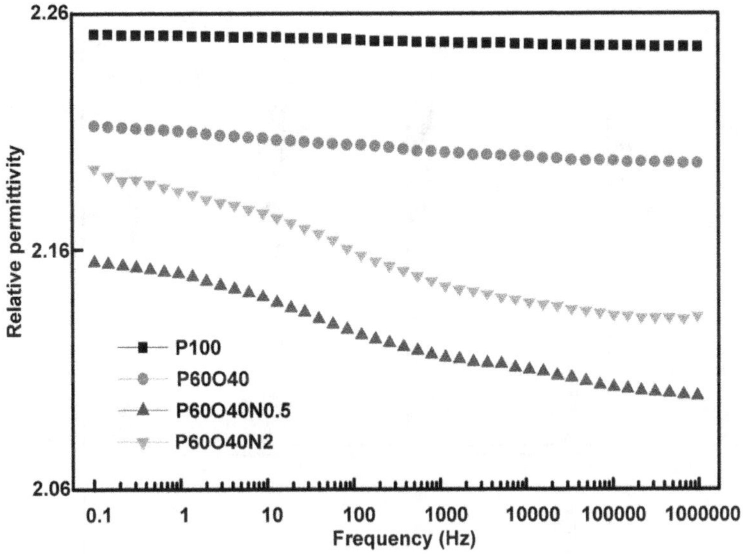

permittivity are introduced, the relative permittivity of the nanocomposites is lower than that of PP/POE blend. This is consistent with the earlier results in the literature that the permittivity values were lower than that of the unfilled polymer. Due to the interaction between the nanoparticle surface and the polymer, the motility of polymer chains as well as the polarization mechanisms in the bulk of materials are restricted, therefore, the nanocomposites show lower relative permittivity than PP/POE blend.

Mechanical property is an important factor for cable insulation application. In order to characterize the mechanical properties, tensile properties for pure PP, PP/POE and its nanocomposites have been obtained. The values of breaking elongation are shown in figure 21a. Due to the rubber elastic properties of POE, the breaking elongation of PP/POE is greatly increased compared to the pure PP, which indicates high mechanical flexibility for cable insulation application. Though the breaking elongation decreases a little with the addition of nano ZnO, it is still much higher than that of pure PP. Figure 21b shows the tensile strength of different specimens, which is an indicator of rigidity. Due to the low tensile strength of POE, the tensile strength of PP/POE is lower than that of PP. With the addition of nano-sized ZnO, the tensile strength is increased due to the formation of interaction between nano-ZnO and the polymer chain, which compensates the decrease of toughness in nanocomposites. It can be seen that the PP/POE/ZnO nanocomposites

Figure 21. Mechanical characteristics of different specimens; (a) breaking elongation; (b) tensile strength

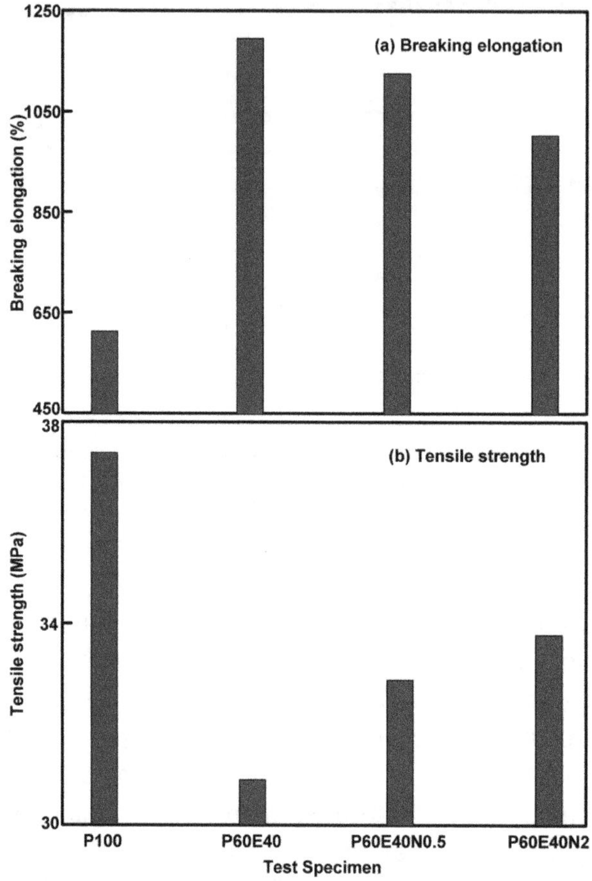

shows high mechanical flexibility and a relatively higher tensile strength than PP/POE blend.

Space Charge Characteristic

Figure 22 shows the dynamic space charge distributions of different specimens under a dc electric field of 70 kV/mm for 3600 s. In order to represent the results clearly, space charge distributions in the thickness range of 35 to 215 μm are also shown in the partly-enlarged graphs. For referencing purpose, Figure 6a shows the space charge behavior of pure PP. Due to the semicrystalline nature of PP, the interfaces between the crystalline and amorphous phases can introduce trapping sites, which would capture the injected charges. With increasing the poling time, homocharges injected from the cathode increase, and heterocharges also appear in the vicinity of anode. It is obvious that the space charge development in the pure PP bulk is minimal.

Comparing the space charge accumulation results in figure 22a and 22b, it is found that much more charges are injected from cathode in the PP/POE blend. With the lapse of time, the injected charges increase and advance deeper into the middle of the specimen, which is related to the microstructure of the PP/POE blend. With the addition of POE, the interfaces between POE and PP introduce many trapping sites, which can easily capture injected charges during the polarization process.

Just by loading the PP/POE blend with a small amount of nano-ZnO, the space charge behavior is improved, as shown in figure 22c and 22d. The charge accumulation in the vicinity of cathode is substantially reduced, and the injected charges move much more slowly compared to the PP/POE blend and accumulate in the vicinity of electrodes. Obvious positive charge injection from anode can be seen in the case of P60E40N2, with low charge density and deep injection distance of about 125 μm. On the whole, the space charge behaviors of nanocomposites are dramatically improved with the addition of nano-ZnO compared to that of PP/POE blend.

The total absolute charge amount for each specimen was calculated according to the following equation:

$$Q(t) = \int_0^L \left|\rho(x,t)\right| S dx \tag{11}$$

where $\rho(x,t)$ is the charge density in depth x (distance from the cathode) at poling time t, S is the electrode area, L is the thickness of the specimen. The calculated total charge of each specimen is shown in figure 23. With the same applied stress, the total charge of pure PP shows the highest start point and is relatively stable. For PP/POE blend, the total charge amount is the highest and keeps increasing with poling time. The total charge amounts of PP/POE/ZnO nanocomposites increase at the initial period and then become stable with poling time. The total charge amount of different specimens after poling for 1 h is in this order: P60E40> P100>P60E40N2>P60E40N0.5.

As the electric field distortion due to space charge formation under DC stress is a great concern, the electric field along the specimen was calculated according to the following equation:

Figure 22. Effect of the poling time at 70 kV/mm DC field on space charge distribution of (a) P100; (b) P60E40; (c) P60E40N0.5; (d) P60E40N2

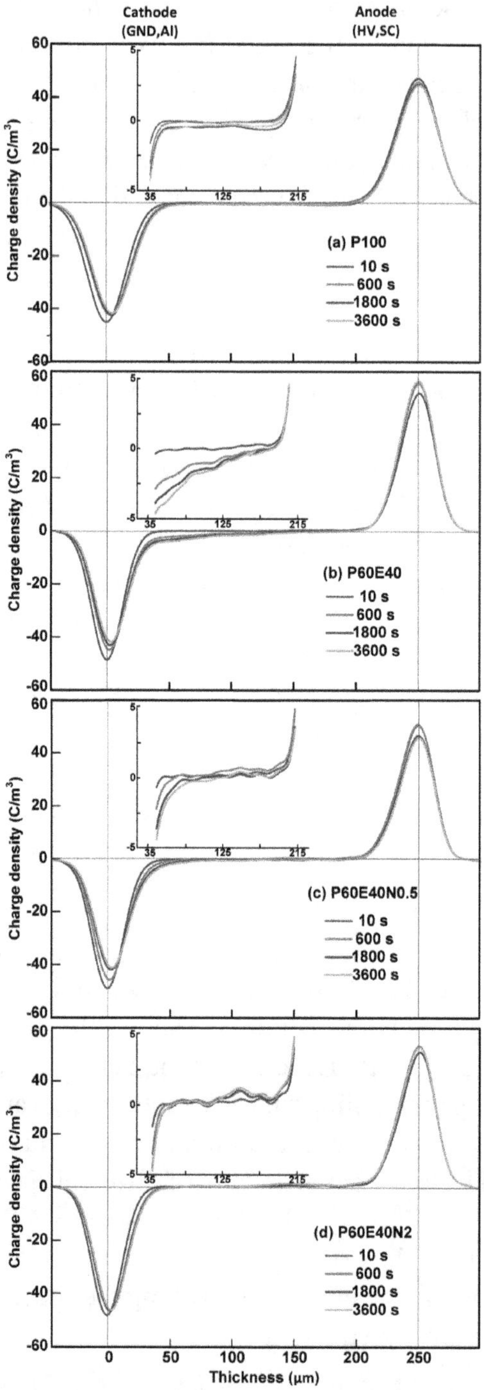

Figure 23. Relationship between the total charge and the poling time

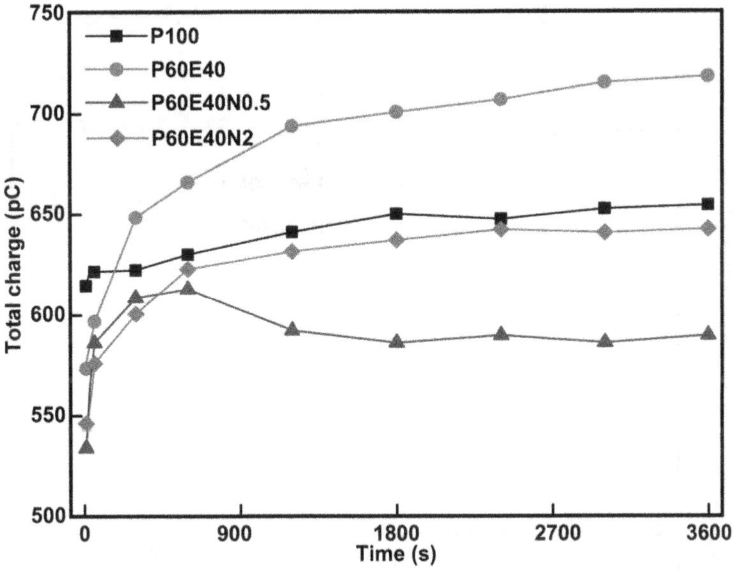

$$E(x) = \frac{1}{\varepsilon_0 \varepsilon_r} \int_0^L \rho(x) dx \tag{12}$$

where $\rho(x)$ is the charge density, ε_0 is the vacuum permittivity, ε_r is the relative permittivity of the specimen.

The electric field distortion results at the polarization time 3600 s are shown in figure 24. It can be seen that the electric field is greatly distorted in PP/POE blend bulk compared to that of pure PP. For PP/POE blend, the electric field is inhibited in the vicinity of cathode, while it is enhanced in the vicinity of anode, which is attributed to the large number of homocharges injected from cathode. Comparing the electric field distribution of PP/POE blend and its nanocomposites, electric field distortion in 0.5 wt% and 2 wt% nanocomposites gets better. It is observed that the electric field of P60E40N0.5 is in close proximity to the ideal electric field of 70 kV/mm, indicating that the electric field distribution is rather uniform in the polymer. In the case of 2 wt% nanocomposites, the electric field in the vicinity of cathode gets enhanced, which is related to the homocharges injection from anode as shown in figure 22d.

The distortion factor of the electric field ΔE can be obtained by the following equation:

Figure 24. Electric field distributions after 1 h of DC poling

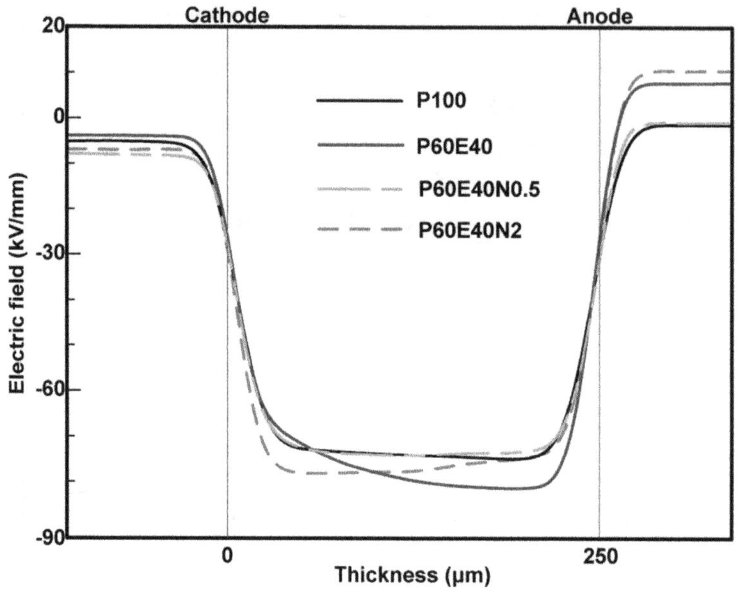

$$\Delta E = \frac{E_{max} - E_{av}}{E_{av}} \tag{13}$$

where the E_{max} is the highest value of the electric field in the specimen bulk after DC poling of 1 h; E_{av} is the ideal electric field in the absence of space charge. The results are shown in figure 25. It is observed that the PP/POE blend shows the highest distortion factor, which is increased by 4 times as compared with pure PP. The severe electric field distortion induced by space charge formation may lead to partial discharge in the micro defects, thus accelerating the insulation aging. Nanoparticles doping can greatly mitigate the electrical field distortion. The distortion factor of specimen with 0.5 wt% nanoparticles is much lower than those of the other specimens, indicating uniform electric field in the polymer.

The comparison of total charge quantity during the depolarization process is described in figure 25. Clearly, the quantity of space charges decays rapidly at the initial period and keeps invariant after about one minute in all specimens. It can be seen that after depolarization for 600 s, the quantity of space charge in pure PP bulk is the least, while large number of space charges still remain in the PP/POE blend. Besides, all PP/POE/ZnO nanocomposites show lower quantity of space charge than the PP/POE blend.

Figure 25. Electric field enhancements after 1 h of DC poling

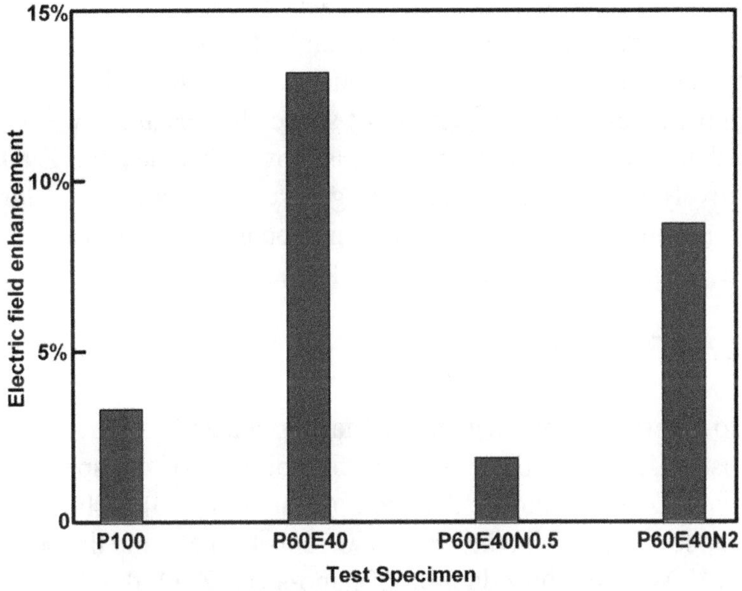

Figure 26. Relationship between total charge and the depolarization time

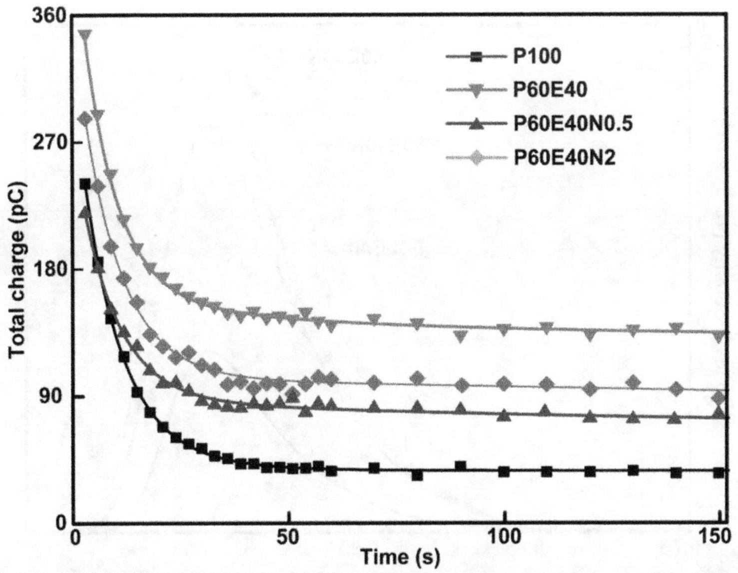

Compared to the results of pure PP, the addition of POE into polypropylene matrix results in more space charge accumulation in the vicinity of cathode and higher electric field distortion factor, which is unexpected for insulating materials in HVDC cable application. While all the PP/POE/ZnO nanocomposites show lower quantity of space charges and lower distortion factor for than those of PP/POE blend. It is proved that the nano-ZnO doping can effectively suppress the space charge accumulation and electric field distortion in polymer, which makes the nanocomposites more suitable for HVDC cable application than PP/POE blend.

Trap Level Distribution

In order to investigate the influence of nano-sized ZnO on the space charge characteristics furtherly, the trap level distributions of the specimens with different nano-ZnO contents calculated by equation (10) is shown in figure 27. The trap levels are all in the range of 0.75 to 1.15 eV, and the peaks locate at ~ 1.03 eV. With the introduction of nano-sized ZnO, the deep trap level

Figure 27. Relationship between the mobility and the depolarization time

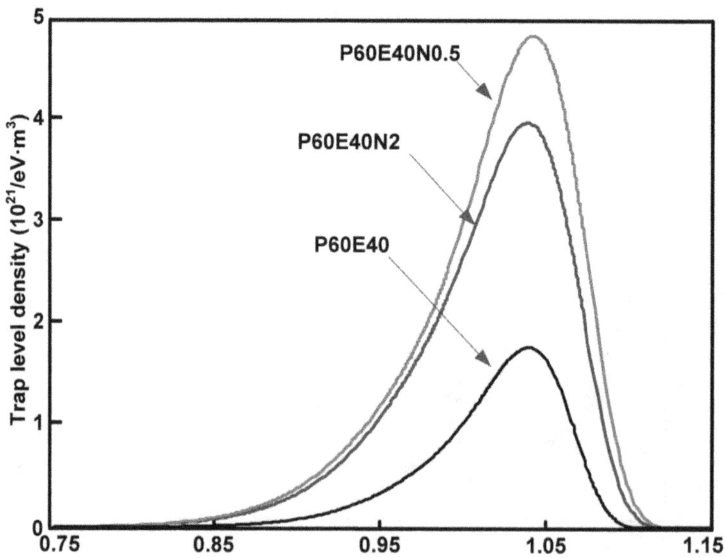

density increases, and the trap density peak of P60E40N0.5 is about 3 times that of PP/POE blend.

The results of trap level distribution clearly indicate that nano-ZnO particles introduce large number of deep trap states into the specimens. Due to the large specific surface area of nanoparticles, extensive interface areas are formed on the boundary between nano-ZnO and the polymer matrix. In interface areas, polymers chains are strongly bound and/or interacted to the nanoparticles, which results in partly immobility of the polymer chains together with the space steric hindrance of the nanoparticles. Therefore, the local properties (chain arrangement, free volume, mobility, etc.) can be changed, which may result in the increase in the trap level density (Dissado et al., 1997; Montanari, 2011; Fu et al., 2008; Tanaka et al., 2011; Calebrese et al., 2011).

The increased trap density is related to the space charge reduction in nanocomposites. The introduced deep trapping states can easily capture a number of electrons or holes. These localized charges in nanocomposites are less mobile, which is confirmed by the slow transport of injected charges shown in figure 22. Moreover, the firstly trapped homocharges accumulate in the vicinity of electrodes, which can result in an independent electric field and decrease the effective electric field between the electrodes and the surface of the specimens, thus forming a blocking layer for further charge injection and suppressing the space charge formation in the specimen bulk. The fact that the nanoparticles doping can result in the increase of trap level density and the suppression of the space charge formation have been confirmed by extensive researches in the literature (Li et al., 1994; Mazzanti et al., 2003).

REFERENCES

Calebrese, C., Le, H., Schadler, L. S., & Nelson, J. K. (2011). A review on the importance of nanocomposite processing to enhance electrical insulation. *IEEE Transactions on Dielectrics and Electrical Insulation*, *18*(4), 938–945. doi:10.1109/TDEI.2011.5976079

Chen, G., Hao, M., Xu, Z., Vaughan, A., Cao, J., & Wang, H. (2015). Review of High Voltage Direct Current Cables. *CSEE J. Power and Energy Systems*, 9-21.

Chen, G., & Tham, C. H. (2009). Electrical Treeing Characteristics in XLPE Power Cable Insulation in Frequency Range Between 20 and 500 Hz. *IEEE Transactions on Dielectrics and Electrical Insulation*, *16*(1), 179–188. doi:10.1109/TDEI.2009.4784566

Crine, J. P. (2005). Influence of Electro-mechanical Stress on Electrical Properties of Dielectric Polymers. *IEEE Transactions on Dielectrics and Electrical Insulation*, *12*(4), 791–800. doi:10.1109/TDEI.2005.1511104

Dakka, M. A., Bulinski, A., & Bamji, S. S. (2010). Space Charge Evolution in Polypropylene Containing Synthetic and Natural Organoclays. *IEEE Conf. Electr. Insul. Dielectr. Phenomena*, 1-4. doi:10.1109/CEIDP.2010.5724020

David, E., Parpal, J. L., & Crine, J. P. (1996). Influence of Internal Mechanical Stress and Strain on Electrical Performance of Polyethylene Electrical Treeing Resistance. *IEEE Transactions on Dielectrics and Electrical Insulation*, *3*(2), 248–257. doi:10.1109/94.486777

Dissado, L. A., Mazzanti, G., & Montanari, G. C. (1997). The Role of Trapped Space Charge in the Electrical Aging of Insulating Materials. *IEEE Transactions on Dielectrics and Electrical Insulation*, *4*(5), 496–506. doi:10.1109/94.625642

Dissado, L. A., Mazzanti, G., & Montanari, G. C. (2002). Elemental Strain and Trapped Space Charge in Thermoelectrical Aging of Insulating Materials Part 1: Elemental Strain Under Thermo-Electrical-Mechanical Stress. *IEEE Transactions on Dielectrics and Electrical Insulation*, *8*(6), 959–965. doi:10.1109/94.971452

Du, B. X., Han, T., & Su, J. G. (2014). Effect of Low Temperature on Tree Characteristics in Silicone Rubber with Different Power Frequency. *IEEE Transactions on Dielectrics and Electrical Insulation*, *21*(4), 1880–1886. doi:10.1109/TDEI.2014.004312

Du, B. X., Han, T., & Su, J. G. (2014). Effect of Low Temperature on Tree Characteristics in Silicone Rubber with Different Power Frequency. *IEEE Transactions on Dielectrics and Electrical Insulation*, *21*(4), 1680–1686.

Du, B. X., Han, T., & Su, J. G. (2015). Electrical Tree Characteristics in Silicone Rubber under Repetitive Pulse Voltage. *IEEE Transactions on Dielectrics and Electrical Insulation*, *22*(2), 720–727. doi:10.1109/TDEI.2015.7076767

Du, B. X., Ma, Z. L., Gao, Y., & Han, T. (2011). Effect of Ambient Temperature on Electrical Treeing Characteristics in Silicone Rubber. *IEEE Transactions on Dielectrics and Electrical Insulation*, *18*(2), 401–407. doi:10.1109/TDEI.2011.5739443

Du, B. X., & Zhu, L. W. (2015). Electrical Tree Characteristics of XLPE under Repetitive Pulse Voltage in Low Temperature. *IEEE Transactions on Dielectrics and Electrical Insulation*, *22*(4), 1801–1808. doi:10.1109/TDEI.2015.005183

Fabiani, D., Montanari, G. C., Cavallini, A., & Mazzanti, G. (2004). Relation Between Space Charge Accumulation and Partial Discharge Activity in Enameled Wires Under PWM-like Voltage Waveforms. *IEEE Transactions on Dielectrics and Electrical Insulation*, *11*(3), 393–405. doi:10.1109/TDEI.2004.1306718

Fu, M., Dissado, L. A., Chen, G., & Fothergill, J. C. (2008). Space charge formation and its modified electric field under applied voltage reversal and temperature gradient in XLPE cable. *IEEE Transactions on Dielectrics and Electrical Insulation*, *15*(3), 851–860. doi:10.1109/TDEI.2008.4543123

Fu, Z., Dai, W. L., Yu, H. M., Zou, X. X., & Chen, B. Q. (2011). Effect of Composition on Fracture Behavior of Polypropylene–wollastonite–polyolefin Elastomer System. *Journal of Materials Science*, *46*(5), 1272–1280. doi:10.1007/s10853-010-4911-8

Gao, Y., Chen, Q. G., Chi, M. H., & Lin, L. (2013). Flow Electrification Characteristics of Oil-Pressboard Insulation under Polarity Reversal Electric Field. *IEEE. Conf. Electr. Insul. Dielectr. Phenomena (CEIDP)*, 1022-1025.

He, Z. M., Li, J., Jiang, T. Y., Bao, L. W., & Cheng, C. K. (2013). Partial Discharge Characteristics Influenced by Different Temperatures under Pulsating DC Voltages. *IEEE. Conf. Solid Dielectr.s (ICSD)*, 1001-1004.

Holto, J., & Ildstad, E. (2010). Electrical Treeing in Extruded Polypropylene High Voltage Cable Insulation. *IEEE. Conf. High Voltage Eng. Application (ICHVE)*, 373-376.

Holto, J., Ildstad, E., & Lunde, I. (2012). Characteristics of Electrical Treeing in Extruded Polypropylene and Cross-Linked Polyethylene Cable Insulation. *IEEE. Conf. High Voltage Eng. Application (ICHVE)*, 63-67.

Jones, J. P., Llewellyn, J. P., & Lewis, T. J. (2005). The Contribution of Field-Induced Morphological Change to the Electrical Aging and Breakdown of Polyethylene. *IEEE Transactions on Dielectrics and Electrical Insulation*, *12*(05), 951–966. doi:10.1109/TDEI.2005.1522189

Kudo, K. (1998). Fractal Analysis of Electrical Trees. *IEEE Transactions on Dielectrics and Electrical Insulation*, *5*(5), 713–727. doi:10.1109/94.729694

Lanca, M. C., Marat-Mendes, J. N., & Dissado, L. A. (2001). The Fractal Analysis of Water Trees - An Estimate of the Fractal Dimension. *IEEE Transactions on Dielectrics and Electrical Insulation*, *8*(5), 838–844. doi:10.1109/94.959714

Lei, Q. Q., Tian, F. Q., Yang, C., He, L. J., & Wang, Y. (2010). Modified Isothermal Discharge Current Theory and its Application in the Determination of Trap Level Distribution in Polyimide Films. *Journal of Electrostatics*, *68*(3), 243–248. doi:10.1016/j.elstat.2010.01.006

Li, J. Y., Zhou, F. S., Min, D. M., Li, S. T., & Xia, R. (2015). The Enerngy Distribution of Trapped Charges in Polymers Based on Isothermal Surface Potential Decay Model. *IEEE Transactions on Dielectrics and Electrical Insulation*, *22*(3), 1723–1732. doi:10.1109/TDEI.2015.7116370

Li, S. T., Yin, G. L., Chen, G. H., Li, J. Y., Bai, S. N., Zhong, L. S., & Lei, Q. Q. et al. (2010). Short-Term Breakdown and Long-Term Failure in Nanodielectrics: A Review. *IEEE Transactions on Dielectrics and Electrical Insulation*, *17*(5), 1523–1535. doi:10.1109/TDEI.2010.5595554

Li, S. T., Zhao, N., Nie, Y. J., Wang, X., Chen, G., & Teyssedre, G. (2015). Space Charge Characteristics of LDPE Nanocomposite/LDPE Insulation System. *IEEE Transactions on Dielectrics and Electrical Insulation*, *22*(1), 92–100. doi:10.1109/TDEI.2014.004524

Li, Y., Yasuda, M., & Takada, T. (1994). Pulsed Electroacoustic Method for Measurement of Charge Accumulation in Solid Dielectrics. *IEEE Transactions on Dielectrics and Electrical Insulation*, *1*(2), 188–195. doi:10.1109/94.300251

Li, Y., Yasuda, M., & Takada, T. (1994). Pulsed electroacoustic method for measurement of charge accumulation in solid dielectrics. *IEEE Transactions on Dielectrics and Electrical Insulation*, *1*(2), 188–195. doi:10.1109/94.300251

Maezawa, T., Taima, J., Hayase, Y., Tanaka, Y., Takada, T., Sekiguchi, Y., & Murata, Y. (2007). Space Charge Formation in LDPE/MgO Nano-composite under High Electric Field at High Temperature. *IEEE Conf. Electr. Insul. Dielectr. Phenomena*, 271-274.

Mazzanti, G., Montanari, G. C., & Alison, J. M. (2003). A Space-charge Based Method for the Estimation of Apparent Mobility and Trap Depth as Markers for Insulation Degradation Theoretical Basis and Experimental Validation. *IEEE Transactions on Dielectrics and Electrical Insulation*, *10*(2), 187–197. doi:10.1109/TDEI.2003.1194099

Mazzanti, G., Montanari, G. C., & Dissado, L. A. (2001). Elemental Strain and Trapped Space Charge in Thermoelectrical Aging of Insulating Materials Life Modeling. *IEEE Transactions on Dielectrics and Electrical Insulation*, *8*(6), 966–971. doi:10.1109/94.971453

Mazzanti, G., Montanari, G. C., & Dissado, L. A. (2011). Elemental Strain and Trapped Space Charge in Thermoelectrical Aging of Insulating Materials Life Modeling. *IEEE Transactions on Dielectrics and Electrical Insulation*, *8*(6), 966–971. doi:10.1109/94.971453

Mita, S., & Yahagi, K. (1975). Effect of Elongation on Dielectric Breakdown Strength in Polyethylene. *Japanese Journal of Applied Physics*, *14*(2), 197–201. doi:10.1143/JJAP.14.197

Montanari, G. C. (2011). Bringing an Insulation to Failure: The Role of Space Charge. *IEEE Transactions on Dielectrics and Electrical Insulation*, *18*(2), 339–364. doi:10.1109/TDEI.2011.5739438

Montanari, G. C., Fabiani, D., Palmieri, F., Kaempfer, D., Thomann, R., & Mulhaupt, R. (2004). Modification of electrical properties and performance of EVA and PP insulation through nanostructure by organophilic silicates. *IEEE Transactions on Dielectrics and Electrical Insulation*, *11*(5), 754–762. doi:10.1109/TDEI.2004.1349780

Murata, Y., & Kanaoka, M. (2006). Development History of HVDC Extruded Cable with Nanocomposite Material. *8th Int'l. Conf. Properties and Applications of Dielectr. Materials*, 460-463. doi:10.1109/ICPADM.2006.284214

Nelson, J. K., & Fothergill, J. C. (2004). Internal Charge Behaviour of Nanocomposites. *Nanotechnology*, *15*(5), 586–595. doi:10.1088/0957-4484/15/5/032

Noto, F., Yoshimura, N., & Ohta, T. (1977). Tree Initiation in Polyethylene by Application of DC and Impulse Voltage. *IEEE Transactions on Electrical Insulation*, *12*(1), 26–30. doi:10.1109/TEI.1977.298003

Roy, M., Nelson, J. K., MacCrone, R. K., & Schadler, L. S. (2007). Candidate Mechanisms Controlling the Electrical Characteristics of Silica/XLPE Nanodielectrics. *Journal of Materials Science*, *42*(11), 3789–3799. doi:10.1007/s10853-006-0413-0

Sjöstedt, H., Gubanski, S. M., & Serdyuk, Y. V. (2009). Charging Characteristics of EPDM and Silicone Rubbers Deduced from Surface Potential Measurements. *IEEE Transactions on Dielectrics and Electrical Insulation*, *16*(3), 696–703. doi:10.1109/TDEI.2009.5128508

Stone, G. C., van Heeswijkb, R. G., & Bartnikas, R. (1992). Electrical Aging and Electroluminescence in Epoxy under Repetitive Voltage Surges. *IEEE Transactions on Electrical Insulation*, *27*(2), 233–244. doi:10.1109/14.135595

Takada, T., Hayase, Y., Tanaka, Y., & Tatsuki, O. (2008). Space Charge Trapping in Electrical Potential Well Caused by Permanent and Induced Dipoles for LDPE/Mgo Nanocomposite. *IEEE Transactions on Dielectrics and Electrical Insulation*, *15*(1), 152–160. doi:10.1109/T-DEI.2008.4446746

Tanaka, T. (2005). Dielectric Nanocomposites with Insulating Properties. *IEEE Transactions on Dielectrics and Electrical Insulation*, *12*(5), 914–928. doi:10.1109/TDEI.2005.1522186

Tanaka, T., Bulinski, A., Castellon, J., Frechette, M., Gubanski, S., Kindersberger, J., & Han, S. J. et al. (2011). Dielectric Properties of XLPE/SiO_2 Nanocomposites Based on CIGRE WG D1.24 Cooperative Test Results. *IEEE Transactions on Dielectrics and Electrical Insulation*, *18*(5), 1484–1517. doi:10.1109/TDEI.2011.6032819

Tanaka, T., Fukuda, T., & Suzuki, S. (1976). Water Tree Formation and Lifetime Estimation in 3.3 kV and 6.6 kV XLPE and PE Power Cables. *IEEE Transactions on Power Apparatus and Systems*, *95*(6), 1892–1900. doi:10.1109/T-PAS.1976.32290

Tanaka, T., Kozako, M., Fuse, N., & Ohki, Y. (2005). Proposal of A Multi-Core Model for Polymer Nanocomposite Dielectrics. *IEEE Transactions on Dielectrics and Electrical Insulation*, *12*(4), 669–681. doi:10.1109/TDEI.2005.1511092

Tanaka, T., Montanari, G. C., & Mulhaupt, R. (2004). Polymer Nanocomposites as Dielectrics and Electrical Insulation - Perspectives For Processing Technologies, Material Characterization and Future Applications. *IEEE Transactions on Dielectrics and Electrical Insulation*, *11*(5), 763–784. doi:10.1109/TDEI.2004.1349782

Tian, F. Q., Lei, Q. Q., Wang, X., & Wang, L. (2012). Investigation of Electrical Properties of LDPE/ZnO Nanocomposite Dielectrics. *IEEE Transactions on Dielectrics and Electrical Insulation*, *19*(3), 763–769. doi:10.1109/TDEI.2012.6215078

Tian, F. Q., Yao, J. Y., Li, P., Wang, Y., Wu, M. L., & Lei, Q. Q. (2015). Stepwise Electric Field Induced Charging Current and Its Correlation with Space Charge Formation in LDPE/ZnO Nanocomposite. *IEEE Transactions on Dielectrics and Electrical Insulation*, *22*(2), 1232–1239. doi:10.1109/TDEI.2015.7076826

Vu, T. T. N., Teyssedre, G., Vissouvanadin, B., Le Roy, S., & Laurent, C. (2015). Correlating Conductivity and Space Charge Measurements in Multi-dielectrics Under Various Electrical and Thermal Stresses. *IEEE Transactions on Dielectrics and Electrical Insulation*, *22*(1), 117–127. doi:10.1109/TDEI.2014.004507

Watson, P. K. (1990). The Energy Distribution of Localized States in Polystyrene, Based on Isothermal Discharge Measurements. *Journal of Physics. D, Applied Physics*, *23*(12), 1479–1484. doi:10.1088/0022-3727/23/12/002

Yahagi, K. (1980). Dielectric Properties and Morphology in Polyethylene. *IEEE Transactions on Electrical Insulation*, *15*(3), 241–250. doi:10.1109/TEI.1980.298316

Chapter 5
HVDC Cable Accessory Insulation

ABSTRACT

HVDC cables have shown great advantages for long-distance, high-power underground or underwater transmission. The performance of cable accessories, as an essential part of the HVDC networks, is a great concern to the reliability of the system. However, Cable accessories made of ethylene-propylene-diene terpolymer (EPDM), which are considered to be the weakest part of HVDC cable system, have to face the problem of space charge accumulation. Nanosized particles have been proved as an effective method to suppress the space charge accumulation in dielectric composites. This chapter presents a study aimed at clarifying the effect of nanoparticles and direct-fluorination on space charge behaviors for HVDC cable accessory insulation. Obtained results show that the dielectric properties and DC conduction for HVDC cable accessory insulation are significantly influenced and the interface charge density can be effectively suppressed by doped fillers.

INTRODUCTION

HVDC cables have shown great advantages as they combine features of the lower energy losses with the same current, lower line costs, unique practical feasibility and the transmission capability for long-distance, high-power underground or underwater transmission. Polymer insulated HVDC cables have demonstrated many advantages as they combine traits of the material

DOI: 10.4018/978-1-5225-2309-3.ch005

and the transmission capability for long-distance, high-power underground or underwater transmission (Y. Zhou et al.,2015; M. Marzinotto and G. Mazzanti,2015; G. Chen et al.,2015; W. Long and S. Nilsson,2007). The performance of cable accessories, as an essential part of the HVDC networks, is of great concern to the reliability of the system, due to the electrical, thermal and mechanical designs of these structures coupled with high risks in environmental pollution and installs errors. However, under high DC electric stress, it is likely to accumulate charges at the interface of materials with different conductivity and permittivity, which will lead to accelerated degradation of insulation systems under various stress conditions, particularly after polarity reversal, leading to material degradation and possibly premature failure (S. Delpino et al.,2008; B. X. Du et al.,2016; M. L. Fu et al.,2008). The interface charge behavior and its impact become much more unpredictable considering the complex geometry, doping of fillers, etc.

Ethylene-propylene-diene Terpolymer (EPDM) composites are used extensively in the cable accessories and provide a better performance compared to traditional materials, which can reduce the physical size of HVDC cable components and the production costs. Besides, EVA composites are also used in the cable accessories for making heat shrinkable insulation, semi-conductive insulation jackets and cable accessory insulation, due to its ability to accept high filler proportions without embrittlement or loss of mechanical integrity, and the ease of crosslinking. Cable accessories can be made by incorporation of specific conductive additives, nonlinear resistive fillers, etc. (L. Donzel et al.,2011; I. Tamayo et al.,2011; L. Gao et al.,2016), which are used in cable accessories to uniform electric field and prevent interface flashover or breakdown. However, the cable accessories have to confront the problem of interface charge accumulation.

Although the formation of interface charge has been extensively investigated today, most of them focus on the interface charge or electric field distribution under multiple stresses. Few researches have studied the interface charge suppression methods, so it is still a major issue for the development of polymeric HVDC cable accessories. Vu et al.(2015) reported the interface charge accumulation in the XLPE/rubber system and the electric field distributions derived from integration over position of space charge profiles. Li et al. (2015) reported the charge transport characteristics and the interface space charge formation as well as the effects of nano-sized TiO_2 concentration in LDPE and doped LDPE films system. Lan et al. (2013) investigated the effects of surface fluorination for LDPE on space charge behaviors at LDPE/EPDM interface and revealed that the property of the interface charge changed a lot after the LDPE layer was surface fluorinated. Delpino et al. (2008) presented the results of space charge and conduction current measurements performed on EVA filled by 5 wt% montmorillonite nanolayer and observed large space charge and conduction current magnitude. Fleming et al. (2008) studied the space charge profiles of low density polyethylene doped with nano and micro-sized ZnO particles by using the laser intensity modulated method. Du et al. (2014) introduced this method to improve the electrical properties of silicone rubber, and pointed out that the fluorinated layer in SiR surface would capture charges and form a shielding layer, which could block further charge injection and suppress the surface charge accumulation.

Nanosized particles have been proved as an effective method to suppress the space charge accumulation in dielectric composites (T. J. Lewis,2014). The mechanism of inorganic particles on the electrical and mechanical properties of polymers is of great interest currently. Many reports revealed that the interfaces between additives and the polymer bulk account for the space charge suppression (T. Tanaka,2005). However, the mechanism of effects of direct-fluorination on space charge behaviors of polymers was still rudimentary, and little attention was paid to the effect of fluorination on the interface charge behavior of cable accessories. So, in this chapter, these methods which can apparently improve the insulation properties for cable accessories are discussed.

INTERFACE CHARGE BEHAVIORS BETWEEN LDPE AND EPDM FILLED WITH CARBON BLACK NANOPARTICLES

Sample Characteristics

In this section, nanocomposites specimens were prepared by dispersing nano-sized carbon black (CB) into the EPDM with various weight ratios of 0, 0.5, 1, 3 and 5 wt%. The relationship among the relative permittivity, AC conductivity, tan (delta) and the carbon black content of the EPDM/carbon black composite is shown in Figure 1, Figure 2 and Figure 3 respectively. The permittivity increases from 2.21 to 4.76 with the carbon black content varying from 0 to 5 wt%. Results show that the AC conductivity of carbon black doped samples is higher than the neat ones at higher frequencies, while an opposite result appears at lower frequencies. The AC conductivity of EPDM presents a weak relation with the frequency at lower frequencies, which is also observed in 0.5 wt% doped sample. Similar results were found in epoxy resin nanoparticles and Lewis has explained such behavior is called "quasi-dc" conduction by the O'Konski's model (J. K. Nelson and J. C Fothergill,2004). The tan(delta) results of the samples has a similar trend with the AC conductivity considering the effect of carbon black doping. Over most of the frequency range, the tan(delta) is basically independent of frequency for the samples containing carbon black particles. Surprisingly, the neat sample shows relatively large increases at the lower frequencies. For the carbon black has ultra-high conductivity and permittivity naturally, the fillers help to increase the volume conductivity and permittivity of the composite at higher frequencies. However, at lower frequencies, the nanoparticle doping introduces interface effects forming local polarization limiting region, which will affect the tan(delta) and conductivity. Or the interaction zone is shorter than the mean free path of carriers, resulting in carrier restraint in the region and lower conductivity. The lossy polarization of dissociative ions or chemical groups is weakened by the traps, accounting for tan(delta) results at lower frequencies. The overlap of the interface effects at higher fillgrade ends with more polarization loss. Besides, the overlap of highly conductive interaction zones will make a mean free path longer and increase the mobility of charge carriers, leading to a rise of the overall conductivity.

The conduction current was also measured at -15 kV/mm, reported in Figure 4. It can be noted that the conductivity of the base EPDM and nanocomposites are generally higher than that of LDPE sample. When the EPDM is doped with 0.5 and 1 wt% carbon black, the DC conduction is decreased than that of base EPDM. The carbon black doping is likely responsible for the change of conduction. Because the nanosized particles doping introduces interface effect forming local traps, which will affect the carrier transport through the sample bulk and decrease the conductivity. The currents of 3 and 5 wt% carbon black doped EPDM are higher than that of the 1 wt% doped EPDM while lower than that of the base EPDM. For the higher doping proportion, the interface effect is weakened by the overlap of interface region.

Interface Charge

The dynamic behavior of interface charge distribution in original LDPE/EPDM composite insulation is presented in Figure 5 (a). Over 1800 s, a mass

Figure 1. Relationship between the carbon black content and the relative permittivity of the EPDM/CB composite

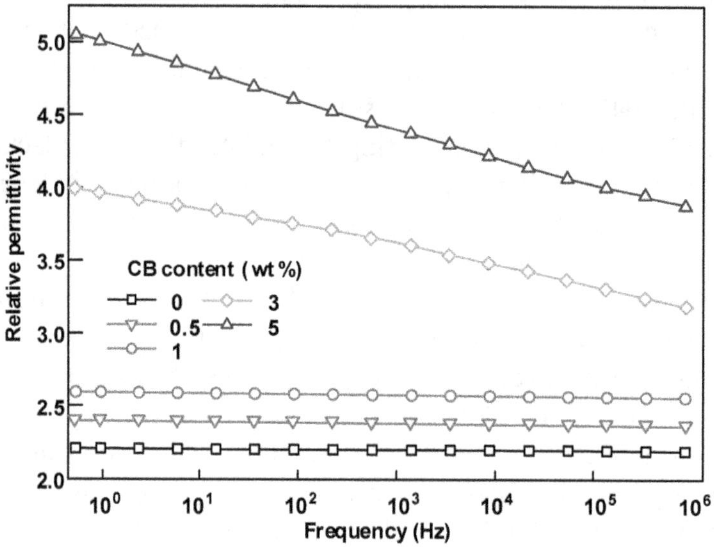

Figure 2. Relationship between the carbon black content and the AC conductivity of the EPDM/CB composite

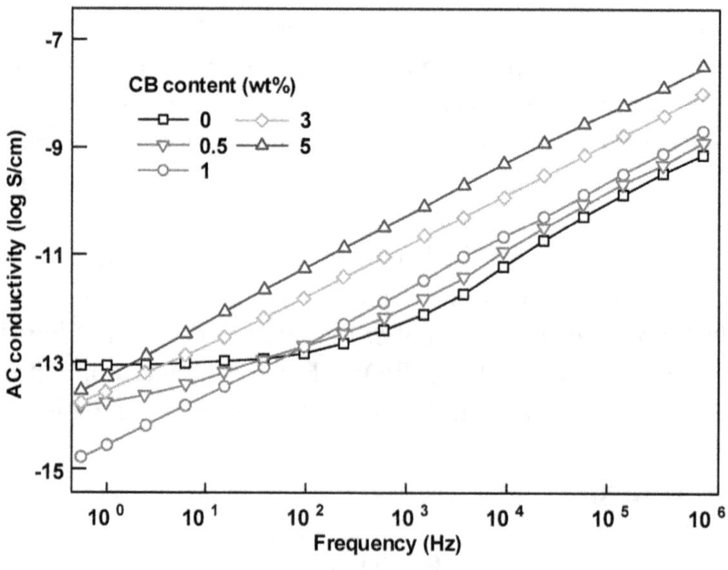

Figure 3. Relationship between the carbon black content and the tan(delta) of the EPDM/CB composite

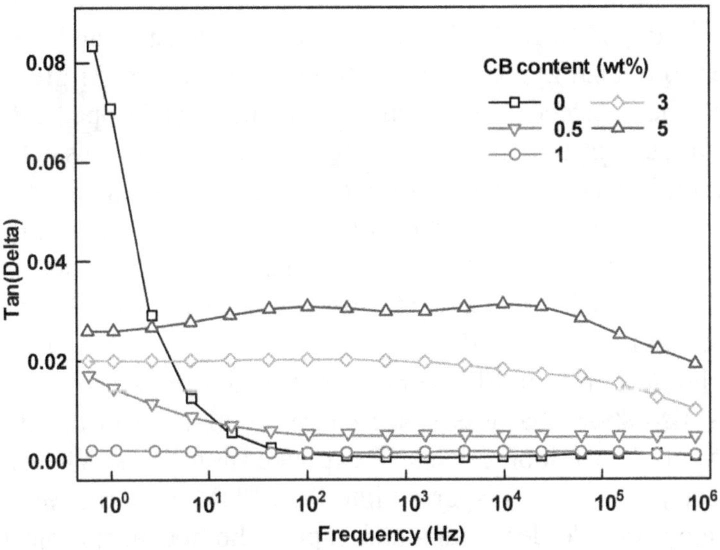

Figure 4. Relationship between the conduction currents and the time with the electric field of 15 kV/mm

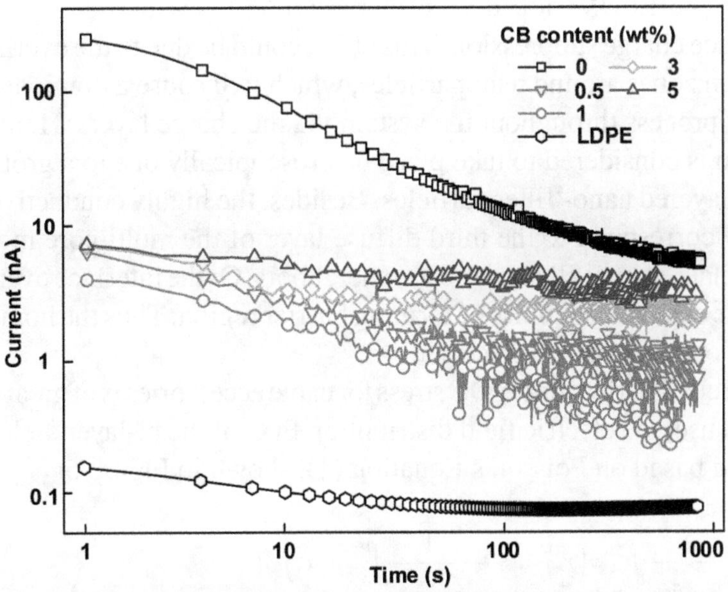

of charges are accumulated at the interface, and the maximum charge density is 1.75 C/m3. The charge polarity was positive with the undoped EPDM. Figure 5 (b) demonstrates the profiles of the interface charge behaviors with 0.5 wt% doped EPDM. The interface charge density can be effectively suppressed to 0.8 C/m3 compared with the undoped EPDM group. Figure 5 (c) shows the interface charge distribution with 1 wt% doped EPDM. The least interface charge builds up at this fillgrade and few charges in the EPDM and LDPE bulk can be measured. In Figure 5 (d), similar trends for interface charge accumulation of 3 wt% doped EPDM can be found with the charge distribution in Figure 5 (b). The value of interface charge density changed into twice or more with increasing the doping ratio to 5 wt%. Heterocharges started to accumulate in EPDM and LDPE layers approaching the interface and a few homocharges in EPDM bulk are tested.

Results also show that less space charge is blocked in the EPDM/CB composite with 1 wt% doping. It can be explained as follows. When nanosized carbon black particles are dispersed into the EPDM bulk, the well bounded interface may provide deep traps and capture homocharges injected from electrodes. These homocharges forming an additional electric field and reduce the effective electric field between the electrode and the surface of sample. Therefore, the homocharges near the electrodes produce a barrier injection, which makes the charge captured by deep traps more difficult to diffuse. However, the EPDM doped with more carbon black particles performs a worse space charge suppression. This effect could be due to the overlap of the interaction zones around nanoparticles, which will cause a dominant charge transport process throughout the system via the charge layers. Homocharge formation is considered to take place macroscopically due to a group effect of multi-layered nano- filler particles. Besides, the highly conductive region here may correspond to the third diffuse layer of the multi-core model that can percolate electrically to its nearest neighbors. Or the interface overlapping has lost its capture capability of forming barrier region. Thus the homocharge injection occurs near the electrode.

As the electric field under DC stress for cable accessories is of great concern for actual use, the electric field distribution E(x) of the bi-layer dielectrics is calculated based on Poisson's Equation (1), shown in Figure 6:

$$E(x) = -\frac{1}{\varepsilon_0 \cdot \varepsilon_{r1}} \int_0^{x_1} \rho(x_1) \cdot dx_1 - \frac{1}{\varepsilon_0 \cdot \varepsilon_{r2}} \int_{d_1}^{x_2} \rho(x_2) \cdot dx_2$$

Figure 5. Interface charge accumulation between LDPE and EPDM/CB nanocomposite under -15 kV/mm

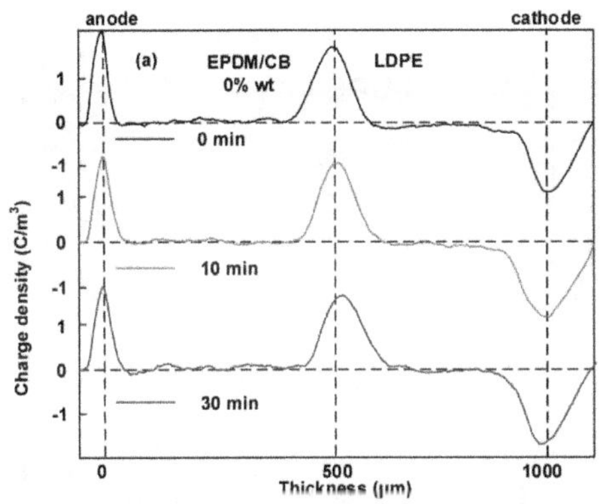

(a) 0 wt% doped EPDM

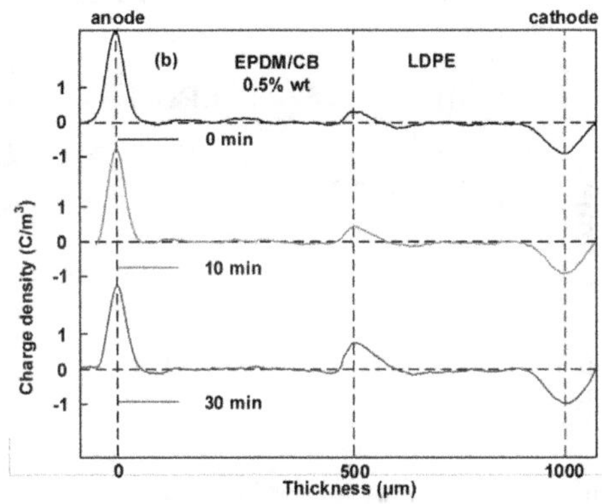

(b) 0.5 wt% doped EPDM

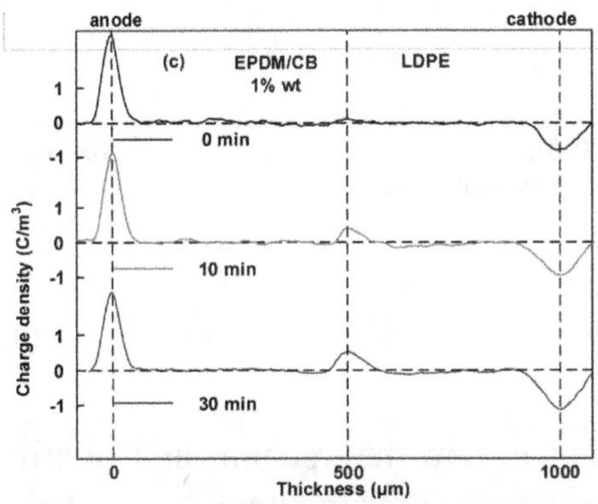

(c) 1 wt% doped EPDM

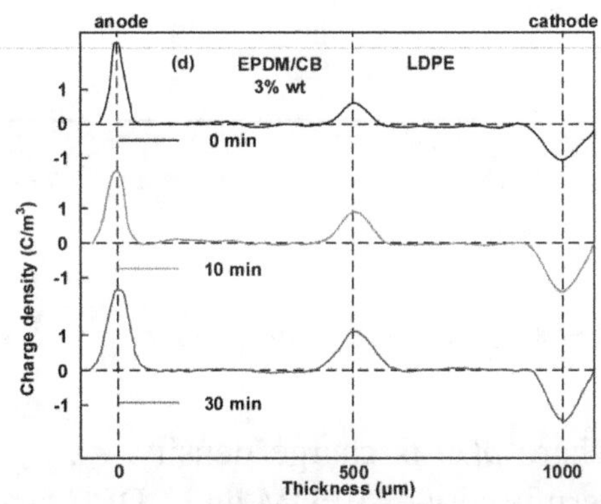

(d) 3 wt% doped EPDM

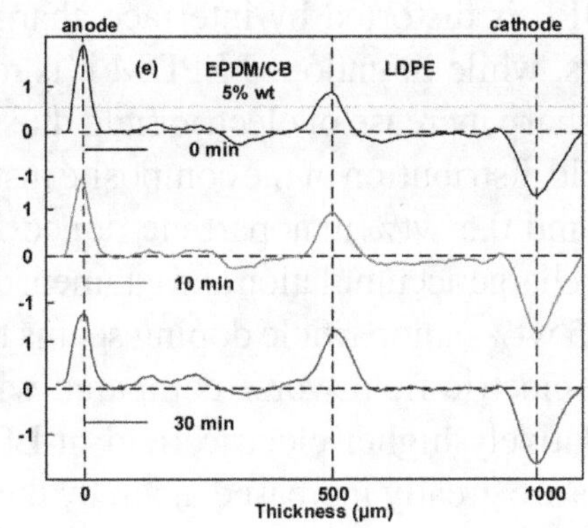

Figure 6. Electric field distribution in LDPE and EPDM/CB composites composite insulation taken after 30 min polarization under -15 kV/mm

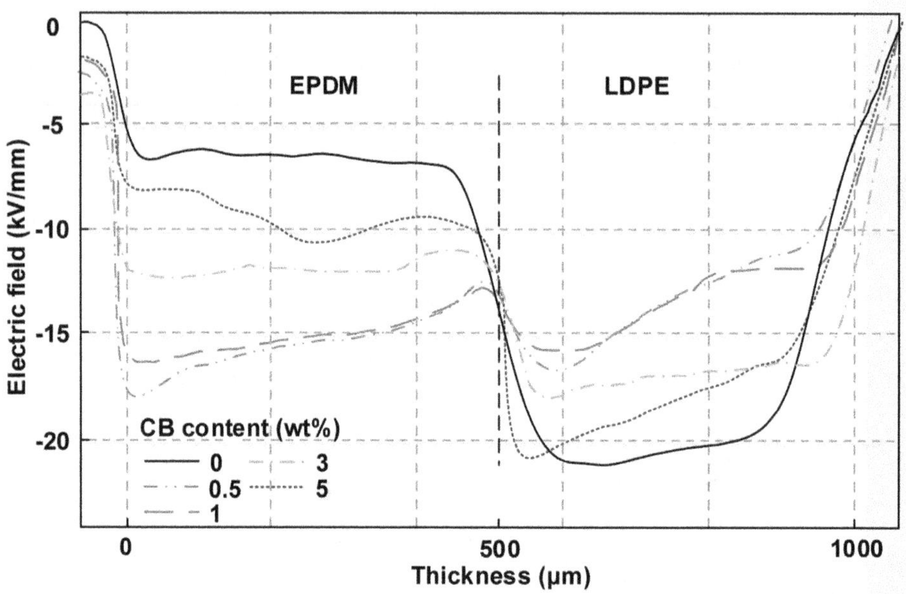

$$0 \le x_1 \le d_1; d_1 \le x_2 \le d_2 \quad (1)$$

where *ρ(x)* is charge density, ε_0, ε_{r1} and ε_{r1} are the vacuum and relative permittivity of EPDM and LDPE, and *d1* and *d2* for the boundary of the *x* are the thickness of the EPDM and LDPE sample (B. X. Du et al.,2013). The electric field in LDPE is distorted by interface charge accumulation more seriously than others, while in undoped EPDM it is relatively lower. At the interface, there is a sharp increase of electric field due to the interface charge build up. Electric field distribution of the composite insulation is significantly improved by the 1 and 0.5 wt% nanoparticle doped EPDM, for which the decrease of interface charge accumulation and augment of permittivity account. However, the 3 and 5 wt% nanoparticle doping seems to get the performance in uniformizing the electric field worse compared with the 1 wt% doping, which results in relatively higher electric field in LDPE insulation. Since the interface charges are greatly increased again by the 5 wt% doped EPDM.

Figure 7 represents interface charge profiles of EPDM/LDPE composite insulation during the depolarization after 30 min stressing. Considering the similar behaviors of 0.5, 1 and 3 wt% doped samples, the interface charge profiles of EPDM with 0, 1 and 5 wt% carbon black are demonstrated here. It can be seen for the original samples in Figure 7 (a), the total charges dissipate fairly fast at the beginning and the EPDM bulk is mostly occupied by positive charge. With the lapse of time, the interface charges are divided into negative and positive charges diffusing towards the anode and cathode respectively, under the force of electric field by the homocharge near the electrodes. Figures 7 (b) and (c) show the interface charge profiles of LDPE and carbon black doped EPDM composite insulation. Similar trends can be found in the interface charge diffusion procedure. Though the initial interface charge of group b is lower than that of the original sample, it dissipates much more slowly. The traps produced by the nanoparticle doping capture the carriers, which slows down the charge transport. While in 5 wt% carbon black doped composite insulation of group c, both the space charge in EPDM and the interface charge turns to dissipate faster. More carbon black nanoparticle may improve the charge carrier mobility during the dissipation process compared with the 1 wt% doped EPDM.

EFFECTS OF ZnO PARTICLES ON SPACE CHARGE OF EVA COPOLYMER FOR HVDC CABLE ACCESSORY INSULATION

Sample Characteristics

Samples with the thickness of ~500 μm were prepared by incorporating ZnO nanoparticles into EVA matrix with fraction of 0, 1, 5 and 10 wt% respectively, with which 5 wt% nano-sized plus 5 wt% micro-sized ZnO doped samples were chosen to be compared. The SEM photograph of 5 wt% nano ZnO doped EVA and its mixed doping composite is shown in Figure 8. In Figure 8 (a), the white spots are the nano fillers. The nano fillers are in homodisperse and no significant aggregation is observed in the samples. The larger particle in Figure 8 (b) represents the micro-sized ZnO. Figure 9 is the results of EDS from the same sample fracture surface in Figure 8 and higher content of zinc element is introduced due to more ZnO doping. The homodisperse of fillers will help to analyze the mechanism for dielectric and space charge properties.

The relationship among the relative permittivity, dissipation factor and the frequency of the EVA composite with different ZnO contents is shown in Figure 10 and 11. The permittivity increases from 2.59 to 2.92 with the ZnO

Figure 7. Interface charge profiles of EPDM/LDPE composite insulation during the depolarization after 30mins stressing

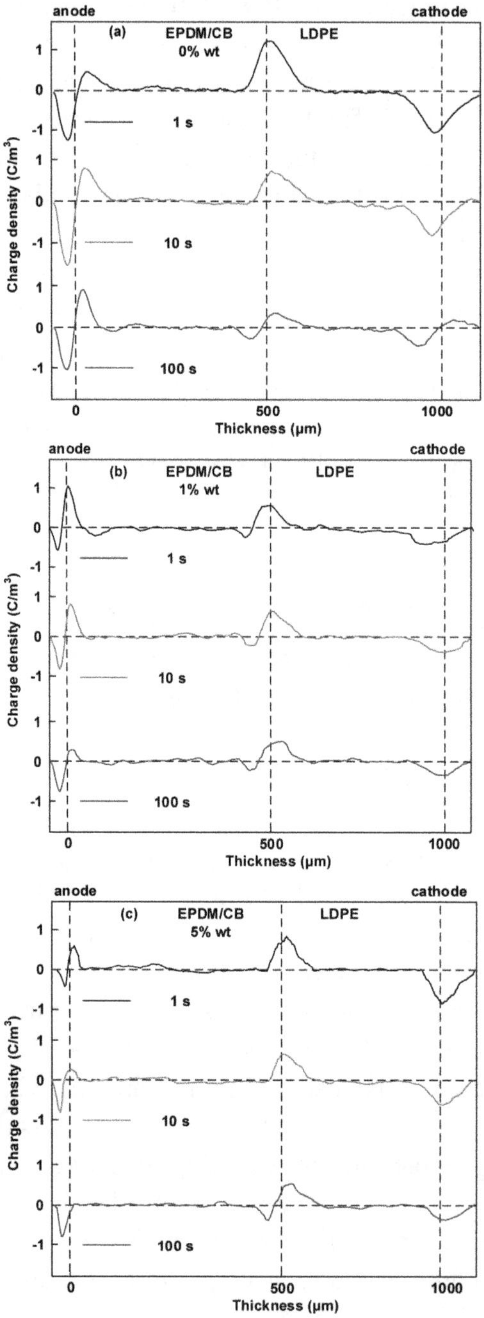

Figure 8. SEM of EVA/ZnO composite fracture surface

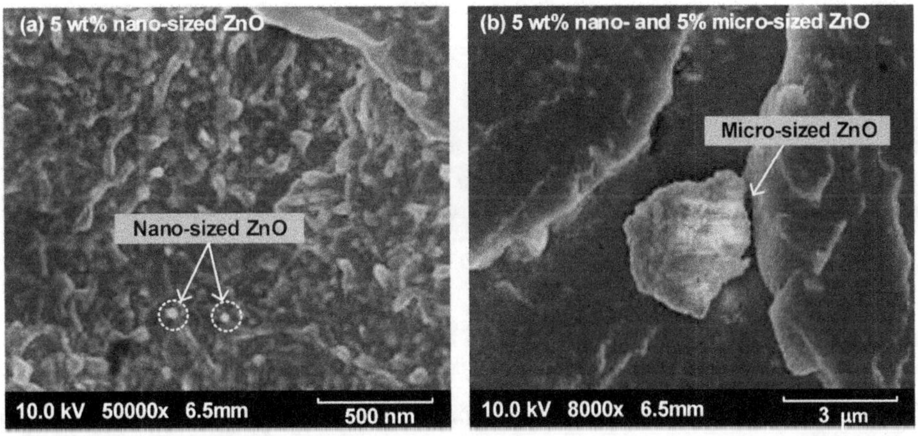

nanoparticles content varying from 0 to 10 wt%. For the sample doped with 5 wt% nanoparticles and 5 wt% microparticles, it is 2.98. The dissipation factor of the samples has a complex trend considering the effect of ZnO doping. Over most of the frequency range, the dissipation factor is largely increasing with the frequency for the samples containing ZnO particles, while the undoped sample shows an opposite trend. At the lower frequencies, dissipation factor of the 1 and 5 wt% nanoparticles doped samples is lower than that of the undoped one. For the higher doping proportion or higher frequency range, the dissipation factor is larger.

The dielectric properties are mainly determined by the dielectric polarization and relaxation mechanisms in the nanocomposites (S. Singha et al.,2008). The polarization associated with the permittivity occurs in EVA, ZnO particles and the matrix-filler interfaces. For the base EVA, the long polymer chain mobility contributes to the polarization and thus the permittivity is low. As the inorganic ZnO particles have relatively higher permittivity, the fillers help to increase the permittivity of the EVA composites with higher fillgrade. However, with less content of particles, the polarization is not increased that much as expected. The likely reason is that nanoparticles doping introduces matrix-filler interfaces, which will restrict the polymer polarization and results in a slow increase of permittivity.

The dissipation factor values varying with the frequencies can be divided into two parts considering different lossy polarization procedures. In the undoped sample, the dissipation is mainly the relaxation loss caused by the polarization of chemical groups or long chains. At lower frequencies, the

Figure 9. EDS of EVA/ZnO composite fracture surface

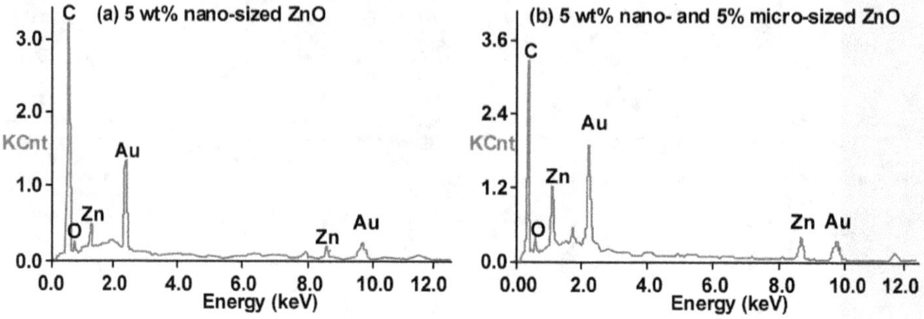

Figure 10. Relationship between the relative permittivity and the frequency with different ZnO contents

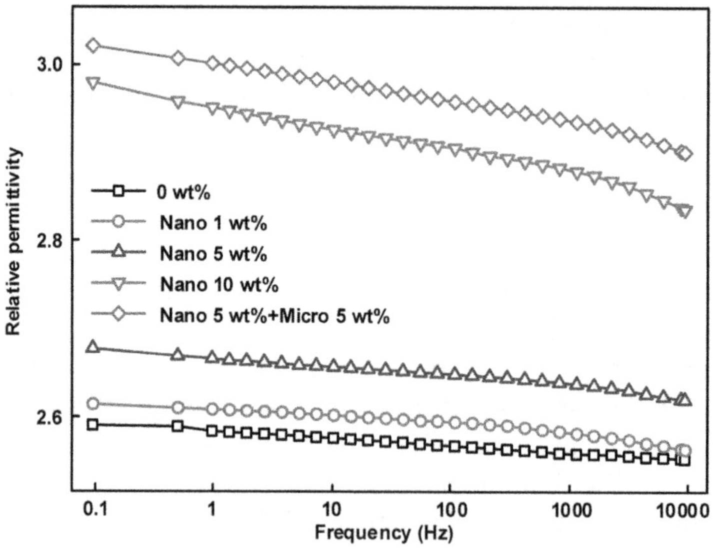

dissipation factor in nanocomposites doped with less particles is lower than the value of unfilled EVA. The nanoparticles doping introduces interface effects forming local polarization limiting region and hindering the carriers transport through different chains and interfaces, which will decrease the dissipation factor. However, over the high frequency range, the polarization will be mainly dominated by lighter electrons provided by the extra ZnO doping, which may be influenced by the overlap of interface at higher fillgrade.

Figure 11. Relationship between the disspation factor and the frequency with different ZnO contents

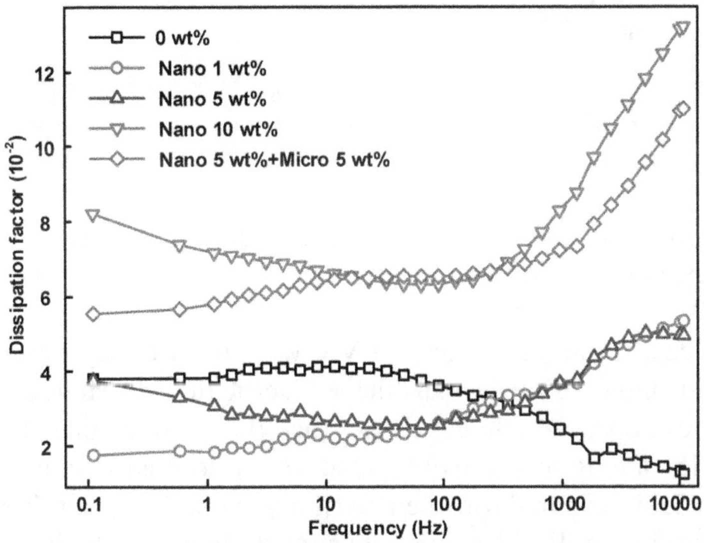

Space Charge Accumulation

Figure 12 shows the space charge profiles of EVA/ZnO composites during the polarization at average electric field of -25 kV/mm. The polarization of the first 600 seconds was recorded. In Figure 12 (a), homocharges in undoped EVA are immediately injected from the anode after the voltage is applied. With the increase of time, the charge package formed by holes continuously moves towards the cathode under theelectric stress. Obvious charge injection is not found in EVA around the cathode. As shown in Figure 12 (b), homocharges including ions are injected into the bulk through cathode, which is opposite with the undoped samples. In Figure 12 (c), the space charges in the form of homocharges distribute close to both sides of electrodes. Noticeably, the EVA composite with 5 wt% ZnO nanoparticles doping demonstrates good performance in suppressing the space charge injection in the polymer bulk. Figure 12 (d) shows that the region of space charge distribution near the cathode of 10 wt% doped sample is larger than that of group c. The space charge patterns reported in Figure 12 (e) show the homocharge can be clearly seen near both electrodes and increases with the time in samples with mixed doping. The homocharge near the cathode and anode gradually moves towards the sample bulk.

As the electric field under DC stress of the material for cable accessories is of great concern for actual use, the electric field distribution E(x) shown in Figure 13 of EVA composite across the sample was calculated based on Poisson's Equation (2):

$$E(x) = -\frac{1}{\varepsilon_0 \cdot \varepsilon_r} \int_0^x \rho(x) dx \tag{2}$$

where $\rho(x)$ is charge density, εo and εr are the vacuum and relative permittivity, and d for the boundary of the x is the thickness of the sample (Y. L. Chong et al.,2007).

The electric field in undoped EVA was distorted by space charge accumulation more seriously than others. Electric field distribution in 1 wt% nanoparticles doped sample gets better than the undoped one. The 5 wt% nanoparticles doping seems to obtain the best performance in uniformizing the electric field compared with the 10 wt% doping, which results in relatively higher electric field. Besides, the field strength in mixed doping sample is the highest with the effect of charge aggregation.

Generally, homocharges often originate from charges injected from electrodes and are determined by the balance of charge injection from the electrode and charge diffusion into the polymer bulk (F. Tian et al.,2012; D. Min et al.,2015). Or in other words, when positive/negative carriers are extracted into the cathode/anode electrodes, the homocharges remain. For the base EVA, the homocharges injected from anode move through the polymer bulk. As the molecular chains in EVA matrix are similar to that in PE containing electrons moving through the interchain space (D. Fabiani et al.,2009), it is easy for the electrons to extract into the anode, resulting in forming positive homocharges and transporting into the polymer bulk.

Since the pure EVA in this experiment has some free chains or other chemical groups, acting as charge carriers, space charge can be easily injected. However, when small amount of ZnO nanoparticles are doped, the homocharges are observed near the cathode. The inorganic particles capture the electrons and prevent them from extracting into the anode. It is prominent that the sample containing 5 wt% nanoparticles suppress the space charge injection to a large extent. Because incorporating of ZnO nanoparticles into EVA matrix can introduce a large number of traps, which can capture enough homopolar charges to weaken the effective electric field between the electrodes and the surfaces of sample and contribute to the suppression

Figure 12. Space charge profiles of EVA/ZnO composites during the polarization at the average applied field of -25 kV/mm

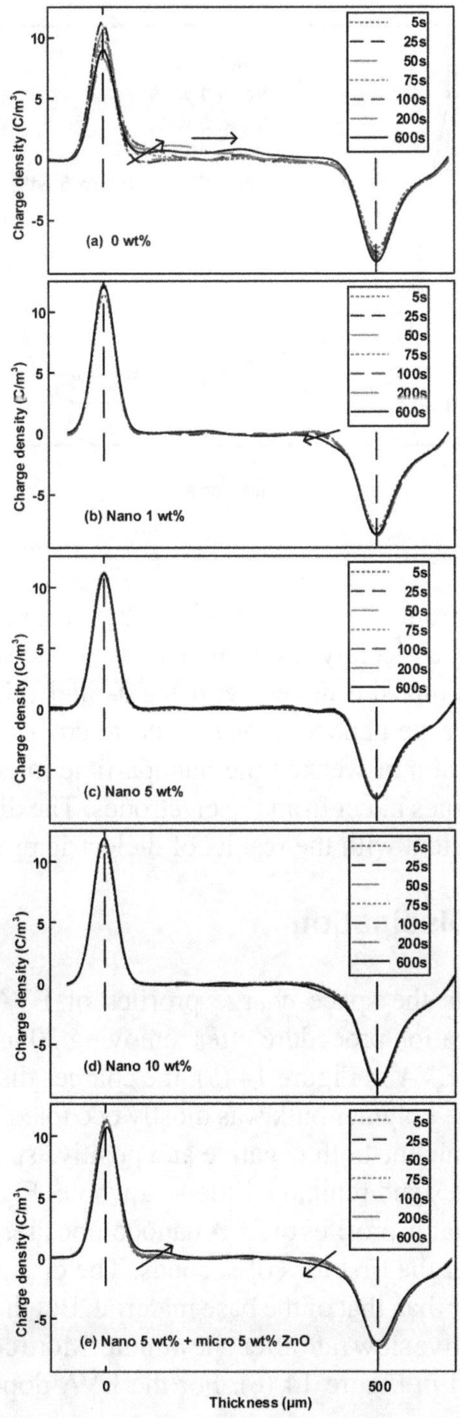

Figure 13. Electric field distribution in EVA/ZnO composites taken after 30 min polarization at the average applied field of -25 kV/mm

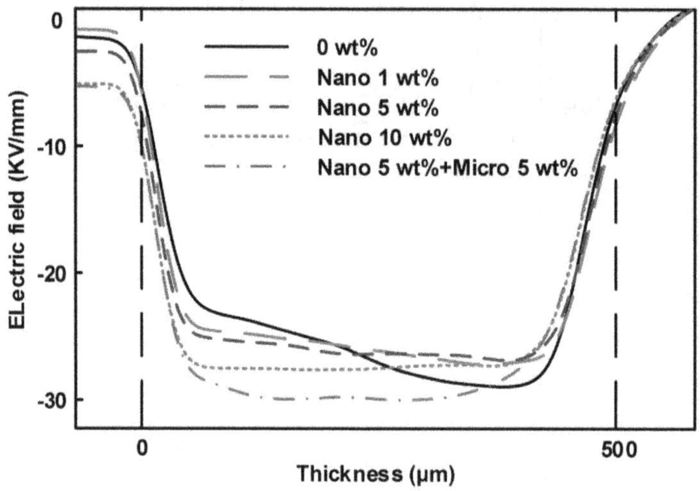

of charge injection from electrodes. With higher fillgrade, the percolation network may be formed locally resulting in a deeper negative injection near the cathode. The composite doped with nano- and micro-sized particles performs worse than the nanocomposite. The microsized fillers can break the main polymer chain or weaken the nanoparticle interface effect, which makes the homocharges inject from the electrodes. The distribution of space charge is well consistent with the results of dielectric properties.

Space Charge Dissipation

Figure 14 represents the space charge profiles of EVA/ZnO composites during the depolarization procedure after removing 30 min stressing. It can be seen for the base EVA in Figure 14 (a), the charges dissipate fairly fast at the beginning and the polymer bulk was mostly occupied by positive charge. With the lapse of time, the both negative and positive space charges diffuse through the electrodes and remain a little as a whole. Figure 14 (b), (c) and (d) show the space charge profiles of EVA nanocomposite. Both homocharges exist in the sample at the first tens of seconds. The dissipation speed of the initial period is faster than that of the base material. But the remaining charge decreases with a relative slow rate after one minute. More complex dissipation process is illustrated in Figure 14 (e). For the EVA doped with nano- and

Figure 14. Space charge profiles of EVA/ZnO composites during the depolarization

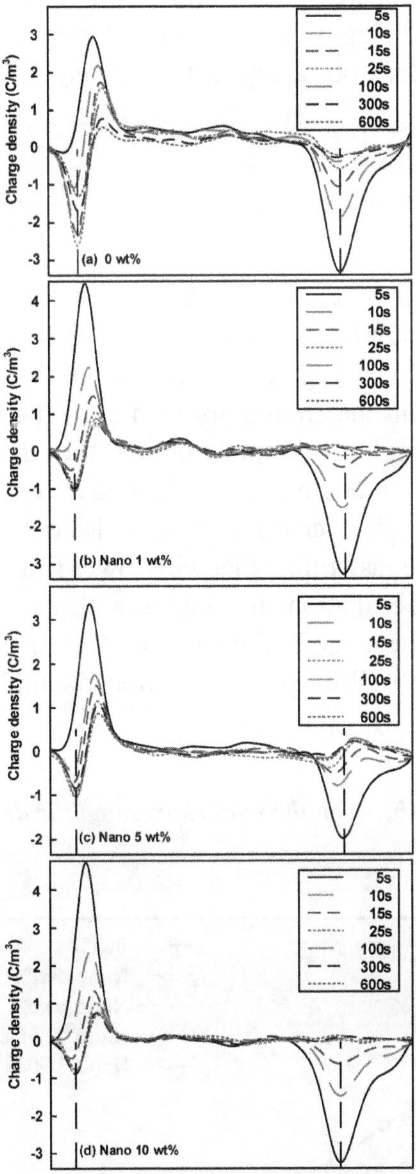

micro-sized ZnO particles, fast dissipation is similar to the nanocomposite at the beginning. Nevertheless, homocharges are injected into deeper region of the bulk instead of being transported into electrodes.

For further discussion, total charge in the bulk is calculated by the Equation (3):

$$Q = \int_0^d \left|\rho(x)\right| \cdot S \cdot dx \tag{3}$$

where $\rho(x)$ is charge density, S is the electrode area, and d is the thickness of the sample. The result of the total charge after being stressed at -25kV/mm for 30 min is shown in Figure 15. Clearly, the five groups dissipate fast at the initial period and the charge dissipation of the base EVA continues with the time. However, for the doped samples, the amount of charges keeps relatively invariant about one minute later. It has been

observed that the total charges of sample with higher contents of nanoparticles dissipate faster than that with lower fillgrade and end in lower values. The charges dissipation in samples with mixed doping shows the trend with generally the same initial value and relatively higher final value compared with base EVA. While the 10 wt% doped sample results in minimum remnant charges.

Figure 15. Relationship between the total charge and the depolarization time with different ZnO contents

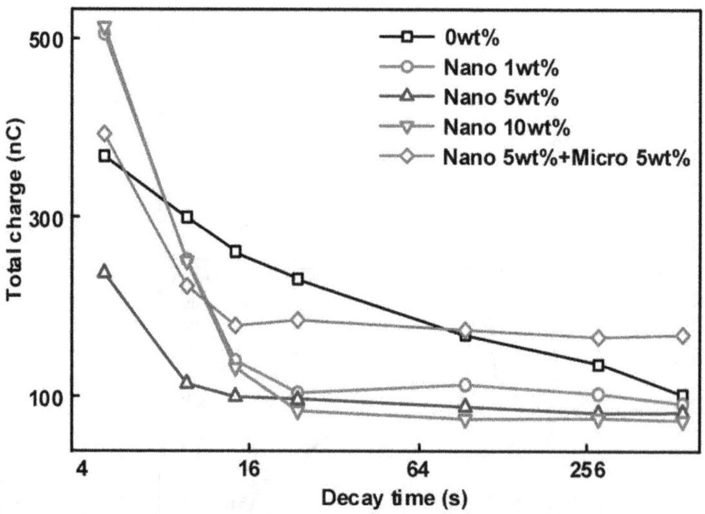

When the applied field is removed from the sample, the trapped charges dissipate with time, mainly due to the detrapping. The difference of the dissipation rate at different time perhaps depends on the detrapping time constant of shallow traps and deep traps. The rapid dissipation at the beginning is corresponding to the detrapping from the shallow traps. Besides, it is harder for the charge carriers captured in the deep traps, resulting in a slow rate for the charge dissipation of doped samples. Nano-sized ZnO particles doping introduces more shallow and deep traps into the polymer. But the micro-sized ZnO particles produce deeper traps compared with the nanocomposite, as the form of new defects, resulting in more remnant charges. For the base EVA, the energy level of the traps is relatively homogeneous. Thus the speed for the base EVA changes little during the dissipation.

Figure 16. Compare with SEM and EDS of the original and fluorinated EPDM fracture surface. (a) original sample; (b) sample fluorinated for 30min

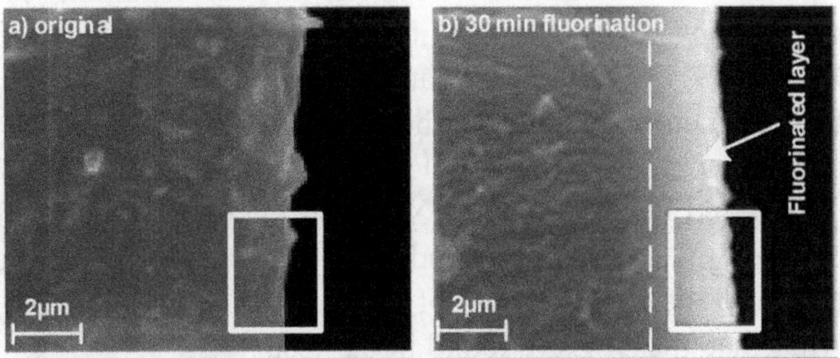

Element	Wt%	At%
CK	78.76	96.16
OK	02.57	02.36
AuL	18. 67	01. 48
Matrix	Correction	ZAF

Element	Wt%	At%
CK	66.91	90.11
OK	03.04	03.07
FK	05.66	04.82
AuL	24.40	02.00
Matrix	Correction	ZAF

SUPPRESSING INTERFACE CHARGE BETWEEN LDPE AND EPDM FOR HVDC CABLE ACCESSORY INSULATION

Sample Characteristics

In this section, The EPDM samples were fluorinated at 25 °C using a F2/N2 mixture gas with 20% F2 by volume at 0.05 MPa for 0, 15, 30 and 60 min, and LDPE for 0 and 30 min, respectively.

The cross-section of EPDM fluorinated for 30 min, as well as the original sample as reference, was observed by scanning electron microscopy (SEM, Hitachi S4800), as shown in Figure 16. It can be seen that the micro-morphology of cross-section was changed apparently after the fluorination process. In addition, it can be observed that the thickness of the fluorinated layer is about ~2 μm. Energy dispersive spectroscopy (EDS) analysis was employed to obtain the quantitative distributions of elements in the scanning area marked with red outlines in Figure 16 (a) and 16 (b). The energy spectra analysis result in Figure 16 (b) presents that the fluorine element (5.66 weight percent) is introduced into the sample surface after fluorination, indicating that a fluorinated layer is formed in the sample surface.

Figure 17 shows the ATR-IR spectra of the original and fluorinated EPDM and LDPE samples. It is clearly indicated that the direct-fluorination gives rise to obvious change in the chemical component of the EPDM surface layer. In the spectrum (a) of the original sample, the absorption bands observed in the regions of 2830-3000 and 1430-1470 cm^{-1} are corresponding to the C-H stretching vibration and the bending vibrations of CH_2 and CH_3, respectively (B. X. Du et al.,2013). The fluorinated sample shows that C-H bond is significantly reduced in strength by the direct-fluorination. A broad strong absorption band appears in the 940-1340 cm^{-1} region, which is associated with C-F, CF_2, and CF_3 groups (B. X. Du et al.,2014). The results show a clear evidence that some C-H bond disappears and fluorine element is incorporated into the surface layer of EPDM by the fluorination. Similar trends of fluorinated LDPE are found in our tests.

The relationship among the relative permittivity and the fluorination time of the EPDM and LDPE is shown in Figure 18. The permittivity of EPDM decreases from 2.36 to 2.28 with the fluorination time varying from 0 to 30 min, while increases to 2.42 for the 60 min fluorination. It can be assumed that the 2s and 2p orbit of fluorine atom are very close to the nucleus, which makes the fluorine element the smallest atom in all chemical elements except

Figure 17. ATR-IR spectra of the original and fluorinated EPDM and LDPE samples. (a) EPDM; (b) LDPE

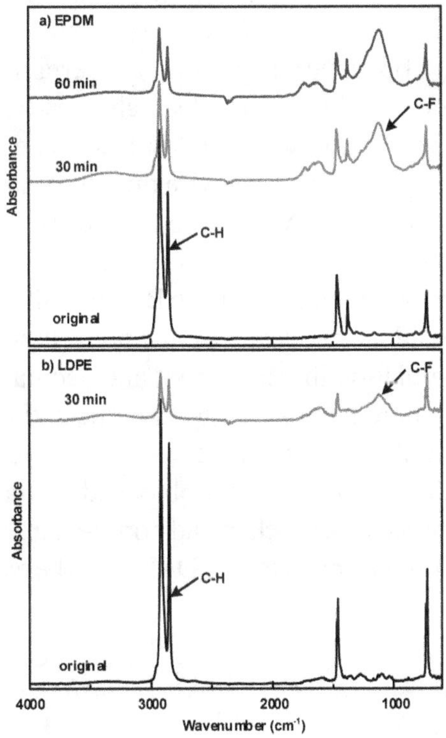

hydrogen atom. As a consequence, the electron polarizability of fluorine element is very low, resulting in a low dielectric permittivity of fluorinated EPDM. When the fluorination time is over 30 min, some scission in polymer chain may occur during the excessive fluorination, resulting in a higher relative permittivity. As for the LDPE sample, the surface fluorination decreases the permittivity, for which the change of replaced C-F bounds may account.

Interface Charge Behaviors

The dynamic behaviors of interface charge distribution in original and fluorinated LDPE/EPDM composite insulation are presented in Figure 19. All the charge polarity of the interface charge was negative, which is consistent with the voltage applied on the EPDM. Over 1800 s in Figure 19 (a), large amounts of charges are accumulated at the interface, and the maximum charge density is 7.55 C/m3. Besides, homo-charges near the interface accumulate gradually with the stressing time and hetero-charges are injected from the cathode. Figure 19 (b) demonstrates the profiles of the interface charge behaviors with the 15 min fluorinated EPDM. The interface charge density is changed to 5.31 C/m3 compared with the undoped EPDM group. This degree of fluorination produces a certain suppression effect and decreases the accumulation of interface charges. Figure 19 (c) shows the interface charge distribution with 30 min fluorinated EPDM. The least interface charge of 3.65 C/m3 builds up with this fluorination time and fewer charges are injected into the EPDM bulk. In Figure 19 (d), similar trends for interface charge accumulation with 60 min fluorinated EPDM can be found with the charge distribution in Figure 19 (b). But the value of interface charge density changed into 6.12 C/m3. While for the both fluorinated samples, the interface charge accumulation increases first and decreases afterwards with time. Simultaneously, more hetero-charges are injected from the cathode.

For the original samples, electrons are easy to injected form the cathode and migrate to the bulk, which are further blocked by the interface. Results show that less space charges are blocked and homo-charges near the interface disappears with the 30 min fluorinated EPDM. It can be explained as follows.

Figure 18. Relationship between the relative permittivity and the frequency with different fluorinated samples

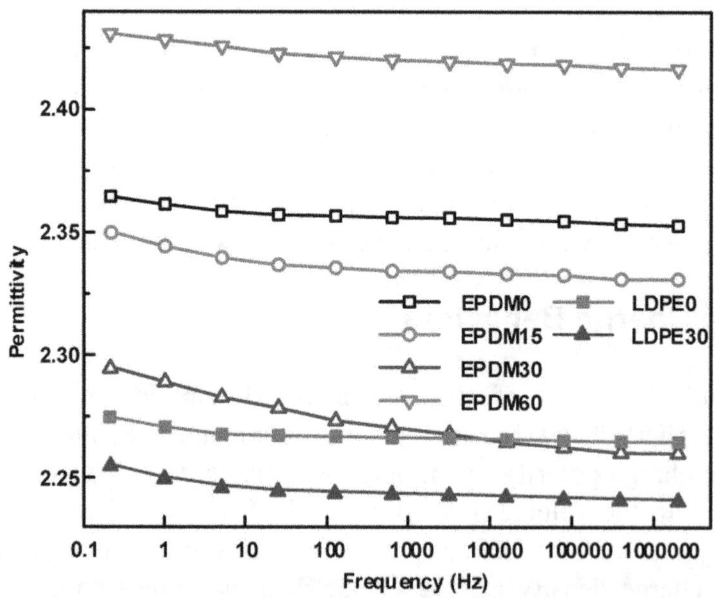

Figure 19. Interface charge accumulation between LDPE and EPDM with different fluorination time under 20 kV/mm

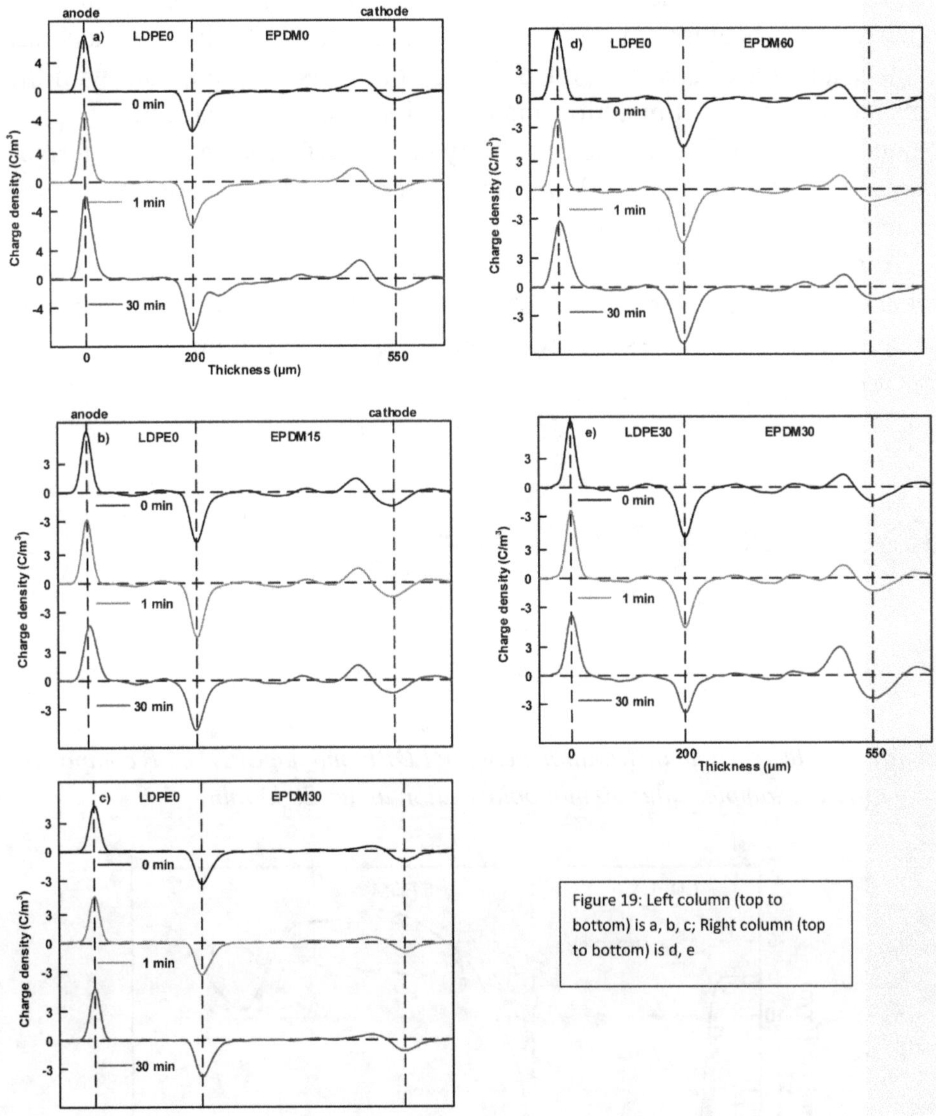

Figure 19: Left column (top to bottom) is a, b, c; Right column (top to bottom) is d, e

It is known that the fluorine element has the largest electro negativity and makes the C-F bond highly polarized (B. X. Du et al.,2013). Therefore, the surface fluorinated layer would capture a quantity of electrons and weaken the effective electric field in the vicinity of cathode, thus resulting in less space charges injection. While some scission in polymer chain may occur in the excessively fluorinated EPDM, which could decrease the charge injection barrier and more space charges are injected. For the both fluorinated LDPE and EPDM samples, the interface charge build process is quite different from others. At the beginning period, the fluorinated layers at the interface may capture amount of electrons once the dc voltage is stressed. And with the lapse of time, the accumulation and migration of carriers reach the dynamic balance. Nevertheless, the compatibility of the surface state at the interface may not be better than the 30 min fluorinated EPDM alone considering the match of permittivity and conductivity.

As the electric field under DC stress for cable accessories is of great concern for actual use, the electric field distribution E(x) of the bi-layer dielectrics is calculated based on Poisson's Equation (4), shown in Figure 20:

$$E(x) = -\frac{1}{\varepsilon_0 \cdot \varepsilon_{r1}} \int_0^{x_1} \rho(x_1) \cdot dx_1 - \frac{1}{\varepsilon_0 \cdot \varepsilon_{r2}} \int_{d_1}^{x_2} \rho(x_2) \cdot dx_2$$

$$0 \le x_1 \le d_1; d_1 \le x_2 \le d_2 \qquad (4)$$

Figure 20. Electric field distribution between LDPE and the EPDM/CB composites with different samples after 30 min polarization under 20 kV/mm

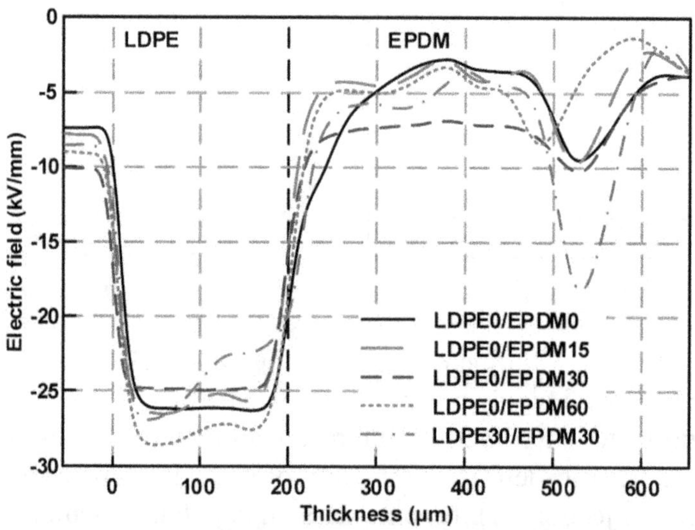

where $\rho(x)$ is charge density, $\varepsilon 0$, $\varepsilon r1$ and $\varepsilon r2$ are the vacuum and relative permittivity of LDPE and EPDM, and $d1$ and $d2$ for the boundary of the x are the thickness of the LDPE and EPDM sample. The electric field in LDPE is higher, while in EPDM it is relatively lower. At the interface, there is a sharp decrease of electric field due to the interface charge build-up. Electric field distribution of the system is significantly improved by the 30 min fluorinated EPDM, for which the decrease of interface charge accumulation and mismatch of permittivity accounts. However, the 60 min fluorination seems to get the performance in uniformizing the electric field worse compared with original samples, which results in relatively higher electric field in LDPE insulation. The high local electric field may lead to the partial discharge at the interface, thus accelerating the insulation aging. For both fluorinated samples, the electric field in LDPE is weakened but strengthened in EPDM due to the hetero-charges near the cathode.

Figure 21 represents interface charge profiles of EPDM/LDPE composite insulation during the depolarization after 30 min stressing. With the lapse of time, the interface charges are divided into negative and positive charges diffusing towards the anode and cathode respectively, under the force of electric field by the homocharge near the electrodes. It can be seen for the original samples in Figure 21 (a), the charges near LDPE dissipate fairly fast at the beginning while both the positive and negative charges near EPDM seem to be remnant with the depolarization time. Because the transportation through LDPE is rather difficult for the charges but easier in the EPDM. Figure 21 (b), (c) and (d) show the interface charge profiles of LDPE and fluorinated EPDM composite insulation. Similar trends can be found in the interface charge diffusion procedure. No speaking of the initial interface charge of the three groups, the charges dissipate much more quickly than that in the original samples. The fluorinated layer at the interface may accelerate the charge transport. For the both fluorinated composite insulation of group (e), an interesting phenomenon that the interface charge changes from negative to positive gradually. Perhaps the interface charges captured by the double fluorinated layers are released towards the anode and cathode after stressing. Besides, there are still a large amount of positive space charges remnant in the EPDM, which is may be influenced by the change of the interface charge polarity.

The energy level of fluorinated samples is lower than that of samples without fluorination. A low energy level represents shallow traps in the sample that cannot obstruct the mobility of the trapped charge. As charges are injected into the bulk only after the traps are filled, the surface potential increases after

Figure 21. Interface charge profiles of EPDM/LDPE composite insulation during the depolarization after 30mins stressing

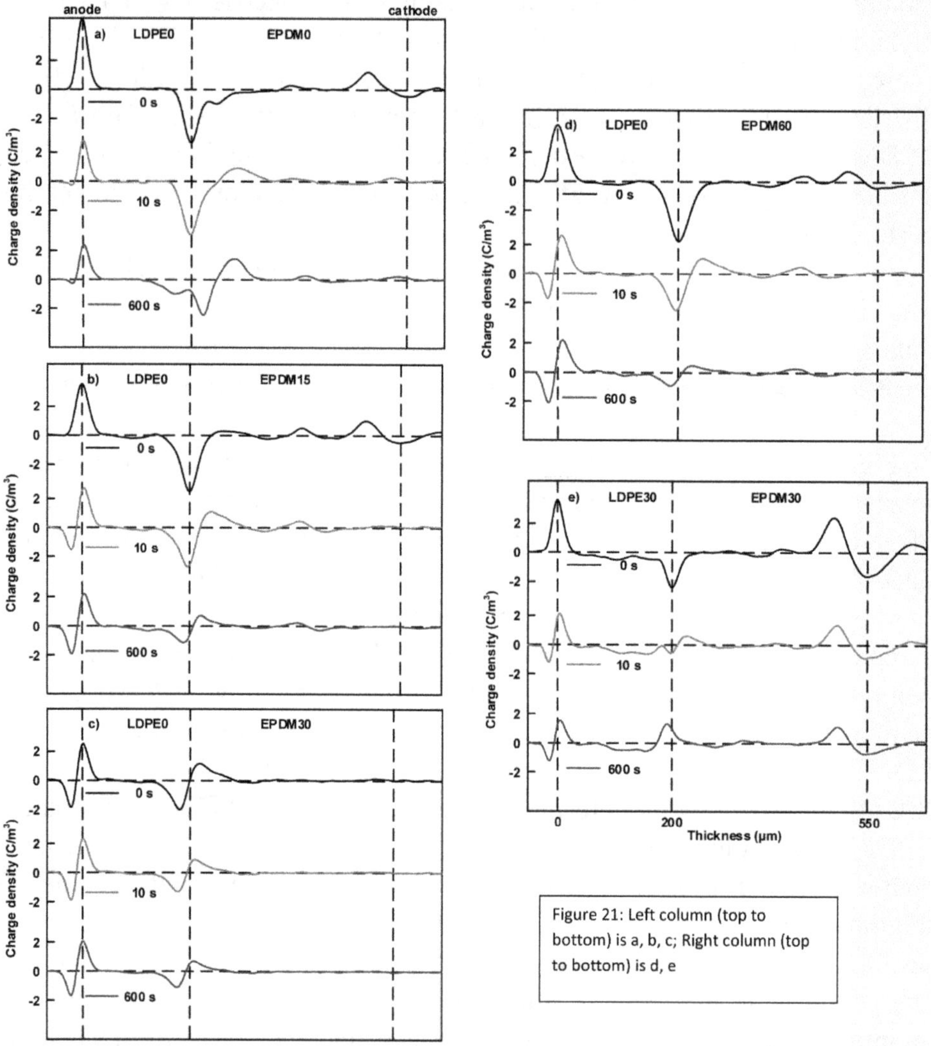

Figure 21: Left column (top to bottom) is a, b, c; Right column (top to bottom) is d, e

the traps are filled. Because of the shallow traps, the charges easily move out of the traps, as a result of which the surface charge of fluorinated samples decays faster than that of samples without fluorination. It can be speculated that the introduction of fluorine to the polyimide chemical structure can lower the traps.

Discussion

For further discussion, total charges at the interface during the polarization and depolarization procedures are calculated by the Equation (5):

$$Q = \int_{d_1-\Delta}^{d_1+\Delta} \rho(x) \cdot S \cdot dx \tag{5}$$

where ρ(x) is interface charge density, S is the electrode area, d1 is the thickness of the LDPE sample, and Δ is 50 μm. Figure 22 (a) shows the accumulation of the interface charge under 20 kV/mm for 30 min. It is observed that the time for interface charge build-up decreases with the fluorination time from

Figure 22. The accumulation and dissipation procedures of the interface charge with the different samples under 20 kV/mm

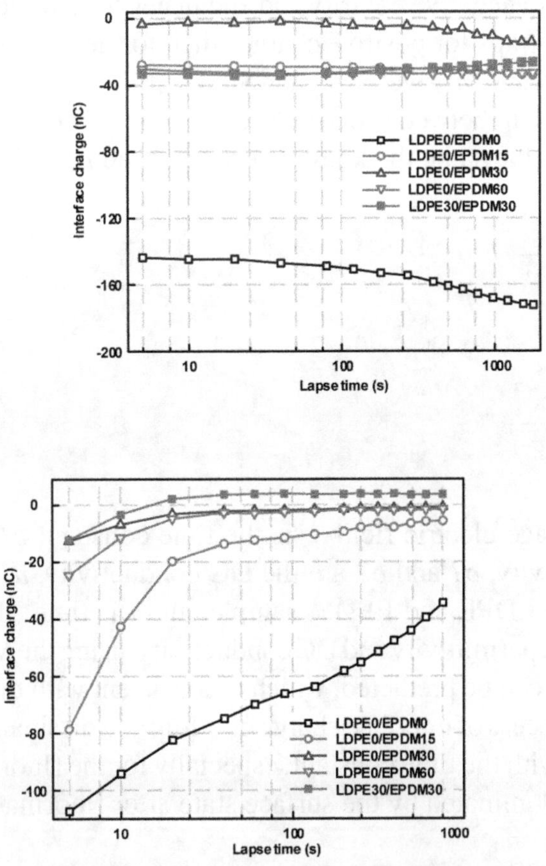

0 to 60 min. The total interface charge in the original composite insulation reaches -175 nC at the time of 30 min and is generally more than the fluorinated ones, of which the total interface charge are 35, 20 and 37 nC for 10, 30 and 60 min fluorinated samples. The interface charge in the both fluorinated samples builds up immediately after applying electric stressing. While with the lapse of time, the amount of interface charge decreases slightly. It may cost less time for the balance of carrier transportation.

The results of the interface charge after being stressed at -20 kV/mm for 30 min are shown in Figure 22 (b) Clearly, the total interface charge of original sample dissipates fast in the initial period and the charge dissipation keeps its speed with time under the logarithmic coordinate. However, for the fluorinated samples, the dissipation is rather rapid at the incipient stage and the balance is completed within decades of seconds. The dissipation speed and remnant charges of 30 min fluorination are in the middle among the five groups. It has been observed that the total charges of sample with higher initial value dissipates faster than that with lower value. The charges of both fluorinated samples start with negative polarity and end in negative values. The interface provides deeper traps for positive charges than for negative ones.

According to the Maxwell-Wagner-Sillars (MWS) theory, the interface charge Qs builds up between two dielectrics with different permittivity and conductivity but the same thickness referring to the Equation (6) and (7):

$$Q_s = \frac{\varepsilon_1 \cdot \sigma_2 - \varepsilon_2 \cdot \sigma_1}{\sigma_1 \cdot d_2 + \sigma_2 \cdot d_1} \cdot U \cdot (1 - e^{-t/t_c}) \tag{6}$$

$$t_c = \frac{(d_1 \cdot \varepsilon_2 + d_2 \cdot \varepsilon_1)}{(d_1 \cdot \sigma_2 + d_2 \cdot \sigma_1)} \tag{7}$$

where E is average electric field, t_c is the time constant, $\varepsilon 1$ and $\varepsilon 2$ are the relative permittivity, $\sigma 1$ and $\sigma 2$ are the DC conductivity, $d1$ and $d2$ are the thickness of the LDPE and EPDM sample, and t is time . From measured values of relative permittivity and DC conductivity, using the MWS theory, the interface charge can be predicted, which is consistent with the actual results. However, the process of interface charge build-up and dissipation seems not to obey Equation with the time constant, especially for the fluorinated samples. Perhaps this is dominated by the surface state after fluorination.

When the applied field is removed from the sample, the trapped charges dissipate with time, mainly due to the thermal detrapping in this condition. An assumption can be proposed that the interface charge dissipation is a process of capacitor discharge, which is also related to the time constant t_c depending on the conductivity and permittivity of the two dielectrics. The difference of the dissipation rate at different time perhaps depends on the detrapping time constant of shallow traps and deep traps. The rapid dissipation at the beginning is corresponding to the detrapping from the shallow traps. Besides, it is harder for the charge carriers captured in the deep traps to detrap, resulting in a slow rate for the charge dissipation of the original samples. Direct- fluorination introduces more shallow traps into the polymer. But excessive fluorination may bring deep traps and limit the charges transporting.

Physical interfaces constituted by the association of two identical dielectrics do not comply with the usual Maxwell-Wagner-Sillars behavior and surface state has to be taken into account. The effects of the fluorinated layer on space charge injection can be explained by using schematic model in Figure 23. For the original sample shown in Figure 23 (a) large quantities of charges are formed at the interface and in the vicinity of cathode in EPDM, which come from the ionization of impurities. These impurities may be introduced by the by-products during the mixing and vulcanization process. The appearance of hetero-charges reduces the potential barrier based on the Schottky injection law, thus causing the further trapping process in the sample bulk. The electrons migrate through the EPDM by the electrical field and accumulate at the interface due to the unbalanced carrier flux. Some of them are neutralized with holes in LDPE.

For the sample fluorinated for 30 min, the fluorinated layer with several microns in width captures large quantities of electrons at the interfaces between electrode-EPDM and LDPE-EPDM. On the cathode side, the accumulation of homo-charges in surface layer and less induced positive charges bring in an equivalent lower electrical field as shown in Figure 23 (b), thereby weakening impurity ionization and decreasing the electrons injection. Besides, a much denser fluorinated layer is formed on the surface of EPDM samples, because of more stable C-F bonds. The fluorine element with the largest electronegativity makes the C-F bond highly polarized, and the fluorinated layer formed on the sample surface would leading to a larger injection barrier for the charges injected from electrodes based on the Schottky injection law. Besides, under the applied electric field, a few electrons are injected from the cathode and move towards the interface. When the electrons come across another fluorinated layer, they are blocked again and result in less accumulation of interface charge. As for the situation of positive voltage applied on EPDM, which is not demonstrated here, the hetero-charges captured by the fluorinated layer and induced charges close to the internal form a local field in the vicinity of anode, which will then weaken the effective field, providing a shield against charge injecting into the bulk, and reduce injection of holes. Due to the combined action of fluorinated layers on both electrons and holes accumulation at electrodes and interface, the charge injection process is evidently suppressed.

Figure 23. Schematic model for illustrating the effect of the fluorinated layer on interface charge build-up

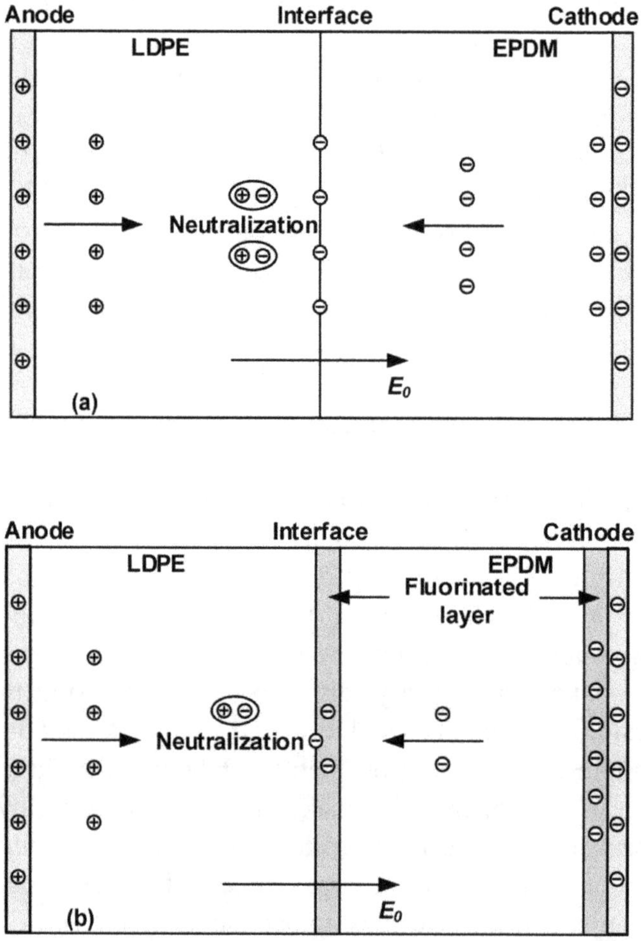

REFERENCES

Chen, G., Hao, M., Xu, Z. Q., Vaughan, A., Cao, J. Z., & Wang, H. T. (2015). Review of high voltage direct current cables. *CSEE J. Power Energy Sys.*, *1*(2), 9–21. doi:10.17775/CSEEJPES.2015.00015

Chong, Y. L., Chen, G., & Ho, Y. F. F. (2007). Temperature effect on space charge dynamics in XLPE insulation. *IEEE Transactions on Dielectrics and Electrical Insulation*, *14*(1), 65–76. doi:10.1109/TDEI.2007.302873

Delpino, S., Fabiani, D., & Montanari, G. C. (2008). High space charge dynamics in EVA-based nanocomposite flat specimens. *IEEE Int. Conf. Electr. Insul. Dielectr. Phenom., (CEIDP)*, 137-140.

Delpino, S., Fabiani, D., Montanari, G. C., Laurent, C., Teyssedre, G., Morshuis, P. H. F., & Dissado, L. A. et al. (2008). Feature Article-Polymeric HVDC Cable Design and Space Charge Accumulation. Part 2. *Insulation Interfaces. IEEE Electr. Insul. Mag.*, *24*(1), 14–24. doi:10.1109/MEI.2008.4455499

Donzel, L., Greuter, F., & Christen, T. (2011). Nonlinear Resistive Electric Field Grading Part 2: Materials and Applications. *IEEE Electr. Insul. Mag.*, *27*(2), 18–29. doi:10.1109/MEI.2011.5739419

Du, B. X., & Li, J. (2016). Effects of Ambient Temperature on Surface Charge and Flashover of Heat-Shrinkable Polymer under Polarity Reversal Voltage. *IEEE Transactions on Dielectrics and Electrical Insulation*, *23*(2), 1190–1197. doi:10.1109/TDEI.2015.005182

Du, B. X., Li, J., & Du, W. (2013). Dynamic Behavior of Surface Charge on Direct-fluorinated Polyimide Films. *IEEE Trans. Dielectr. Electr. Insul.*, *20*(3), 947-954.

Du, B. X., Li, Z. L., & Li, J. (2014). Surface Charge Accumulation and Decay of Direct-fluorinated RTV Silicone Rubber. *IEEE Trans. Dielec. Electr. Insul.*, *21*(5), 2338-2342.

Fabiani, D., Montanari, G. C., Dissado, L., Laurent, C., & Teyssedre, G. (2009). Fast and slow charge packets in polymeric materials under DC stress. *IEEE Transactions on Dielectrics and Electrical Insulation*, *16*(1), 241–250. doi:10.1109/TDEI.2009.4784573

Fleming, R. J., Ammala, A., Lang, S. B., & Casey, P. S. (2008). Conductivity and space charge in LDPE containing nano- and micro-sized ZnO particles. *IEEE Transactions on Dielectrics and Electrical Insulation*, *15*(1), 118–126. doi:10.1109/T-DEI.2008.4446742

Fu, M. L., Dissado, L. A., Chen, G., & Fothergill, J. C. (2008). Space Charge Formation and Its Modified Electric Field under Applied Voltage Reversal and Temperature Gradient in XLPE Cable. *IEEE Transactions on Dielectrics and Electrical Insulation*, *15*(3), 851–860. doi:10.1109/TDEI.2008.4543123

Gao, L., Yang, X., Hu, J., & He, J. (2016). ZnO Microvaristors Doped Polymer Composites with Electrical Field Dependent Nonlinear Conductive and Dielectric Characteristics. *Materials Letters*, *171*, 1–4. doi:10.1016/j.matlet.2016.02.016

Lan, L., Zhong, Q., Yin, Y., & Li, X. (2013). Effect of Surface Fluorination on Space Charge Behavior at LDPE/EPDM Interface. *IEEE Int. Conf. Solid Dielectr. (ICSD),* 444-447.

Lewis, T. J. (2014). Charge transport in polyethylene nano dielectrics. *IEEE Transactions on Dielectrics and Electrical Insulation*, *21*(2), 497–502. doi:10.1109/TDEI.2013.004173

Li, S. T., Zhao, N., Nie, Y. J., Wang, X., Chen, G., & Teyssedre, G. (2015). Space Charge Characteristics of LDPE Nanocomposite/LDPE Insulation System. *IEEE Transactions on Dielectrics and Electrical Insulation*, *22*(1), 92–100. doi:10.1109/TDEI.2014.004524

Long, W., & Nilsson, S. (2007). HVDC transmission: Yesterday and today. *IEEE Power Energy Mag.*, *5*(2), 22–31. doi:10.1109/MPAE.2007.329175

Marzinotto, M., & Mazzanti, G. (2015). The statistical enlargement law for HVDC cable lines part 1: Theory and application to the enlargement in length. *IEEE Transactions on Dielectrics and Electrical Insulation*, *22*(1), 192–201. doi:10.1109/TDEI.2014.004626

Min, D., Wang, W., & Li, S. (2015). Numerical analysis of space charge accumulation and conduction properties in LDPE nanodielectrics. *IEEE Transactions on Dielectrics and Electrical Insulation*, *22*(3), 1483–1491. doi:10.1109/TDEI.2015.7116341

Nelson, J. K., & Fothergill, J. C. (2004). Internal Charge Behavior of Nanocomposites. *Nanotechnology*, *15*(5), 586–595. doi:10.1088/0957-4484/15/5/032

Singha, S., & Thomas, M. J. (2008). Dielectric properties of epoxy nanocomposites. *IEEE Transactions on Dielectrics and Electrical Insulation*, *15*(1), 12–23. doi:10.1109/T-DEI.2008.4446732

Tamayo, I., Òrrit, J., Diego, J. A., Belana, J., & Cañadas, J. C. (2011). Influence of EVA-Carbon Black Based Sc-shield in Free Charge in LDPE. *IEEE Int. Symp. Electrets (ISE)*, 97-98. doi:10.1109/ISE.2011.6085000

Tanaka, T. (2005). Dielectric nanocomposites with insulating properties. *IEEE Transactions on Dielectrics and Electrical Insulation*, *12*(5), 914–928. doi:10.1109/TDEI.2005.1522186

Tian, F., Lei, Q., Wang, X., & Wang, Y. (2012). Investigation of electrical properties of LDPE/ZnO nanocomposite dielectrics. *IEEE Transactions on Dielectrics and Electrical Insulation*, *19*(3), 763–769. doi:10.1109/TDEI.2012.6215078

Vu, T. N., Teyssedre, G., Vissouvanadin, B., Roy, S., & Laurent, C. (2015). Correlating Conductivity and Space Charge Measurements in Multi-dielectrics under Various Electrical and Thermal Stresses. *IEEE Transactions on Dielectrics and Electrical Insulation*, *22*(1), 117–127. doi:10.1109/TDEI.2014.004507

Chapter 6
Oil and Paper Insulation for DC Converter Transformer

ABSTRACT

Transformer oil and oil-impregnated paper, serve as the essential parts of converter transformer, suffering various electric fields. The accumulation of surface charge on the paper would lead to flashover. When the power flow of the HVDC system is reversed, the charge field will easily lead to discharge. Direct-fluorination is a method which could affect the material property without alternating the bulk property. Besides, a new type of nano-modified transformer oil is a method to improve properties. This chapter presents a study of the effect of fluorination on surface charge behavior, the effect of polarity reversal voltages on interface charge behavior and the effect of Boron nitride (BN) nanoparticles on the high thermal conductivity of transformer oil. Results show that fluorination had an influence on the chemical property of the paper and BN nanoparticles has improvements in heat transfer process. In the polarity reversal test, the dissipation rate becomes smaller as the reversal time gets longer.

INTRODUCTION

In recent year, with the development of the worldwide power industry and voltage levels, the capacity and level of power transformers are increasing. The HVDC system is becoming very popular and its safe operation of high capacity converter transformers is related to the security of entire grid.

DOI: 10.4018/978-1-5225-2309-3.ch006

Meanwhile, transformer oil and paper are vital essential parts of EHV and UHV transformers, and oil–paper insulation is also commonly used in large power transformers. As a bridge between AC and DC transmission systems, the valve side of the main insulation in the converter transformer under normal conditions should have considerable ability to withstand the AC and DC voltages. Due to the imbalance of permittivity and resistivity in mass dielectrics such as the transformer oil, pressboard, and others, the interface between the dielectrics will accumulate charge (Wada et al., 2006; Yang et al., 1997; Xu et al., 2016; Hanaoka et al., 2002). When the power flow of the system is reversed or a sudden outrage occurs, the local electric strength of the dielectrics will have a serious distortion, most likely leading to the occurrence of discharge or insulation damage. Owing to the fact that the flashover breakdown field strength along the oil–paper interface direction is lower than the bulk field strength which is vertical to the interface, surface discharge can occur easily, and will cause power failures. So the effect of the interface charge on oil–paper insulation becomes the topic of focus (Huang et al., 2014).

Oil–paper insulation dielectrics are complex combinations of solid–liquid two–phase morphology, and the injection, transport, accumulation, attenuation characteristics and mechanism of space charge are not entirely clear yet. These characteristics are closely associated with aging and damage mechanisms and directly affect the variation in electrical performance and insulation life of oil–paper insulating dielectrics. Direct-fluorination is a method to modify the chemical constituents of the material which possess the advantage of alternating the surface property only and leaving little modification on the essential property of the matrix, and thus, has been widely applied in many areas (Yi et al., 2015; Saulnier et al., 2013; Charlet et al., 2015; Prorokova et al., 2015; Lee et al., 2009). In recent years, researches have focused on the paper applied in transformers for improved dielectric properties (Chen et al., 2014; Zhuge et al., 2015). However, most studies on direct-fluorination have concentrated on the chemical stability, wettability, mechanical strength, etc. Few researches discussed the effect on electrical property after the treatment. In recent research, efforts have been paid on the surface modification of insulation material for improvement on the dielectric property.

Moreover, the oil applied in transformers is highly refined mineral oil which performs dual functions of insulating and cooling materials. The oil fills up the pores in the fibrous insulation thus affecting the dielectric property of oil-paper insulation system and increasing the breakdown strength. For thermal property the oil carries away the heat generated during normal operation by conduction and convention (IEEE, 1995). The miniaturization and higher voltage of electric devices have brought new challenge to traditional transformer oil. New type of transformer oil with enhanced dielectric property and high thermal conductivity is in urgent need. In recent years, many researchers have been concentrated on preparing nano-oil by dispersing nanoparticles into transformer oil and measuring the electrical property (Mansour et al., 2012; Atiya et al., 2015; Du et al., 2012; Zhong et al., 2013; Liu et al., 2011; Lv et al., 2014). However, most studies on nano-modified transformer oil have been limited to the dielectric property or thermal property only. Very few researchers have considered developing a type of nano-oil with improved properties of both breakdown strength and thermal conductivity, and investigating the relation between the thermal conductivity and the dielectric property.

Therefore, this chapter will focus on the studying the effect of fluorination on surface charge of oil-impregnated paper and the effect of boron nitride (BN) nanoparticles on the high thermal conductivity of transformer oil. Furthermore, the surface charge accumulation and decay of multi-layer oil impregnated paper under DC and polarity reversal voltages will also be discussed in the last part.

SURFACE CHARGE ACCUMULATION AND DECAY ON DIRECT-FLUORINATED OIL-IMPREGNATED PAPER

Fluorination Effect on Sample Characteristics

The EDAX analysis of the fluorinated paper without impregnated in the oil is illustrated in the Figure 1. The peak of hydrogen was missed in the spectrum due to its light weight of atom which is beyond the measurement accuracy of EDAX technique (Nowell et al., 2004). It is obvious that a little amount of fluorine with the atomic percent of 3.4% appears in the sample as shown in the spectrum between 0.6 to 0.7 keV indicating the fluorination in the cellulose matrix.

In addition, Figure 2 illustrates the attenuated total reflection infrared (ATR-IR) spectrums of the paper before and after fluorination as a method to analysis the alternation of functional group of the paper. As shown in the spectrum of the untreated paper, the absorption peaks between 3200-3500

Figure 1. EDAX analysis of the fluorinated paper

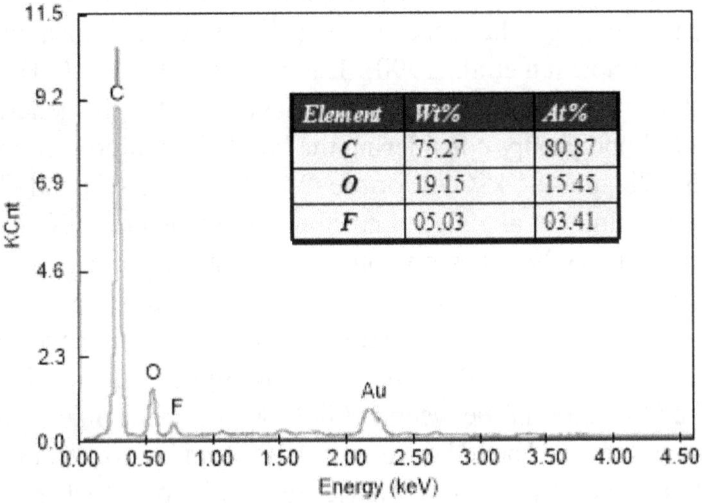

Figure 2. ATR-IR spectra of the untreated and fluorinated paper

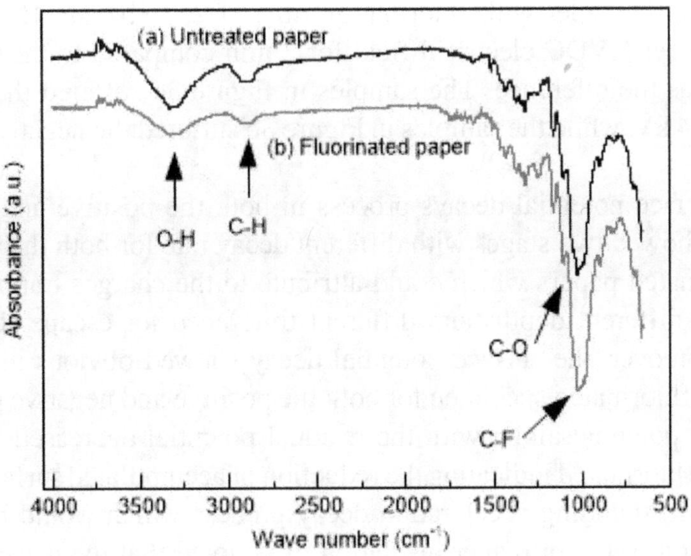

cm^{-1} and 2800-3000 cm^{-1} represents the O-H bond and C-H bond respectively. And the absorption peak between 1000-1300 cm^{-1} is related to the C-O bond which is confused by another absorption peak of C-H bond located between 800-1000 cm^{-1} (Norman et al., 1990). It is reported that the C-H bond and C-OH bond would be broken in fluorination and result in the generated C-F, C-F_2 and C-F_3 bond groups considering the bond dissociation energy (BDE) of 336, 445, 378 and 486 kJ/mol for the C-H, CH-H, C-OH and C-F bond, respectively (Saulnier et al., 2013; Lee et al., 2009; Du et al., 2014; Benson, 1965; Dean et al., 1999). It is obvious that the absorption peaks between 3200-3500 cm^{-1} and 2800-3000 cm^{-1} of the fluorinated paper were reduced indicating the substitution reaction in fluorination. Meanwhile, the absorption peak of C-F bond between 900-1100 cm^{-1} was slightly enhanced due to the facts that the C-O-C bond (between 1000-1300 cm^{-1}) was barely influenced by the F_2 gas and the C-H bond between 800-1000 cm^{-1} was overlapped by the generated C-F bond. As shown in the figure, no obvious double absorption peaks appear in the spectrum of the fluorinated paper indicating the little amount of C-F_2 and C-F_3 bond groups which were hard to be produced.

Effect of Fluorination Time

The surface charge of the impregnated paper was injected by corona charge. Meanwhile, the paper with fluorination time of 2, 5, 10 and 15 min were stressed under HVDC electrical field for 2 min compared to the untreated specimen as the reference. The samples in Figure 3a suffered the positive voltage of 4 kV, while the samples in Figure 3b suffered the negative voltage of -4 kV.

The surface potential decays process in both the positive and negative potential showed two stages with different decay rate for both the untreated and fluorinated papers which could attribute to the charges trapped in the traps with different depths and different time need for escape (Du et al., 2015). Moreover, the surface potential decay showed obvious high decay rate in the fluorinated specimen for both the positive and negative potential. The initial potential along with the residual potential decreased when the paper was fluorinated indicating the reduction of accumulated surface charge and the corresponding acceleration decay process which would benefit in the operation of the oil-paper insulation. It is noted that the decay process accelerated when the fluorination time increased from 0 to 5 min while slowed down if the fluorination continued resulting the paper fluorinated for 5 min

Figure 3. Relation between the surface potential and the decay time with different fluorination time. (a)charging with positive voltage. (b)charging with negative voltage

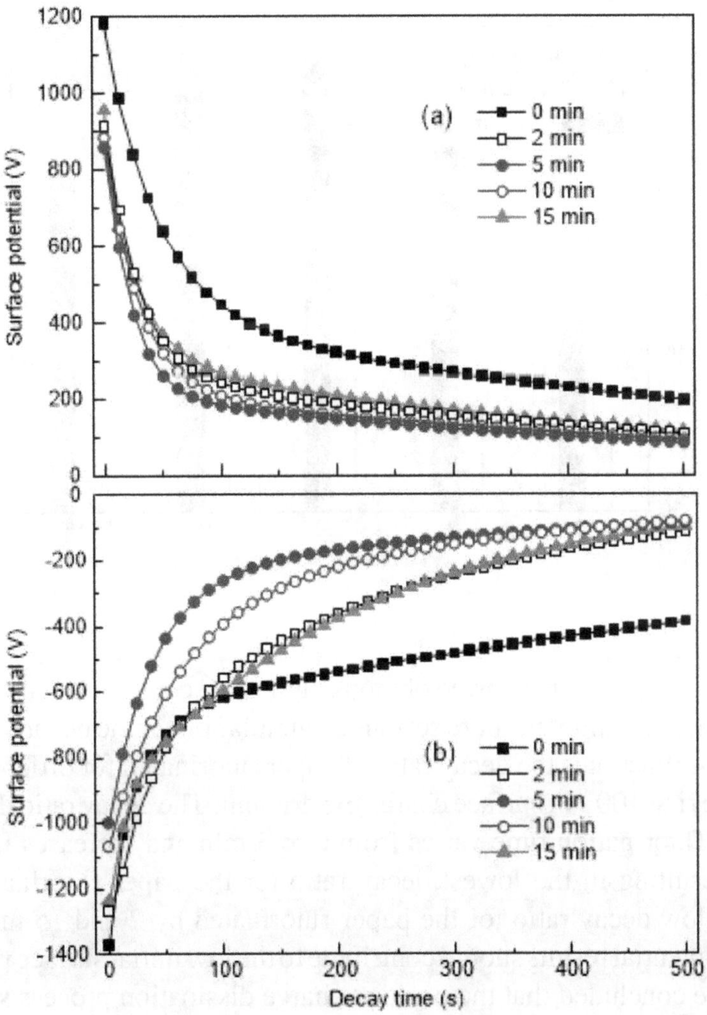

the excellent decay behavior, and the inconspicuous difference between the papers fluorinated for 2 min and 15 min in the negative experiment. The similar tendency with varying the fluorination time is shown in the positive experiment which could attribute to the alternation in chemical and physical structure. With the evolution of fluorination, the domination of the two kinds of influence on trap characteristics changes leading to the various behaviors in decay process (Du et al., 2014; An et al., 2015). The initial potential in

Figure 4. Relation between the decay ratio and the fluorination time with the different charging polarities

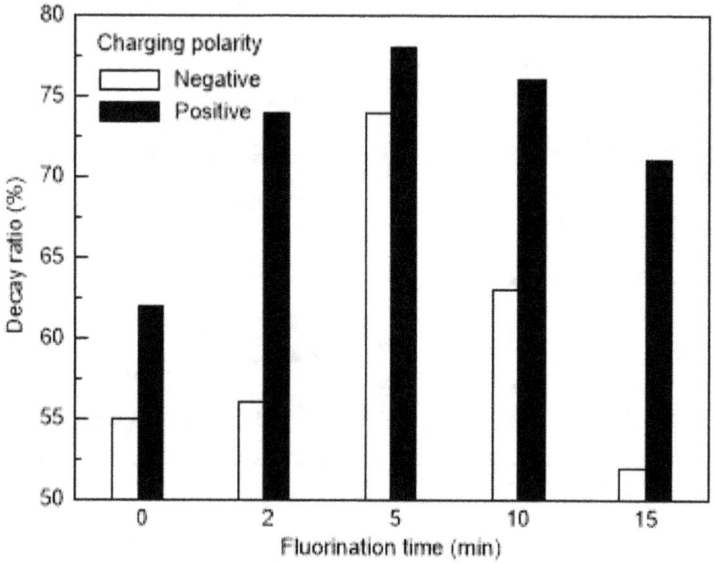

the negative experiment showed obvious high value compared to the positive experiments revealing the more serious accumulation for the negative charge.

Figure 4 illustrates the decay ratio of paper fluorinated for different times during the first 100 s in surface charge experiment. The decay ratio decreases when the fluorination time varies from 0 to 5 min and increases from 5 to 15 min resulting in the lowest decay ratio for the paper fluorinated by 5 min. The low decay ratio for the paper fluorinated by 2 and 15 min in the negative potential results should contribute to the low initial surface potential. It is can be concluded that the positive charge dissipation process showed a fast tendency to the stable stage and the fluorination showed more obvious influence on the negative charge dissipation process.

Figure 5 shows the trap distribution of the impregnated paper with various fluorination time. The distribution showed two obvious peaks located at different energy levels related to the charge captured in the shallow and deep traps, respectively. The charges captured in shallow traps de-trapped in short time after the electrical field was moved off and dominated the first stage in the surface potential decay process with high decay rate as shown in Figure 3 (Du et al., 2015). Meanwhile, the charges in the deep traps need more time

Figure 5. The trap distribution at various fluorination times after applying a negative voltage

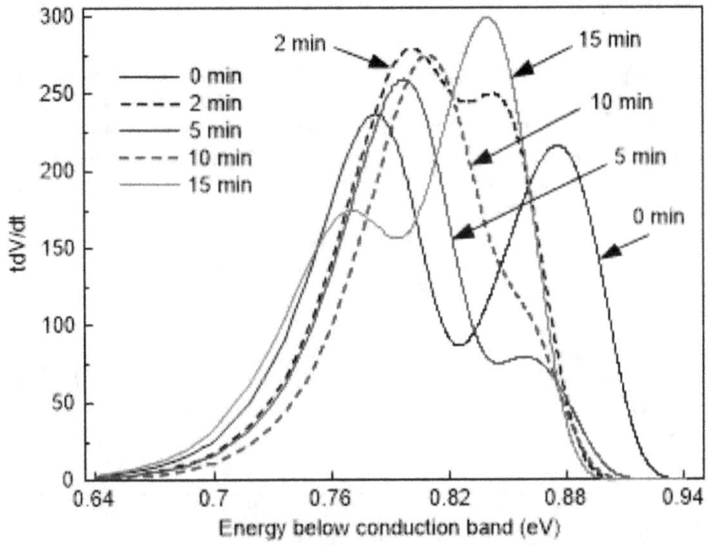

to escape leading to the relative gentle decay tendency with low decay rate in the whole process. With the treatment of fluorination, the peak related to the deep trap moved to the energy band with low level indicating the reduction of charge captured in the deep traps. Moreover, with the fluorination time varying from 2 to 5 min, the peaks decreased, especially for the peak related to the deep trap. However, with the evolution of fluorination, the charge captured in deep traps increased resulting in the minimum deep traps for the impregnated paper fluorinated by 5 min and leading to the excellent behavior in the surface potential decay process shown in Figure 3.

Effect of Charging Method

Figure 6 shows the surface potential decay rate of the untreated and fluorinated impregnated paper when they were corona charged with positive and negative voltage at identical magnitude and the same charging time, respectively. The treatment of fluorination showed different influence on the decay rate behavior for the positive and negative potential. The higher decay rate at the initial stage for the untreated paper when charged with negative voltage should attribute to the higher initial surface potential shown in Figure 3 leading to

Figure 6. Relation between the decay rate and the decay time with different charging polarities

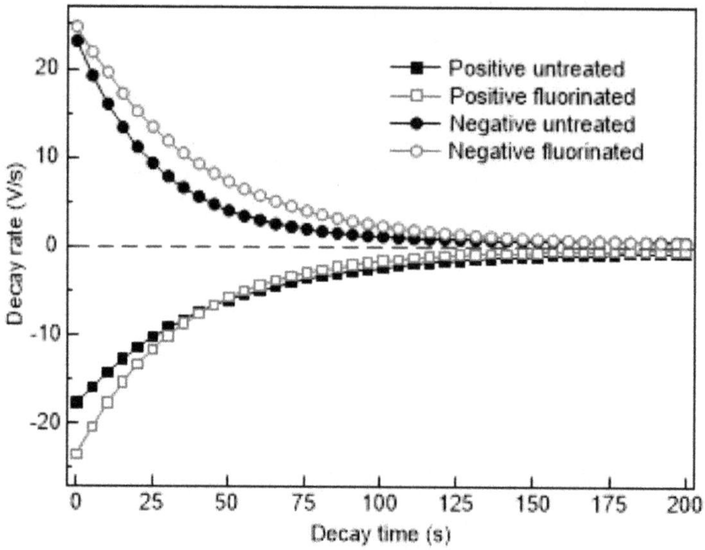

Figure 7. The trap distribution of the paper after applying the positive and negative voltage

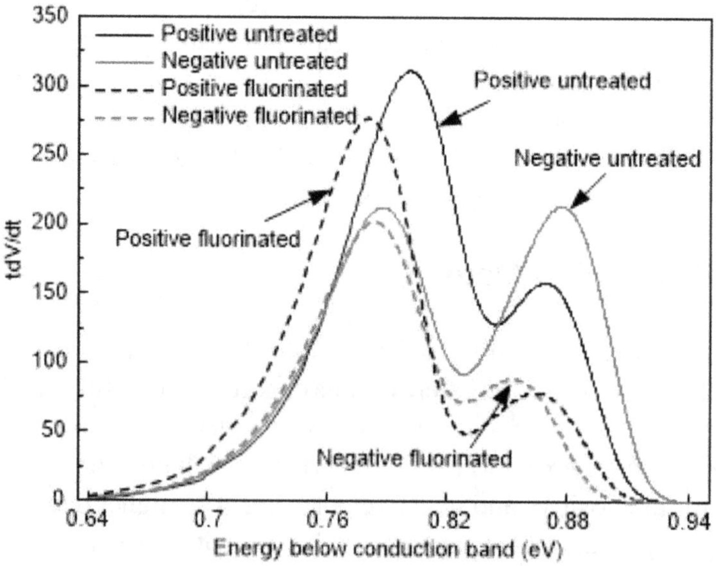

the higher vertical electrical field and the probable horizontal field built by the accumulated charge. The decay rate of the fluorinated paper showed no obvious difference at the beginning of the process in the negative experiment, while exhibited relative high decay rate in the whole decay process leading to the delayed inflection point in Figure 3b compared to the untreated paper.

However, the acceleration in decay process due to fluorination significantly affected the decay behavior from the initiate stage and slowed down with the evolution of decay resulting in the cross point of the decay rate in the positive experiment. The different influence of fluorination on decay behavior would attribute to the different decay process of the untreated paper at positive and negative condition that the positive potential showed more smooth decay process with low initial decay rate and slow decrease of the rate while the negative potential decay fast in the first stage and the rate slowed down quickly compared to the positive potential.

The surface charge dissipation pattern varied with the charging polarity and the correlative trap distribution is shown in Figure 7. It is obvious that more charge captured in shallow traps instead of the deep traps in positive experiment for the untreated paper which leads to the relative higher decay rate before the surface potential moved to the stage with a gentle decay rate as shown in Figure 8. When treated with fluorination, the charges trapped in the deep traps decreased due to the shielding effect of the fluorination layer resulting in the reduction of the peaks related to the shallow and deep traps. Meanwhile, the movement of the correlative peak of the shallow trap to the energy band with low level in positive experiment may contribute to the high decay rate at the initial stage shown in Figure 6 as different from the effect of fluorination in the negative experiment.

Figure 8 illustrates the different decay rate behavior of the fluorinated and untreated paper under different charging conditions. The decay rate of the impregnated paper charged with the same negative voltage at the time of 30 s and 4 min is shown in Figure 8a. It is obvious that the potential decay fast when charged with short time for both kinds of papers due to the more charges trapped in shallow traps.

Moreover, the fluorination treatment exhibited significant influence on the decay behavior at different charging times that the fluorinated paper charged for 30 s showed remarkable high decay rate especially at the initial stage of the decay process compared to the fluorinated paper charged for 4 min. Meanwhile, the fluorinated paper showed obvious high decay rate compared to the untreated paper and slowed down to the same rate in the experiment charging for 30 s, while the fluorinated paper exhibited no obvious difference

Figure 8. Relation between the decay rate and the decay time with different charging conditions. (a)charging with positive voltage. (b)charging with negative voltage

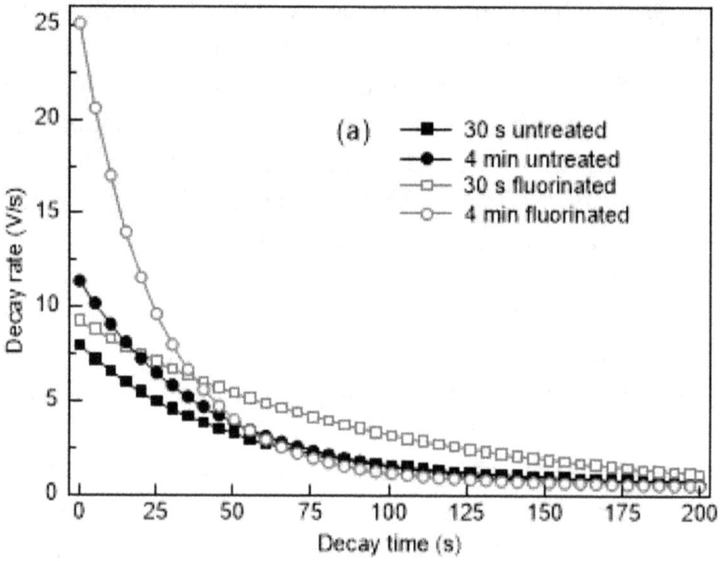

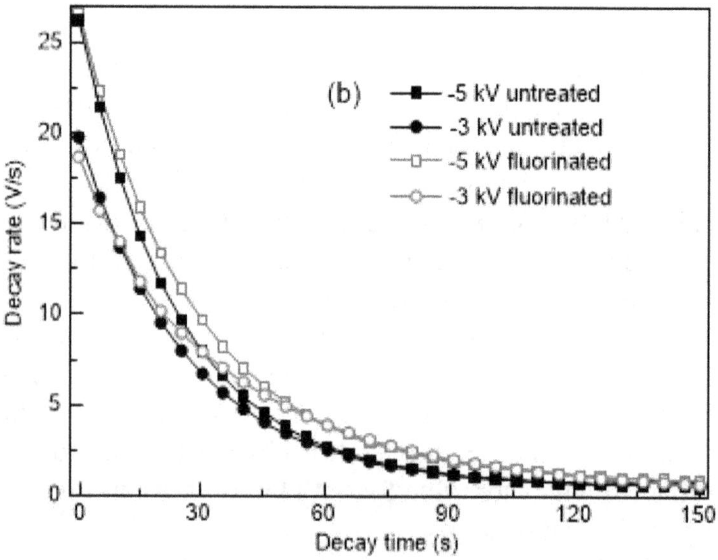

at the beginning but held a relative high decay rate to the untreated paper in the process indicating the considerable reduction of the time need for the fluorinated paper to the stable stage when charged in short time.

The surface potential decay rate of the paper before and after fluorination when stressed with different charging voltage is shown in Figure 8b. When increasing the charging voltage, the initial decay rate increased correspondingly for both the untreated and fluorinated paper due to the high initial surface potential. The decay rate showed no obvious difference at the initial stage

Figure 9. The trap distribution of the untreated and fluorinated paper with different charging conditions. (a)charging with positive voltage. (b)charging with negative voltage

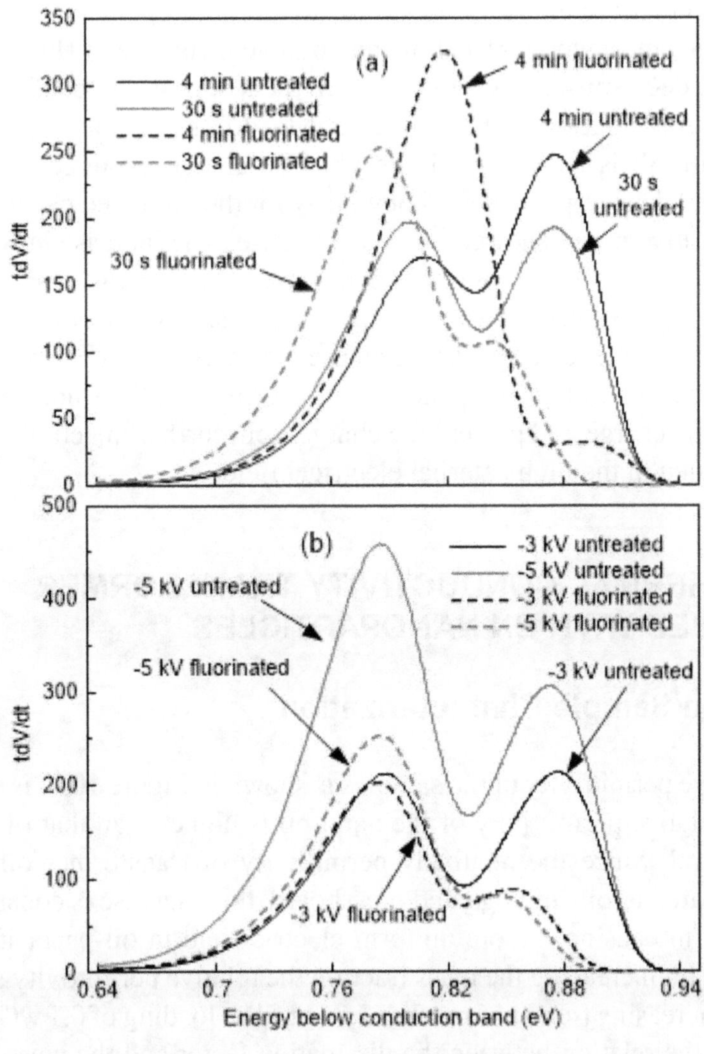

considering the two kinds of paper and the rate slowed down in a moderate pace for the fluorinated paper leading to the relative high decay rate to the untreated paper for the experiment charged at different voltages indicating no obvious influence of fluorination on the charging voltage which is different from the result in Figure 8a.

The charge captured in deep traps increased while the charge trapped in the shallow traps decreased when the charging time increased for both the untreated and the fluorinated paper as shown in Figure 9a. Moreover, the peak related to the shallow trap moved to the energy with high level obviously and showed increase in trap energy density for the fluorinated paper. Thereby, the alternation in trap distribution could illuminate the high decay rate for the fluorinated paper charged for 30 s compared to the paper charged for 4 min in Figure 8a, meanwhile, reveal the remarkable influence of fluorination on the charge decay process when applied in the fast charging condition.

The trap distribution for the paper charged at different voltage in accordance with Figure 8b is shown in Figure 9b. The charge captured in both the shallow and deep traps increased obviously for the untreated paper since the two correlative peaks showed obvious increase. The increase in the charge captured by the shallow traps would contribute to the increase in decay rate at the initial stage. However, the trap energy density showed no obvious increase especially for the deep trap when increase the charging voltage which could attribute to the shielding layer generated in fluorination. The fluorination layer captured the charge and prevent the charge from further injection especially when applied in the high external electrical field.

HIGH THERMAL CONDUCTIVITY TRANSFORMER OIL FILLED WITH BN NANOPARTICLES

Effect on Sample Characterization

The relative permittivity of the samples is shown in Figure 10. It is observed that the relative permittivity of the nano-oil is higher than that of the non-modified oil. Since the relatively permittivity of transformer oil is quite low compared to oil-impregnated pressboard, this increase is considered be beneficial in causing a more uniform electric field in oil-paper insulation system. With increasing the mass fraction the relative permittivity shows an obvious increasing trend and reaches 2.34 at filler loading of 0.1 wt%. Figure 11 shows the relation between the dissipation factor and the mass fraction.

Figure 10. Relation between the relative permittivity and the mass fraction

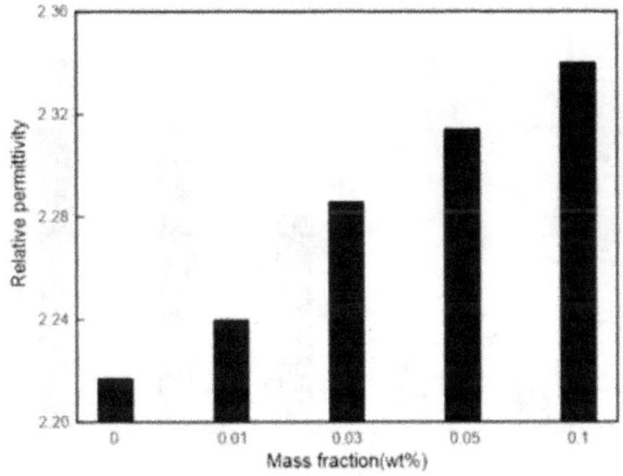

Figure 11. Relation between the dissipation factor and the mass fraction

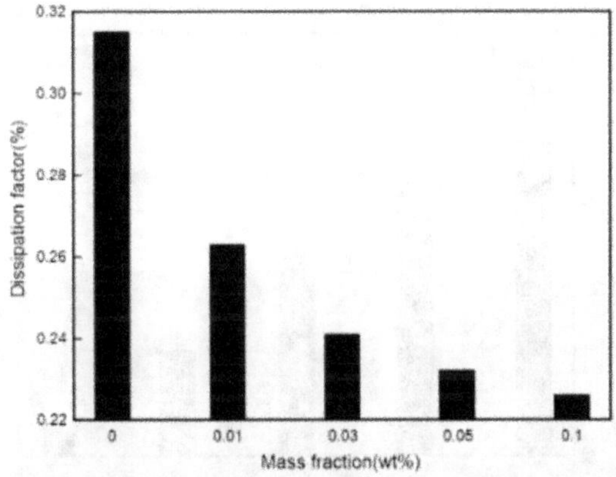

It is noticed that the dissipation factor decreases with the addition of the BN nanoparticles.

As the mass fraction increases, the dissipation factor decreases from 0.315% to 0.226%. This indicates that with the presence of the nanoparticles, the nano-oil shows improved dielectric property thereby contributes to restricting the insulation ageing process and enhancing the insulating strength in transformer oil.

Figure 12. Relation between the breakdown strength and the mass fraction

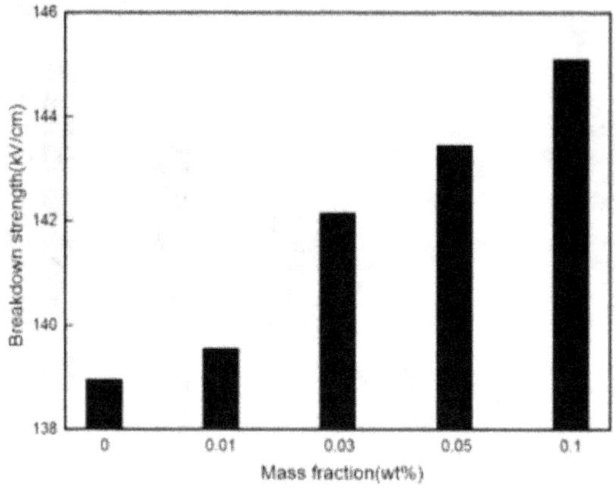

Figure 13. Relation between the discharge magnitude and the mass fraction

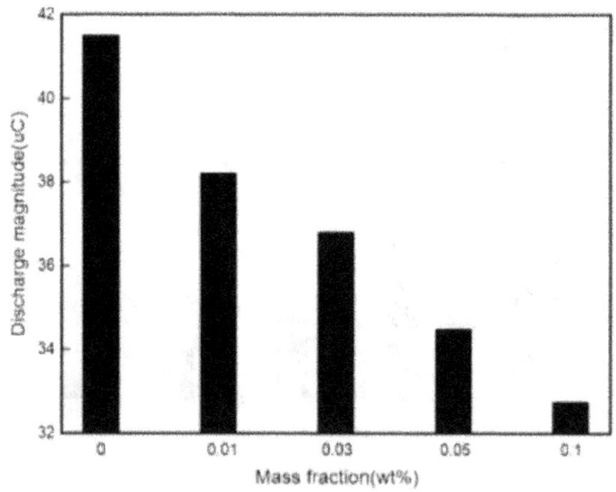

Effect on Dielectric Breakdown

The relation between the breakdown strength and the mass fraction is shown Figure 12. It is interesting to notice that the breakdown strength of the nano-oil is larger than that of non-modified oil. With the increase of mass fraction, the breakdown strength of nano-oil increases obviously confirming that BN nanoparticles in transformer oil effectively inhibit the process of dielectric breakdown.

Figure 13 shows the discharge magnitude cumulated in the first 1 ms of the breakdown process for the samples. It is noticed that the discharge magnitude shows an obvious decrease trend when increasing the mass fraction of the nanoparticles, which reveals the decline of the intensity in discharge process. The small discharge magnitude indicates the difficulty in breaking of the chemical bond, which attributes to the inhibitory effect of BN nanoparticles in breakdown process. Figure 14 shows the discharge procedure along the image width direction during the breakdown process with different mass fractions. The image frames are proportional to the time in breakdown process, and the figure which contains 100 frames reflects the first 5 seconds in breakdown process of the samples. With increasing the mass fraction, the discharge intensity in the samples shows a significant decrease trend. It is noticed that the discharge phenomenon in the non-modified oil is the strongest. In the sample with the mass fraction of 0.1 wt%, the discharge process shows discontinuous phenomenon. The obtained results indicate that

Figure 14. Relation between the discharge intensity and the breakdown process with the different mass fractions

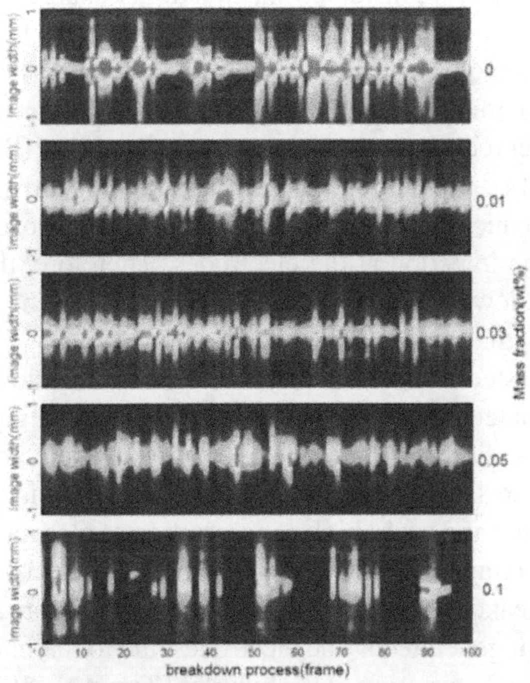

BN nanoparticles can effectively restrain the discharge process in nano-oil to decease the discharge intensity.

Hwang et al have found that the charge relaxation time constant can reveal the influence of the nano-filter on the electrodynamic process in streamer propagation and explain the considerable enhancement in electric property of the oil (Hwang et al., 2010; Lv et al., 2014). However, this theory showed some limitations in recent researches. For these reasons, the enhancement in electrical property can be considered as the result of the nanoscale of materials. The major influence of the additional nano-materials is the high surface to volume of the nanoparticles. The volume atoms are much more than the surface atoms in macro materials. However, in nano-scale the surface atoms increase observably relative to volume atoms and become a dominant role in material property which makes the nano-material show different properties from the homologous macro material. Furthermore, in nano-oil the interface between the nanoparticle and the oil matrix can affect the properties of the surface atoms (Sugumaran et al., 2012).

According to the multi-core model by Tanaka, the interfacial region in nano-dielectrics has an effect on the charge behavior and causes change in dielectric properties (Tanaka, 2005; Tanaka et al., 2005; Zhang et al., 2014). With decreasing the size of the additional nanoparticles, the interfacial effect gradually dominates. It is considered that the surface charge of the nanoparticle forms a charge layer at the interfacial region, and the layer is altered along with the alternation of the charge distribution at the interface, consequently may cause change in density and depth of trap sites. In Smith's research this change can further reduce the carrier energy and mobility (Smith et al., 2008; Saleem et al., 2014). When the carrier is trapped frequently, the homocharge due to trapping reduces the electric field intensity and causes increase in the voltage for charge injection at the electrodes. Therefore, this increases the short-term breakdown voltage and eventually leads to the enhancement in breakdown strength.

Kao has suggested a bubble theory for electric breakdown in liquid, that breakdown is initiated by bubbles due to electrons accelerating. And then these bubbles elongate in liquid and provide free path for the electrons to accelerate in the field. These free electrons with high energy induce the ionization process in the liquid and cause high concentration of plasma generated which forms the conducting channel in liquid gap and eventually results in the occurrence of breakdown (Kao et al., 1970). Furthermore, it is considered that the Joule heat generated at the electrodes due to the injection of carries may cause the initial formation of the bubbles (Kao et al., 1976). This reveals the influence of the improved thermal conductivity on the enhancement of breakdown strength. It can be concluded that in the oil with better thermal property the heat disperses to the vicinity in a faster way which results in the inhibition of bubble formation. Therefore, the breakdown process is blocked due to the enhancement of thermal property. Based on the analysis above, the enhancement of the breakdown strength is achieved by the nanoscale of materials in cooperation with the increased thermal property.

Effect on Thermal Conductivity

The temperature rise in heating process of the samples is given in Figure 15. The maximum temperature of the samples was measured by an infrared thermal imager for 50 minutes. It is observed that the maximum temperature of the nano-oil rises rapidly compared with non-modified oil. The temperature rise shows obvious acceleration tendency with increasing the mass fraction. When the oil was heated for 50 minutes the maximum temperature of the 0.1 wt% sample is 102 °C, while only 79 °C of the non-modified sample. The results indicate that the nano-oil has superior heat absorption property due to the addition of BN nanoparticles by conducting heat generated in the

Figure 15. Relation between the oil temperature and the heating time with the different mass fractions

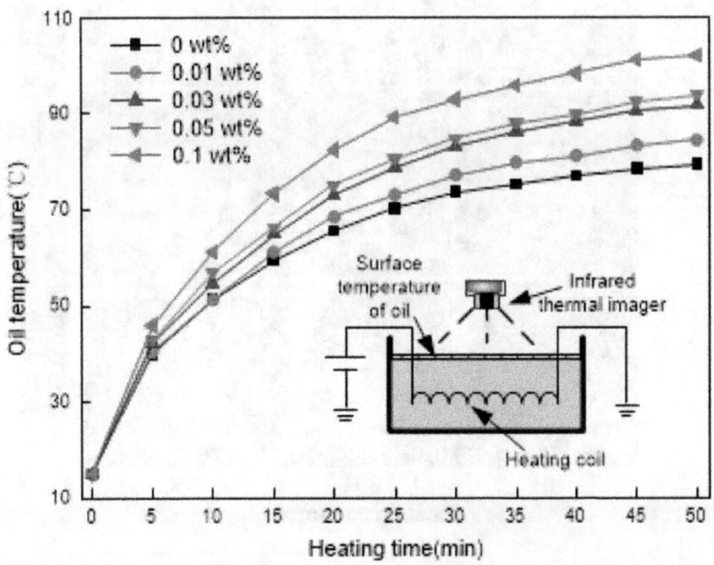

source to the oil in a rapid way. Figure 16 shows the temperature distribution in heating process at the time of 10, 30 and 50 minutes. It is showed that the high temperature area of the samples extends obviously with increasing the mass fraction. Furthermore, the maximum temperature area shows a significant increase trend when the mass fraction increases which is in accordance with the result in Figure 15. Both the thermographs and the temperature curve show remarkable enhancement in thermal property due to the addition of BN

Figure 16. Relation between the mass fraction and the heating time

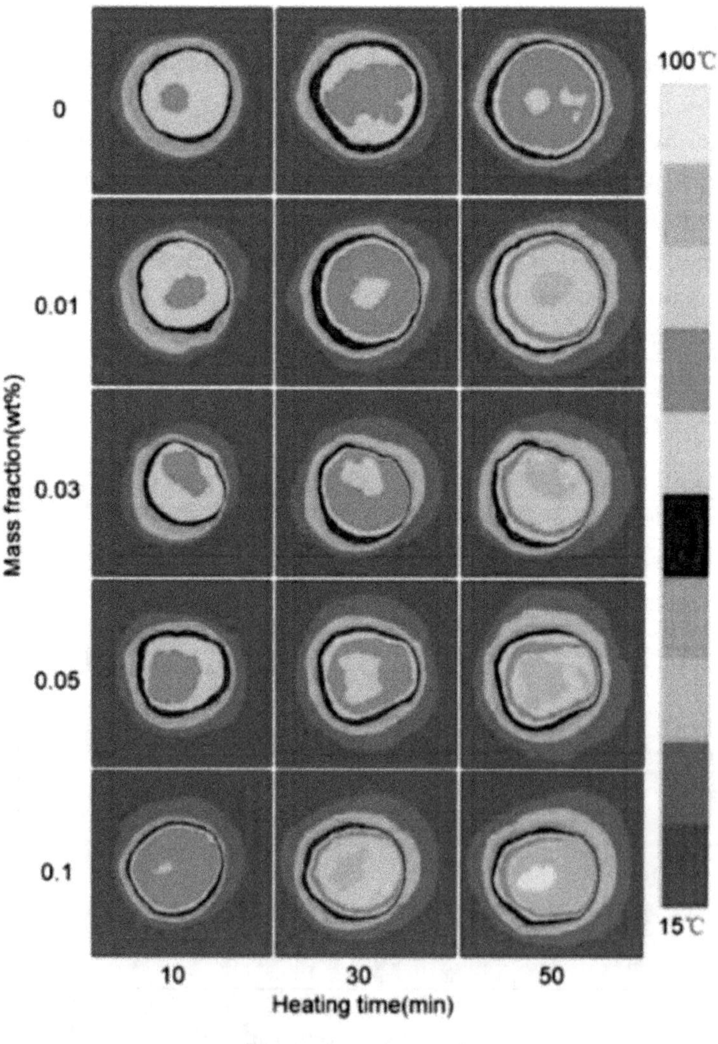

**For a more accurate representation of this figure, please see the electronic version.*

Figure 17. Relation between the coil temperature and the decay time with the different mass fractions

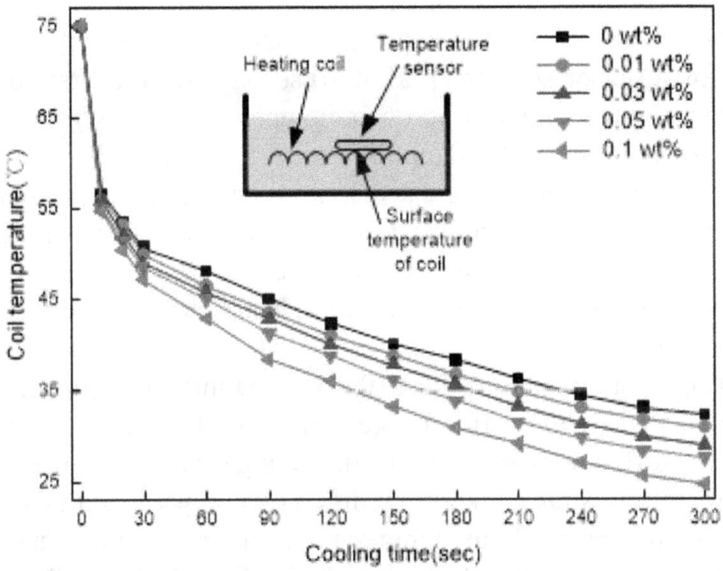

Figure 18. Relation between the oil temperature and the lapse time with the different mass factions

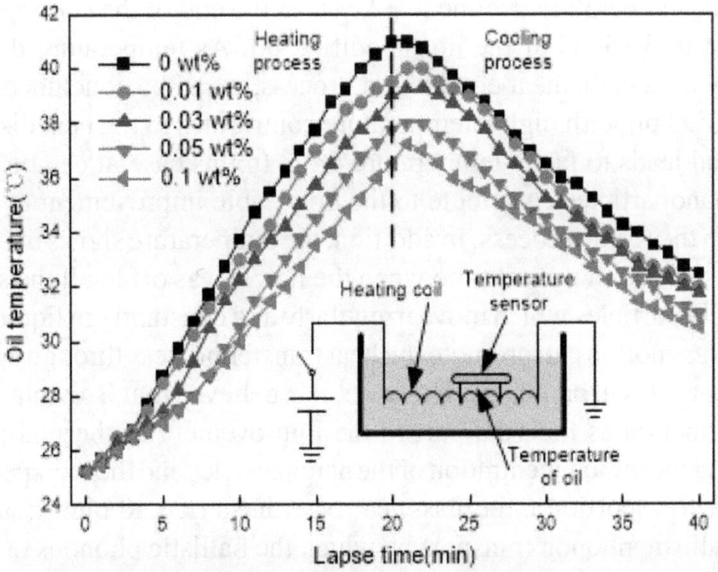

nanoparticles which indicates the heat generated by the coil is conducted to the oil rapidly instead of accumulated in the heat source.

The experiment for the cooling process was carried out and surface temperature of the heating coil is given in Figure 17. The temperature was measured by a sensor at an interval of 10 seconds for the first 30 seconds in order to reflect the rapid drop of the temperature at beginning. For the following measurements the sampling time was fixed at 30 seconds. It is observed that the cooling process accelerates obviously with increasing the mass fraction of BN nanoparticles. Furthermore, the surface temperature of the coil reaches to 32 °C at the cooling time of 300 s in non-modified oil while only 25 °C in the sample of 0.1 wt%. The obtained results indicate a considerable improvement in heat conduction which contributes to reducing the probability of operation failure due to the overheating in electric transformer.

Figure 18 shows the relation between the oil temperature and the lapse time with the different mass fractions. In heating process, it is observed that the temperature rises faster in non-modified oil compared with the nano-oil and reaches a higher maximum temperature. The maximum temperature of the non-modified oil is 40.5 °C while only 36 °C of the 0.1 wt% sample.

It can be confirmed that the heat generated by the source is conducted and dissipates to the entire oil in a faster way due to the addition of BN nanoparticles. Moreover, the improvement of thermal property shows an evident increase trend with increasing the mass fraction. When the concentration of the nano-oil is 0.1 wt% the temperature decline to 32.5 °C at the end of the cooling process in contrast to 30.5 °C of the non-modified oil. As temperature difference effectively affects the heat conduction process, it can be concluded that the non-modified oil with higher temperature contributes to the heat dissipation process and leads to faster temperature drop. In this case, it can be inferred that BN nanoparticles contribute to the noticeable improvement of thermal property in the whole process. In addition, the temperature shows unexpected delayed rise trend in a short time when the power was off in all the samples.

The solid particles which move irregularly and constantly in liquid caused by Brownian motion can promote the heat transfer process through collisions between particles in base fluid. However, it is believed that Brownian motion may not function as the vital part in the improvement of thermal property considering the low concentration of the nanoparticles and the low speed of the movement in proportion as the mass heat to be transferred. Keblinski suggested that the ballistic phonon transport by which the ballistic phonons originated in one particle can spread to an adjacent particle attributes to the remarkable enhancement much compared to Brownian motion. The nanoparticles get close to each other even at low concentration and may be much closer affected by Brownian motion. Thereby the ballistic phonon transport operating only in short distance dominates in the increase of thermal property of the nano-oil even at extremely low mass fraction (Keblinski et al., 2001; Suganthi et al., 2014; Henderson et al., 1984; Joshi et al., 1993; Ye et al., 1993).

Based on the analysis above, it is considered that the notable improvement of thermal property in nano-oil could be mainly attributed to ballistic phonon transport among nanoparticles. Brownian motion appears not to be important, but it may have an indirect positive effect on the ballistic phonon transport and promote the thermal property in auxiliary pattern.

INTERFACE CHARGE BEHAVIOR OF OIL-IMPREGNATED PAPER UNDER DC AND POLARITY REVERSAL VOLTAGES

Effect of Numbers of Paper Layers

The curves of the charge decay are shown in Figure 19. The interface charge density decreases with time. The decay is fast in the initial period and then becomes slower with time. The decay behavior of negative charge polarity is similar to that of positive polarity. The decay curves of paper charged at 4 kV voltages show distinguished features. As can be seen in Figure 19a and Figure 19b, the curves show the relations among different paper layers from one to four. Four curves decrease following the same trend. This is mainly due to the reduction of the initial electric field. With the lapse of time, the decay curves tend to become flat, revealing that the interface charge density becomes stable over time. This is mainly because the trapping of injected charge is enhanced during this period, and the implanted charge on the surface in turn inhibits charge from transferring into the bulk, thus prolonging the decay. Both the positive polarity and the negative polarity continue to follow the same trend, meaning that the absolute value of density decreases very fast at the beginning, then the decay rate becomes a little slower than before, and finally the density falls to a specific number. In addition, the interface charge density of the single-layer paper is much larger than in other situations. Accordingly, it reflects the fact that the insulation of a single layer is weaker than that of more layers. The density of four layers is a little larger than that of three layers, and three layers is sufficient to withstand the high voltage.

Figure 19. Relationship between the interface charge density and the decay time, after a 5-min 4 kV charging voltage for different paper layers with a thickness of 0.13 mm. (a)charging with positive voltage. (b)charging with negative voltage

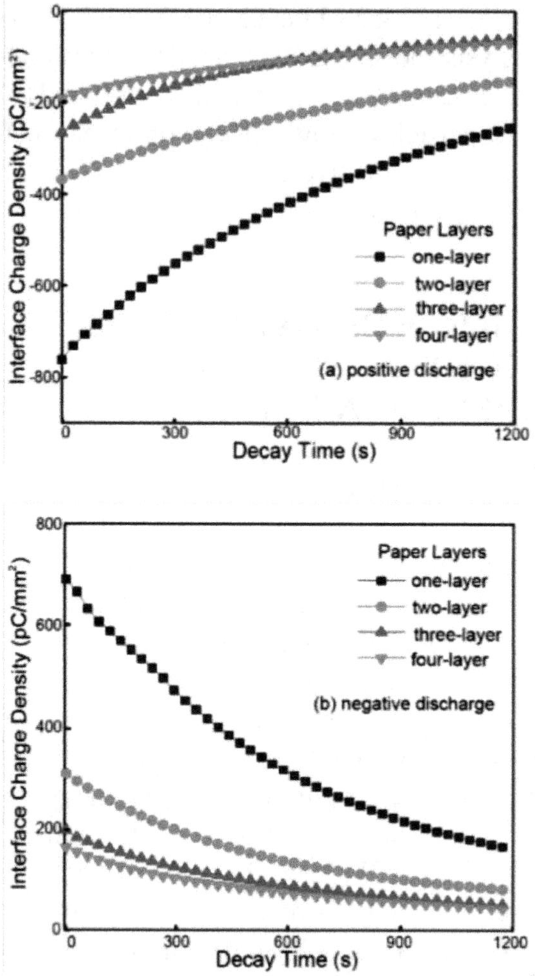

From Figure 20, the paper layers can be compared with the dissipation rate of interface charge. From the figure it can be seen that, for the thickness of 0.08 mm, the dissipation rates of the positive interface charge for one, two, three, and four layers are 46.67, 54.28, 59.06, and 68.50%, while the corresponding figures are 65.57, 68.98, 77.34, and 84.05% for the thickness of 0.13 mm. It can be concluded that with the increase in the number of paper layers, the dissipation rate rises. The reason for this must be that the more

Figure 20. Relationship between the dissipation rate of interface charge and the paper layers, after a 5-min 4 kV charging voltage for different thicknesses of 0.08 and 0.13 mm

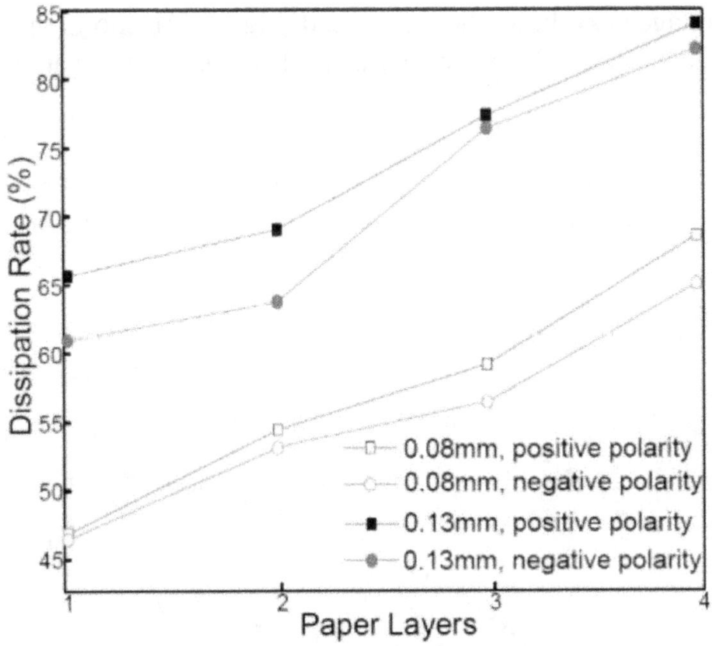

paper there is, the more easily the interface charges accumulate on the surface, and the more quickly the interface charge decreases. Meanwhile, according to the above data, it can be found that for the same number of paper layers, the dissipation rate is higher for 0.13-mm-thick paper compared to 0.08-mm-thick paper. This trend is the same as that of the paper layers, meaning that increasing the thickness of the paper is equivalent to adding paper layers.

The negative discharge accords with the same law. The dissipation rates of the negative interface charge for one, two, three, and four layers are 46.25, 52.95, 56.27, and 64.97%, for the thickness of 0.08 mm, respectively, and 60.84, 63.67, 76.38, and 82.18% for the thickness of 0.13 mm. The trend is almost the same.

Effect of Unipolarity Reversal Voltage

Figure 21 a shows the relationship between the interface charge density and the decay time under the positive-negative voltage. From the figure it can be seen that the dissipation trend of the interface charge under the polarity reversal voltage is the same as that under DC voltage. Also, according to the experiments, different thicknesses and polarity reversal times show the same trend. Under these conditions the dissipation is the same. The negative-positive voltage also shows the same trend. Figure 21b shows the situation in which the interface charge density is measured under a negative-positive

Figure 21. Relationship between the interface charge density and the decay time, after a 4 kV charging polarity reversal voltage with the reversal time of 25 s with a thickness of 0.08 mm. (a)charging with positive voltage. (b)charging with negative voltage

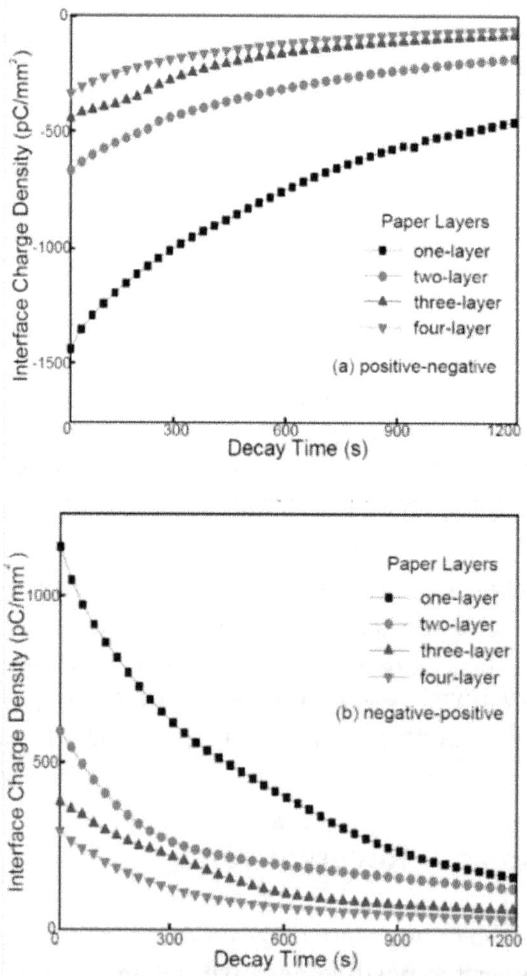

Figure 22. Relationship between the surplus ratio and the decay time, after a 4 kV charging polarity reversal voltage with the reversal time of 2 min with a thickness of 0.08 mm: (a)charging with negative voltage (b)charging with positive voltage

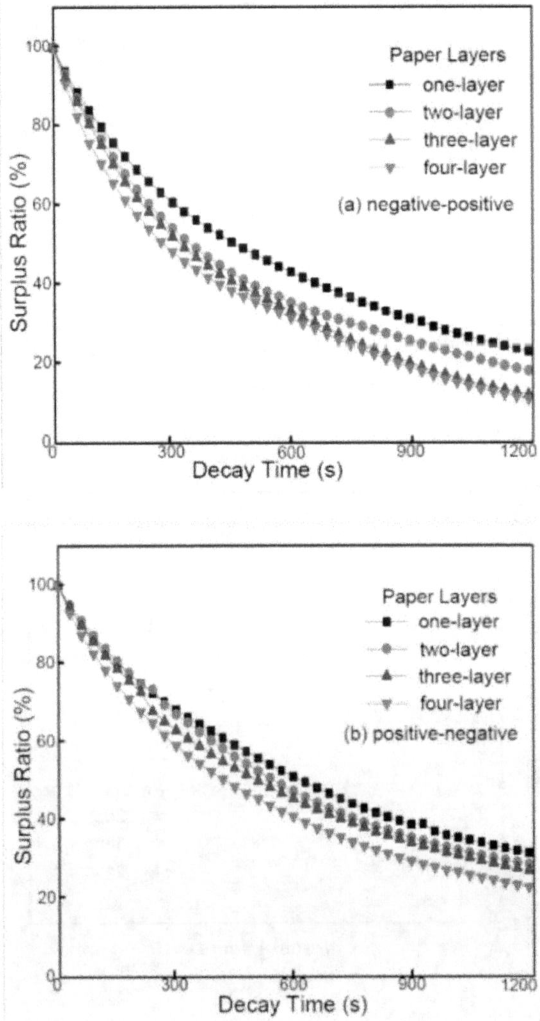

voltage. In that figure, because the final voltage applied in the experiment is positive, all the values are positive. But the absolute value of the numbers in Figure 21b is obviously a little smaller than that in Figure 21a.

Figure 22 shows the relationship between the surplus ratio and the decay time. In the figures, with the increase in the number of paper layers, the surplus ratio decreases. This means that the interface charge dissipates more quickly as the number of paper layer increases. The same conclusion is drawn

Figure 23. Relationship between the dissipation rate and the paper layers, after 4 kV charging positive–negative polarity reversal voltage for different thicknesses. (a)the thickness of 0.08 mm. (b)the thickness of 0.13mm

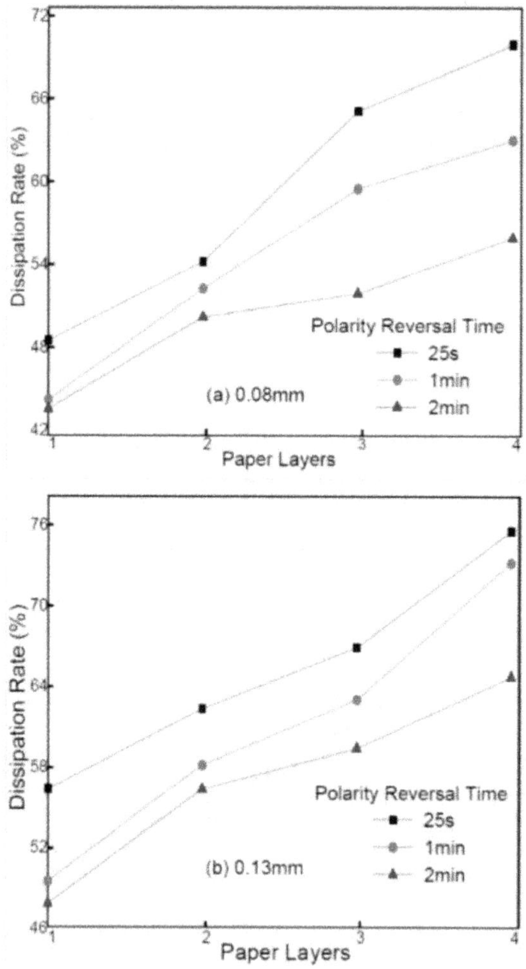

from Figure 23. Now, take the data of 25 s of polarity reversal time under the positive-negative polarity reversal voltage for the thickness of 0.13 mm as an example. The dissipation rates for one, two, three, and four layers are 56.28, 62.25, 66.82, and 75.49% respectively. Meanwhile, the dissipation rates for 1 min of polarity reversal time for the thickness of 0.08 mm are 44.59, 52.66, 59.89 and 63.41% respectively. It is the same in situations of negative-positive polarity reversal voltage.

Figure 24. Relationship between the surplus ratio and the decay time, after a 4 kV charging positive–negative polarity reversal voltage with the reversal time of 25 s for the two-layer paper

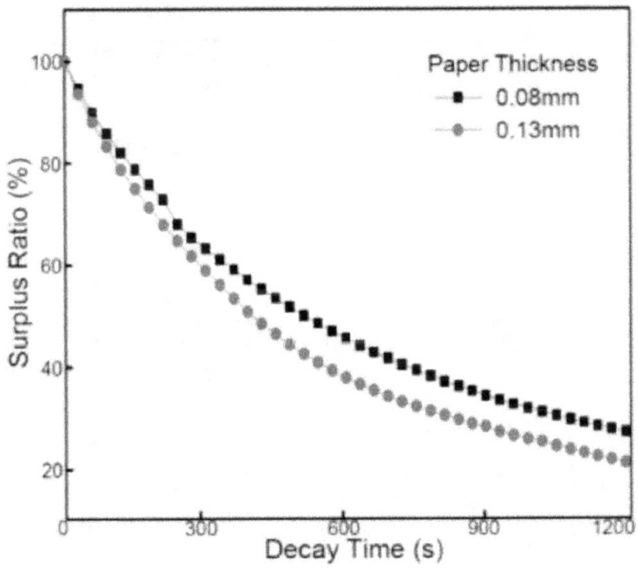

Figure 25. Relationship between the surplus ratio and the decay time, after a 4 kV charging polarity reversal voltage with a thickness of 0.08 mm and the polarity reversal time of 1 min for the four-layer paper

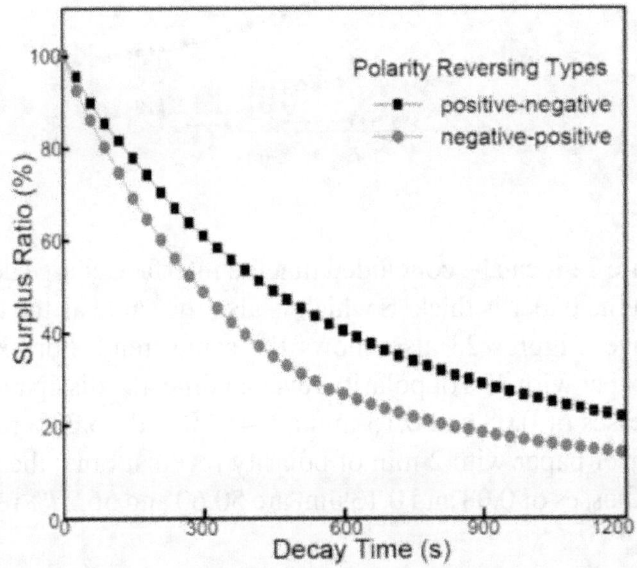

Figure 26. Relationship between the interface charge density and the decay time, after a 4 kV charging polarity reversal voltage with the reversal time of 25 s: (a) charging with negative-positive-negative voltage (b) charging with positive-negative-positive voltage

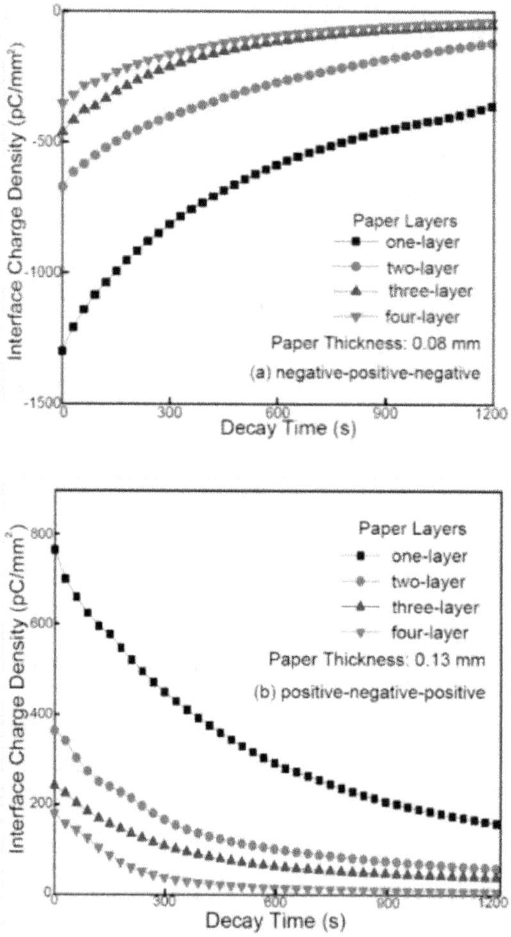

From Figure 24 it can be concluded that the interface charge decays more quickly when the paper is thicker, which is also the same as the trend under the DC voltage. Figures 23 also shows the same trend. For example, for single-layer paper with 25 s of polarity reversal time, the dissipation rates for paper thicknesses of 0.08 and 0.13 mm are 48.90 and 56.28% respectively. For double-layer paper with 2 min of polarity reversal time, the dissipation rates for thicknesses of 0.08 and 0.13 mm are 50.60 and 56.27% respectively.

Figure 27. Relationship between the surplus ratio and the decay time, after a 4 kV charging positive–negative–positive polarity reversal voltage with the reversal time of 25 s with a thickness of 0.08 mm

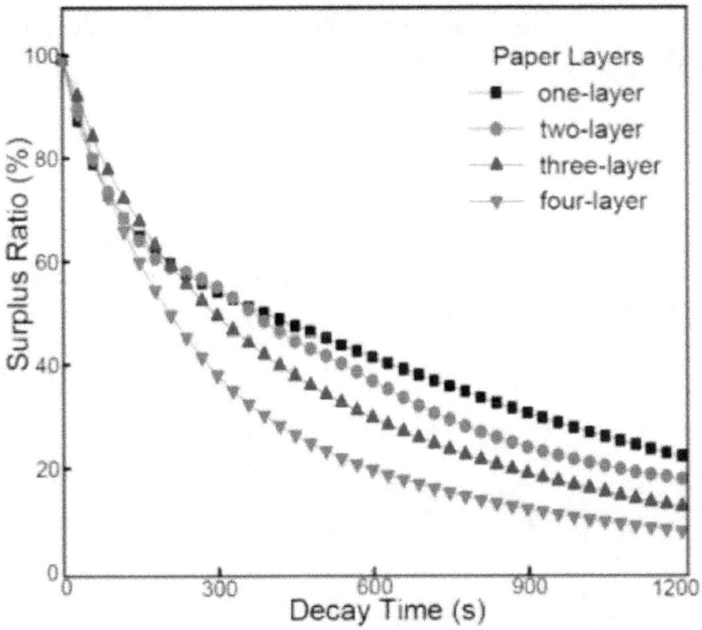

The other groups of data show the same trend. This means that increasing the thickness of the paper is equivalent to adding more paper layers.

The situations of the negative–positive voltage also follow the same rules. It is true that the different polarity reversal times have different dissipation rates. It is easily found from Figure 23 that a longer polarity reversal time causes a lower dissipation rate. From Figure 23 it can be seen that in the positive–negative reversal test, for single-layer paper with a thickness of 0.08 mm, the dissipation rates with reversal times of 25 s and 1 and 2 min are 48.90, 44.59, and 43.93% respectively, while for four-layer paper with a thickness of 0.13 mm, the dissipation rates with reversal times of 25 s and 1 and 2 min are 75.49, 73.11, and 64.64% respectively.

Meanwhile, the trend in the negative–positive reversal test is the same as that in the positive–negative case. Figure 21b shows the relationship between the dissipation rate and the number of paper layers, after applying a 4 kV negative–positive polarity reversal voltage. Through comparison with Figure 21a, it is easily found that the dissipation rate of the negative–positive situation is larger than that of the positive–negative situation. Figure 25 reflects this trend more significantly. In Figure 25, the two curves represent the situation in which the thickness of paper is 0.08 mm for four layers and the polarity reversal time is 1 min. But the polarity reversal types are opposite to one another. Obviously, the dissipation of the negative–positive one is much faster than that of the positive–negative one.

Figure 28. Relationship between the dissipation rate and the paper layers, after 4 kV charging positive–negative–negative polarity reversal voltage (a)the thickness of 0.08 mm. (b)the thickness of 0.13mm

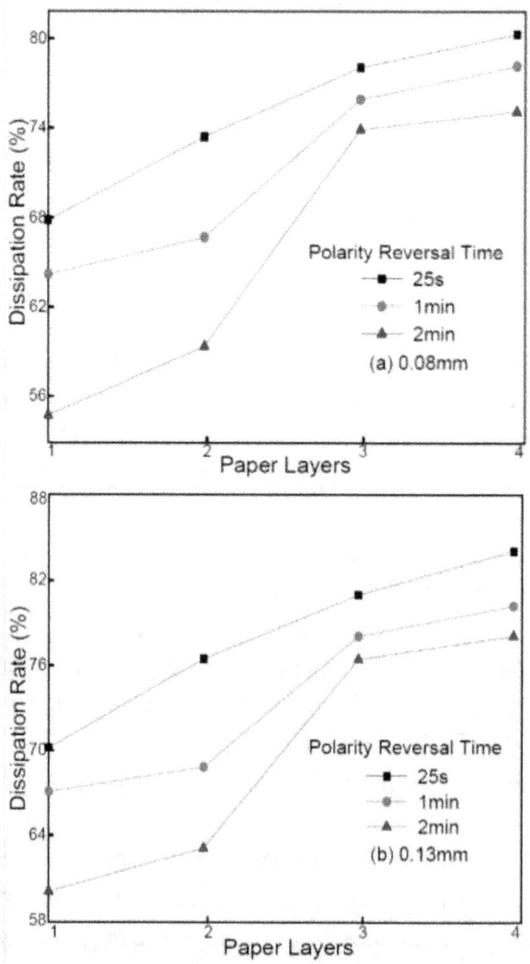

Effect of Bipolarity Reversal Voltage

The relationship between the interface charge density and the decay time is shown in Figure 26. The dissipation of interface charge under the bipolarity reversal voltage is the same as that under the DC and unipolarity reversal voltages. In these figures, it is clearly found out that as the number of paper layer increases, the absolute value of the interface charge density decreases. This means that single-layer paper keeps the most interface charge. Three-layer and four-layer paper have almost the same amount of interface charge.

From Figure 27, it is obvious that the more paper layers there are, the more severely the interface charge dissipates. Now take the data of 25 s of polarity reversal time for the thickness of 0.08 mm as an example. From Figure 28, it can be seen that the dissipation rates for one, two, three, and four paper layers are 67.97, 73.57, 78.27, and 80.45%, respectively. Meanwhile the dissipation rates for 1 min of polarity reversal time for the thickness of 0.13 mm are 67.23, 68.97, 78.20, and 80.34% respectively. The other four groups of data confirm the same rule. From Figure 29, the conclusion can be drawn that the thicker the paper is, the more severely the interface charge decays. For example, for

Figure 29. Relationship between the surplus ratio and the decay time, after a 4 kV charging positive–negative–positive polarity reversal voltage with the reversal time of 2 min for the two-layer paper

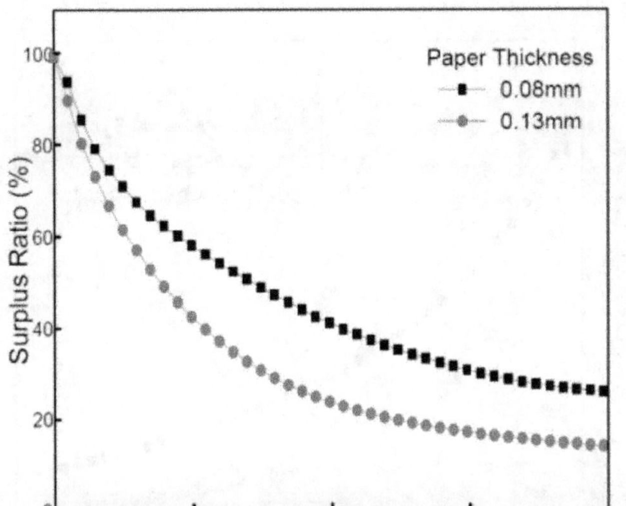

Figure 30. Relationship between the surplus ratio and the decay time, after a 4 kV charging positive–negative–positive polarity reversal voltage with a thickness of 0.13 mm for the two-layer paper

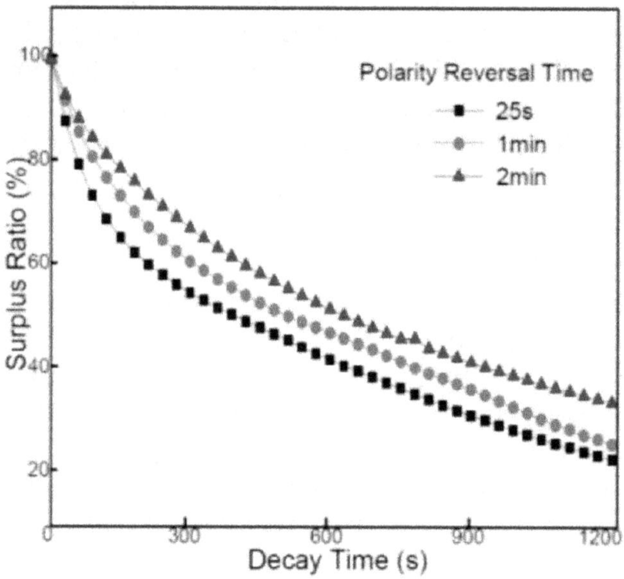

Figure 31. Relationship between the surplus ratio and the decay time, after a 4 kV charging polarity reversal voltage with a thickness of 0.08 mm and the polarity reversal time of 1 min for the three-layer paper

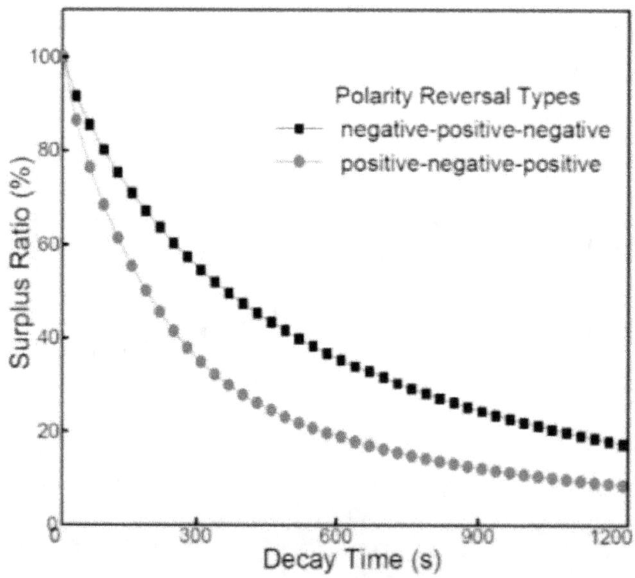

single-layer paper with 25 s of polarity reversal time, it can be seen from Figures 30a and 6.30b that the dissipation rates for paper thicknesses of 0.08 and 0.13 mm are 67.97 and 70.32%, respectively. For double-layer paper with 2 min of polarity reversal time, the dissipation rates for paper thicknesses of 0.08 and 0.13 mm are 59.48 and 63.24%, respectively. The other groups of data show the same trend. This means that increasing the thickness of the paper is equivalent to adding the paper layers.

It is easily found from Figure 30 that longer polarity reversal times cause lower dissipation rates. From Figure 28, it can be seen that for single-layer paper with a thickness of 0.08 mm, the dissipation rates with reversal times of 25 s and 1 and 2 min are 67.97, 64.30, and 54.84% respectively, while for four-layer paper with a thickness of 0.13 mm, the dissipation rates with reversal times of 25 s and 1 and 2 min are 84.19, 80.34, and 78.23% respectively. In Figure 31, the two curves represent the situation in which the thickness of paper is 0.08 mm for three layers and the polarity reversal time is 1 min. The only difference is between the polarity reversal types. Obviously, the dissipation of the positive–negative–positive one is much faster than that of the negative–positive–negative one. The other data show the same trend.

REFERENCES

An, Z., Yin, Q., Liu, Y., Zheng, F., Lei, Q., & Zhang, Y. (2015). Modulation of surface electrical properties of epoxy resin insulator by changing fluorination temperature and time. *IEEE Transactions on Dielectrics and Electrical Insulation*, *22*(1), 526–534. doi:10.1109/TDEI.2014.004551

Atiya, E. G., Mansour, D. E. A., Khattab, R. M., & Azmy, A. M. (2015). Dispersion behavior and breakdown strength of transformer oil filled with TiO_2 nanoparticles. *IEEE Transactions on Dielectrics and Electrical Insulation*, *22*(5), 2463–2472.

Benson, S. W. (1965). Bond energies. *Journal of Chemical Education*, *42*(9), 502–518. doi:10.1021/ed042p502

Charlet, K., Saulnier, F., Dubois, M., & Béakou, A. (2015). Improvement of wood polymer composite mechanical properties by direct fluorination. *Materials & Design*, *74*(1), 61–66. doi:10.1016/j.matdes.2015.02.012

Chen, Q. G., Liu, H. Q., Zhuge, X. L., Chi, M. H., & Wei, X. L. (2014). Analysis of electric field homogenization of converter transformer barrier system based on nano-modification of pressboard. *Adv. Mater. Res.*, *981*(1), 940–945. doi:10.4028/www.scientific.net/AMR.981.940

Dean, J. A. (1999). *Lange's Handbook of Chemistry* (15th ed.). New York: McGraw-Hill Companies.

Du, B. X., & Guo, Y. G. (2015). Effects of surface fluorination on dielectric properties and surface charge behavior of water absorbed polyimide film. *IEEE Transactions on Dielectrics and Electrical Insulation*, *22*(3), 1369–1375. doi:10.1109/TDEI.2015.7116324

Du, B. X., Li, Z. L., & Li, J. (2014). Surface charge accumulation and decay of direct-fluorinated RTV silicone rubber. *IEEE Transactions on Dielectrics and Electrical Insulation*, *21*(5), 2338–2342. doi:10.1109/TDEI.2014.004079

Du, Y. F., Lv, Y. Z., Li, C. R., Chen, M. T., Zhong, Y. X., & Zhou, J. Q. (2012). Effect of semiconductive nanoparticles on insulating performances of transformer oil. *IEEE Transactions on Dielectrics and Electrical Insulation*, *19*(3), 770–776. doi:10.1109/TDEI.2012.6215079

Hanaoka, R., Kohrin, T., Miyagawa, T., & Nishi, T. (2002). Creepage discharge characteristics over solid-liquid interfaces with grounded side electrode. *IEEE Transactions on Dielectrics and Electrical Insulation*, *9*(2), 308–315. doi:10.1109/94.993749

Henderson, J. R., & Swol, F. V. (1984). On the interface between a fluid and a planar wall theory and simulations of a hard sphere fluid at a hard wall. *Molecular Physics*, *51*(4), 991–1010. doi:10.1080/00268978400100651

Huang, M., Zhou, Y. X., Chen, W. J., Sha, Y. C., & Wang, Z. D. (2014). Influence of voltage reversal on space charge behavior in oil-paper insulation. *IEEE Transactions on Dielectrics and Electrical Insulation*, *21*(1), 1625–1634. doi:10.1109/TDEI.2013.004010

Hwang, J. G., Zahn, M., O'Sullivan, F. M., Pettersson, L. A. A., Hjortstam, O., & Liu, R. S. (2010). Effects of Nanoparticle Charging on Streamer Development in Transformer Oil-Based Nano-fluids. *Journal of Applied Physics*, *107*(014310).

IEEE. (1995). *Guide for loading mineral-oil-immersed transformers, annex I: transformer insulation life*. IEEE Standard C57.91.

Joshi, A. A., & Majumdar, A. (1993). Transient ballistic and diffusive phonon heat transport in thin films. *Journal of Applied Physics*, *74*(1), 31–39. doi:10.1063/1.354111

Kao, K. C. (1976). Theory of high-field electric conduction and breakdown in dielectric liquids. *IEEE Transactions on Electrical Insulation*, *11*(4), 121–128. doi:10.1109/TEI.1976.297919

Kao, K. C., & McMath, J. P. C. (1970). Time-dependent pressure effect in liquid dielectrics. *IEEE Transactions on Electrical Insulation*, *5*(3), 64–68. doi:10.1109/TEI.1970.299097

Keblinski, P., Phillpot, S. R., Choi, S. U. S., & Eastman, J. A. (2001). Mechanisms of heat flow in suspensions of nano-sized particles (nanofluids). Intl. *J. Heat Mass Transf.*, *45*(4), 855–863. doi:10.1016/S0017-9310(01)00175-2

Lee, S. A., Oh, S. H., & Lee, W. (2009). The effect of direct fluorination of polydimethylsiloxane films on their surface properties. *Journal of Colloid and Interface Science*, *332*(2), 461–466. doi:10.1016/j.jcis.2008.12.062 PMID:19147155

Liu, R. S., Pettersson, L. A. A., Auletta, T., & Hjortstam, O. (2011). Fundamental research on the application of nano dielectrics to transformers. *IEEE Conf. Electr. Insul. Dielectr. Phenom. (CEIDP)*, 423-427. doi:10.1109/CEIDP.2011.6232685

Lv, Y. Z., Zhou, Y., Li, C. R., Ma, K. B., Wang, Q., Wang, W., & Jin, Z. et al. (2014). Nanoparticle effects on creeping flashover characteristics of oil/pressboard interface. *IEEE Transactions on Dielectrics and Electrical Insulation*, *21*(2), 556–562. doi:10.1109/TDEI.2013.004151

Lv, Y. Z., Zhou, Y., Li, C. R., Wang, Q., & Qi, B. (2014). Recent progress in nanofluids based on transformer oil: Preparation and electrical insulation properties. *IEEE Electr. Insul. Mag.*, *30*(5), 23–32. doi:10.1109/MEI.2014.6882597

Mansour, D., Atiya, E., Khattab, R., & Azmy, A. (2012). Effect of titania nanoparticles on the dielectric properties of transformer oil-based nanofluids. *IEEE Conf. Electr. Insul. Dielectr. Phenom. (CEIDP)*, 295-298. doi:10.1109/CEIDP.2012.6378779

Norman, B. C., Lawrence, H. D., & Stephen, E. W. (1990). *Introduction to infrared and raman spectroscopy* (3rd ed.). New York: Academic Press.

Nowell, M. M., & Wright, S. I. (2004). Phase differentiation via combined EBSD and XEDS. *Journal of Microscopy*, *213*(3), 2628–2638. doi:10.1111/j.0022-2720.2004.01299.x PMID:15009697

Prorokova, N. P., Istratkin, V. A., Kumeeva, T. Y., Vavilova, S. Y., Kharitonov, A. P., & Bouznik, V. M. (2015). Improvement of polypropylene nonwoven fabric antibacterial properties by the direct fluorination. *RSC Advances*, *5*(55), 44545–44549. doi:10.1039/C5RA05112F

Saleem, H., Thunga, M., Kollosche, M., Kessler, M. R., & Laflamme, S. (2014). Interfacial treatment effects on behavior of soft nano-composites for highly stretchable dielectrics. *Polymer*, *55*(17), 4531–4537. doi:10.1016/j.polymer.2014.06.054

Saulnier, F., Dubois, M., Charlet, K., Frezet, L., & Beakou, A. (2013). Direct fluorination applied to wood flour used as a reinforcement for polymers. *Carbohydrate Polymers*, *94*(1), 642–646. doi:10.1016/j.carbpol.2013.01.060 PMID:23544585

Smith, R. C., Liang, C., Landry, M., Nelson, J. K., & Schadler, L. S. (2008). The mechanisms leading to the useful electrical properties of polymer nanodielectrics. *IEEE Transactions on Dielectrics and Electrical Insulation*, *15*(1), 187–196. doi:10.1109/T-DEI.2008.4446750

Suganthi, K. S., Vinodhan, V. L., & Rajan, K. S. (2014). Heat transfer performance and transport properties of ZnO–ethylene glycol and ZnO–ethylene glycol–water nanofluid coolants. *Applied Energy*, *135*, 548–559. doi:10.1016/j.apenergy.2014.09.023

Sugumaran, C. P. (2012). Experimental evaluation on dielectric and thermal characteristics of nano filler added transformer oil. *IEEE Int'l. Conf. High Voltage Eng. Appl. (ICHVE)*, 207-210.

Tanaka, T. (2005). Dielectric nanocomposites with insulating properties. *IEEE Transactions on Dielectrics and Electrical Insulation*, *12*(2), 914–928. doi:10.1109/TDEI.2005.1522186

Tanaka, T., Kozako, M., Fuse, N., & Ohki, Y. (2005). Proposal of a multi-core model for polymer nanocomposite dielectrics. *IEEE Transactions on Dielectrics and Electrical Insulation*, *12*(4), 669–681. doi:10.1109/TDEI.2005.1511092

Wada, J., Nakajima, A., Miyahara, H., Takuma, T., Yanabu, S., Okabe, S., & Kohtoh, M. (2006). Surface breakdown characteristics of silicone oil for electric power apparatus. *IEEE Transactions on Dielectrics and Electrical Insulation*, *13*(4), 830–837. doi:10.1109/TDEI.2006.1667742

Xu, Z. Y., Li, J. Z., & Ke, W. (2016). Surface partial discharge and breakdown characteristics of oil-paper insulation under -40°C condition. *Proceedings of CSEE*, *36*(10), 2827–2835.

Yang, J. X., Chi, X. C., Ding, L. J., & Liu, J. (1997). Creeping discharge performance of oil-paper insulation with streaming electrification. *IEEE Transactions on Dielectrics and Electrical Insulation*, *4*(6), 780–784. doi:10.1109/94.654705

Ye, L., Liu, J., Sheng, P., Huang, J. S., & Weitz, D. A. (1993). Sound propagation in colloidal systems. *Journal of Physics*, *3*(1), 183–196.

Yi, X., & Wang, Z. (2015). The influences of solid surface on the propagation of creepage discharge in insulating liquids. *IEEE Transactions on Dielectrics and Electrical Insulation*, *22*(1), 303–312. doi:10.1109/TDEI.2014.004195

Zhang, L., Zhou, Y., Huang, M., Sha, Y., Tian, J., & Ye, Q. (2014). Effect of nanoparticle surface modification on charge transport characteristics in XLPE/SiO_2 nanocomposites. *IEEE Transactions on Dielectrics and Electrical Insulation*, *21*(2), 424–433. doi:10.1109/TDEI.2013.004145

Zhong, Y. X., Lv, Y. Z., Li, C. R., Du, Y. F., Chen, M., Zhang, S., & Chen, L. et al. (2013). Insulating properties and charge characteristics of natural ester fluid modified by TiO_2 semiconductive nanoparticles. *IEEE Transactions on Dielectrics and Electrical Insulation*, *20*(1), 135–140. doi:10.1109/TDEI.2013.6451351

Zhuge, X., Chi, M., Wang, X., & Liu, H. (2015). Electric field homogenization for converter transformer with nano-modified pressboard. *IEEE Int. Conf. Prop. Appl. Dielectr. Mater. (ICPADM)*, 44-47.

Chapter 7
Epoxy Resin for GIS Disc-Type Insulator

ABSTRACT

Epoxy resin is widely used in disc-type spacers of gas insulated switchgear (GIS), such as disc-type insulator, post insulator and so on. While surface flashover related to the existence of surface charge and tracking failure often occur on the spacer, which is a great threat to the application of GIS and the safety of entire grid. This chapter presents a study aimed at clarifying surface charge behaviors, surface flashover and tracking resistance of EX and EX/ Al_2O_3 composites. It is indicated that surface charge behaviors of EX charged by AC and positive pulse voltage are totally different from that charged by AC and negative pulse. And it can be concluded that the characteristics of surface charge behaviors are changed by adding nanoparticles. With increasing the filler content from 0 to 5 wt%, the relative discharge intensity, weight loss and pattern area of the specimens decrease, the last time of the tracking process increases.

INTRODUCTION

EPOXY resin (EX) has attracted increasing attention because of its excellent insulation properties and low cost (Iwata, 2016; An et al., 2016; Li et al., 2015; Nguyen et al., 2015; Ning et al., 2015). The addition of nanoparticles into epoxy can change the molecular structure and introduce some traps at the interfaces. The effects of nanoparticles on electrical properties have

DOI: 10.4018/978-1-5225-2309-3.ch007

been investigated. It was found that the charge behavior can be affected by nanoparticles (Lewis, 2004; Tagami et al., 2008; Sarathi et al., 2016). For example, the surface flashover characteristics on solid dielectrics were found to be closely related to the charge behavior (Kumara et al, 2012; Kimura et al., 2002; He et al., 2016).

Many investgators have investigated the effect of nanoparticles on charge behavior (Hayase et al, 2006; Du et al., 2012; Perlman et al., 1987; Nath et al., 1990; Takada et al., 2008). Hayase et al reported that low density polyethylene (LDPE)/MgO nanocomposite at loadings higher than 0.5 per hundred (phr) of the base polymer showed a superior property to prevent the injection of space charge into the bulk (Hayase et al, 2006). Du studied the effect of TiO_2 nanoparticles on the dynamic behavior of surface charge accumulation and charge decay of Epoxy/TiO_2 nanocomposites. The obtained results showed that the charge accumulation and charge decay were restricted by nanoparticles (Du et al., 2012). Perlman et al showed that new deep traps were introduced by the titanium dioxide at all concentrations in polyethylene (Perlman et al., 1987). Using the same method, Nath et al concluded that charge storage in polyarylate increased with increasing dopant concentration to an optimum at 10 wt% titanium dioxide (Nath et al., 1990). Takada reported that trap depth and the effective range of the induced dipole due to nanoparticles depended on the relative dielectric constant, size, and shape of the nanoparticles (Takada et al., 2008).

For the purpose of obtaining more excellent properties, a lot of work has been done by adding nano fillers into epoxy including nano-Al_2O_3. Thomas et al have discovered that the adding of Al_2O_3 reduces the PD magnitude in LDPE (Alapati et al., 2012). Edin et al have found the dimension of the nano-particle has great influence on the magnitude of tan δ (Jäverberg et al., 2012). Tanaka et al have investigated that nano particles impact the initiation treeing time (Tanaka et al., 2008; Chen et al., 2010). They have also suggested that the nano-Al_2O_3 particles increase the dielectric breakdown strength of epoxy composites (Li et al., 2010). Meanwhile, Moraveji et al have studied that the thermal conductivity increases when adding nano-Al_2O_3 into a fluid (Moraveji et al., 2012). It is shown that the adding content of nano particles has a great effect on the dielectric performance of polymer nanocomposites (Du et al., 2012).

EFFECTS OF AC AND PULSE VOLTAGE COMBINATION ON SURFACE CHARGE ACCUMULATION AND DECAY OF EPOXY RESIN

Effect of Voltage Amplitudes

Under normal conditions, because GIS operates under AC voltage, charges will accumulate on the surface of spacer. When switching overvoltage occurs, pulse voltage subsequently appears. So it is necessary to ascertain the charge coupling behavior. Actually, the surface charge density has a linear relationship with the surface potential. Hence, the change in surface potential can be regarded as the accumulation and decay of surface charge.

The relationship between the initial surface potential and the pulse voltage amplitudes is shown in Figures 1a and 1b. According to these two Figures, the initial potential increases with the AC voltage level without applying any pulse. When the AC voltages are 0, 10, 12 and 14 kV, the initial potential are 0, -1045, -1312 and -1634 V, respectively. Negative charge is easier to move than positive charge, so the negative charge is easier to be injected in. After each period, the amount of negative charge injected into the sample should be greater than that of positive charge. If AC voltage is applied on sample for a period time, there will be a lot of negative charges accumulating on the surface of sample. The higher AC voltage level is, the larger surface negative charge amount is. When only pulse voltage is applied on the needle electrode, the needle will inject charges into the sample. More surface charges accumulate with the pulse amplitude increasing. As shown in Figures 1a and 1b, when the negative pulse amplitude rises from 0 V to -2.5 kV, the initial potential increases from 0 V to -1832 V. When the positive pulse voltage rises from 0 to +3 kV the initial potential increases from 0 V to +1531 V.

The negative pulse voltages with different amplitude are applied to the sample after AC voltage was applied. The pulse number is 500 and the frequency is 500Hz. With increasing the amplitude of negative pulse voltage, the initial potential increases and then decreases. The curves show a clear inflection point. As shown in Figure 1a, AC voltages are 10, 12 and 14 kV. The initial potentials reach the minimum when the negative pulse amplitudes are -1000, -1500 and -2000 V, respectively. Figure 1b shows relationship between the initial surface potential and the pulse voltage amplitudes after AC and positive pulse voltage applied. The curves show a single trend.

Figure 1. Relationship between the initial surface potential and the pulse voltage amplitudes, after an application of AC and 500 Hz pulse voltage with pulse number of 500

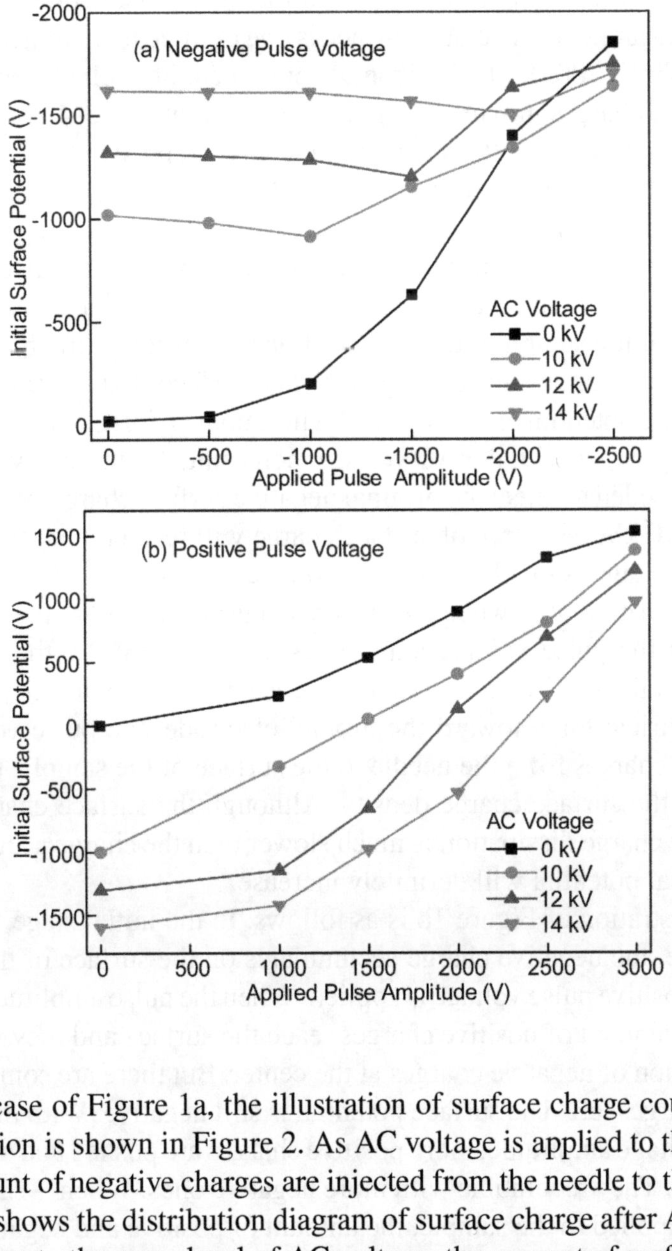

In the case of Figure 1a, the illustration of surface charge coupling and accumulation is shown in Figure 2. As AC voltage is applied to the sample, large amount of negative charges are injected from the needle to the sample. Figure 2a shows the distribution diagram of surface charge after AC voltage applied. Due to the same level of AC voltage, the amount of surface charge is certain. Because of the rejection of the homo-charge on the surface, the charge from the needle caused by pulse voltage will be difficultly injected into the sample. In addition, when the pulse voltages are small, the original surface charges will also be influenced by the Coulombic forces caused by pulse voltages. The repulsive force produced by smaller pulse voltage makes a few surface charges go through the bulk. Meanwhile, that charges transporting along the insulator surface is one of three main routes for charge decay (Molinie, 2005). The horizontal component of applied electrical field can enhance charges moving along the gas-solid interface. Therefore, a few surface charges move to the surrounding area by repulsive force component paralleling the surface as shown in Figure 2b. The arrow in the Figures is the main motion direction of charges. Hence, a small pulse voltage amplitude has little influence on the initial surface potential. When the pulse amplitude increases, as shown in Figure 2c, the repulsive force increases as well. More charges from the sample interface will transport through the bulk or move from the center to the surrounding area. The surface charges reduce and the initial surface potential reduces as well. The amount of original surface charge will become larger with the AC voltage increasing. So more powerful pulse voltage is needed to reject the original negative surface charge. When the AC voltage is 10 kV, the curve of surface charge will reach the minimum when the pulse voltage is -1 kV. If the AC voltage increases to 14 kV, the curve will reach the minimum when the pulse voltage is -2 kV. With the continuous increase in the pulse voltage amplitudes, as in Figure 2d, the amount of charge from the needle continues to increase more. At that time, due to the strong electrical force toward the ground electrode, the pulse voltage will "push" the charges from the needle to the surface of the sample, thus easily increasing the surface charge density. Although the surface charges decay as well, the charge dissipation is much slower than the charge accumulation. So the initial potential will definitely increase.

The illustration of Figure 1b is as follows. In the initial stage, due to the AC voltage, the negative charge accumulates on the surface of the sample. Then the positive pulse voltage is applied. When the pulse amplitude is small, only small amount of positive charges reach the surface and they recombine with a portion of negative charges at the center. But there are some negative surface charges left. The surface potential is still negative. With the increase in pulse voltage amplitude, more positive charges are pushed and appealed to the surface. They recombine with more negative ones. When +1.5 kV pulse voltage is applied to the sample, the amount of positive and negative charge caused by AC 10 kV are basically equal. So the initial surface potential is almost zero. As the pulse amplitude continues rising, more positive charges will accumulate on the surface. So the surface potential tends to become

Figure 2. Illustration of surface charge accumulation, after AC voltage and negative pulse voltage applied in different amplitudes

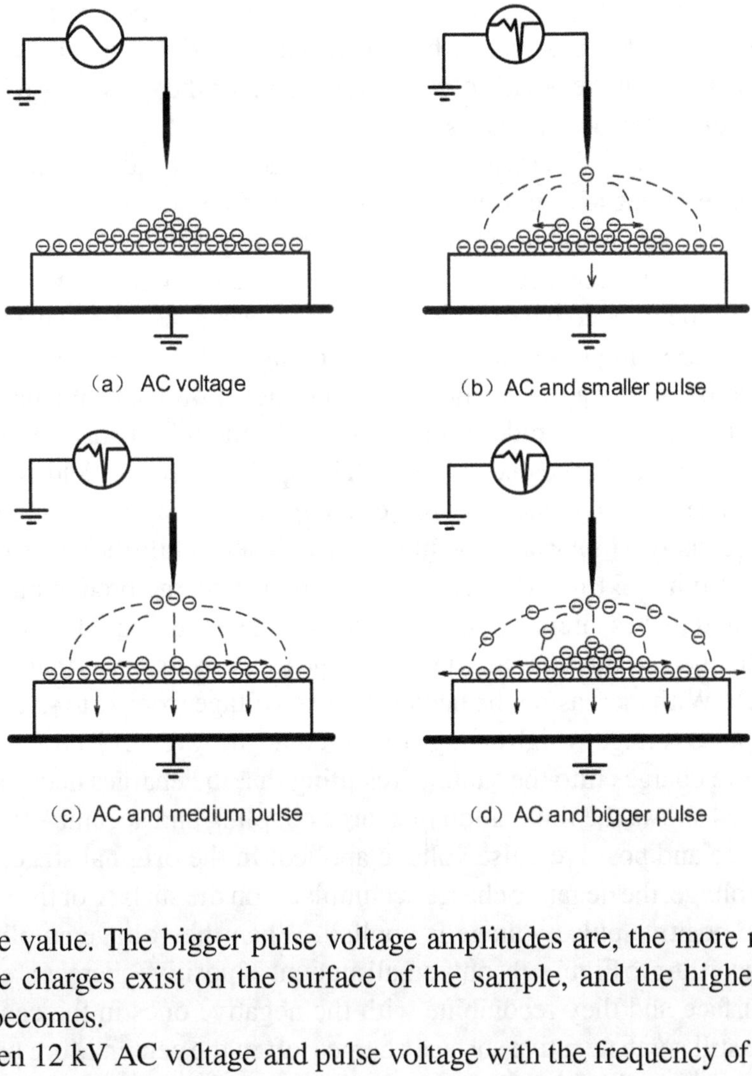

positive value. The bigger pulse voltage amplitudes are, the more residual positive charges exist on the surface of the sample, and the higher initial value becomes.

When 12 kV AC voltage and pulse voltage with the frequency of 500 Hz and the pulse number of 500 have been applied, the relationship between the surface potential and the decay time under different pulse amplitude is shown in Figure 3. Figure 3a shows the curves of AC voltage and negative pulse voltage. The potential decreases with the increase of time. The decay process is fast at the beginning, and then becomes slower. Figure 3b shows the relationship between the surface potential and the decay time after applying 12 kV AC voltage and positive pulse voltage with different amplitude. While the positive pulse amplitude is +1 kV. The surface potential is negative, increasing from -657 to -858 V and then becoming smaller. While the pulse voltage increases to +2 kV, the surface potential is positive value at the beginning, decreasing fast from +128 to 0 V and then becoming negative value. The +2.5 kV case has a higher initial potential and shows a faster decay process than the +2 kV case. The potential of the +3 kV case is entire positive during the decay process.

The explanation of charges decay process shown in Figure 3 is as follows. For the case of negative pulses, actually, the surface potential dissipation is relevant to the release of electrons from the trapping state, indicating that it is the detrapping of these electrons that leads to the decline in surface potential with time. The trapped charges in shallow traps easily escape, causing rapid initial decay. During next period, the detrapping of the charges in deep traps occurs, thus the decay in surface potential slows down. With increasing the negative pulse amplitude from 0 to -1.5 kV, the initial surface potential decreases and the charge decays faster. The applied electric field caused by smaller pulse voltage makes the surface charges get energy and escape from traps more easily. The surface potential decays faster. While the negative pulse increases from -1.5 to -2 kV, the existing negative charge producing reverse electric field makes later charges only accumulate on the sample surface and cannot be injected into the sample. So charges still decay faster than the case of -2.5 kV. With increasing the negative pulse voltage from -2 to -2.5 kV, the applied pulse voltage is high enough to overcome the reverse electric field and inject more charges into the sample, resulting that the charges decay slower.

Figure 4 shows the illustration of charge coupling and accumulation after AC voltage and positive pulse voltage applied. In the original stage, due to the AC voltage, the negative charge accumulates on the surface of the sample. Then the positive pulse voltage is applied. When the voltage amplitude is small, according to Figure 4b, only small amount of positive charges are driven to the surface and they recombine with the negative ones in the center. So there are still some negative surface charges left in the center. More negative charges are recombined under the needle electrode than other positions of the sample. Therefore, the potential under the needle is different with the surrounding potential, resulting in the negative charges detrapping and moving from the surrounding area to the center. Thus the negative potential in the center area increases at the beginning and then decreases. With the increase in pulse voltage amplitude, more positive charges are pushed to the surface and they recombine with the negative charges, indicated from Figure 4c. When +2 kV pulse voltage is applied to the sample, the negative charges in the center are recombined and some positive charges remain on the surface,

Figure 3. Relationship between the surface potential and the decay time, after the application of 12 kV AC voltage and 500 Hz pulse voltage with the pulse number of 500, and the different amplitudes

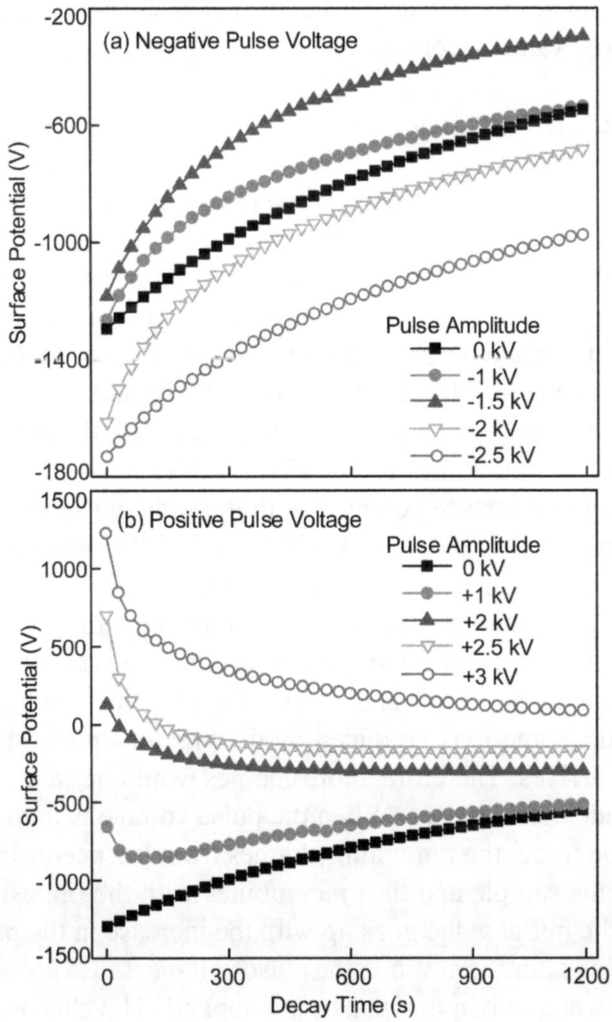

resulting in the potential becoming positive. Then the surrounding negative charges move to the center and recombine with the positive charges, leading to the potential changing from positive value to negative value. After applying positive pulse, the decay rate at beginning is bigger than that without applying positive pulse. It is because that the bigger potential difference will appear between the center and surrounding area due to the positive pulse. It enhances the local electric field and accelerates the charge recombining process. From

Figure 4d it can be seen that as the pulse voltage continues rising, all negative charges are recombined and positive charges will accumulate on the surface, so the surface potential becomes positive. The larger pulse voltage amplitude is, the more residual positive charges exist on the surface of the sample, and the higher initial value becomes.

Effect of Pulse Numbers

In order to study the effect of pulse numbers on epoxy resin insulation, the relationship between the initial surface potential and the pulse numbers is shown in Figure 5. The pulse voltage with different numbers were applied to the sample after the corona charging under 12 kV AC voltage for 20 min. The initial values of surface potential with the pulse numbers of 0, 50, 500 and 1000 for a +2 kV pulse voltage are -1354, -264, 125 and 371 V, respectively. Meanwhile, the initial values of surface potential with those pulse numbers for a -1.5 kV pulse voltage are -1354, -1242, -1183 and -1133 V, respectively. The initial values of surface potential with the same pulse numbers for a -2 kV pulse voltage are -1354, -1395, -1613 and -1797 V, respectively. The similarity of these three curves is that the initial values change more with the increase of pulse numbers. As the pulse number continues increasing, the influence on initial value becomes weaker and becomes gradually saturated. Actually, for curves of +2 kV and -2 kV, when the pulse voltage is applied to the electrode, some ions produced by the previous pulse still exist until the next pulse arrives. Therefore, more charges would be easily injected into the surface under more pulses. When the pulse voltage is high enough, due to the repulsive force, the remaining charges from the needle can no longer go straight to the sample and thus recombines with the opposite polar ions in the air. So the initial value goes up with the increase in the pulse number, and then reaches saturation. While the pulse voltage is -1.5 kV, some charges have decayed when only a few pulses are applied. The charges which were captured by shallow traps detrapped more easily due to the applied electric field at the beginning of pulse application time, while the charges in the deep trap detrap hardly because the energy provided by the pulses is not enough under the same voltage amplitude, causing that the potential decreases only a little even though the number of pulses increases.

The curves in Figure 6 present relationship between the surface potential and the decay time, after 12 kV AC voltage and 500 Hz pulse voltage with the different pulse numbers. As shown in Figure 6a, the decay rate increases

Figure 4. Illustration of surface charge accumulation, after an AC voltage and a positive pulse voltage applied in different amplitudes

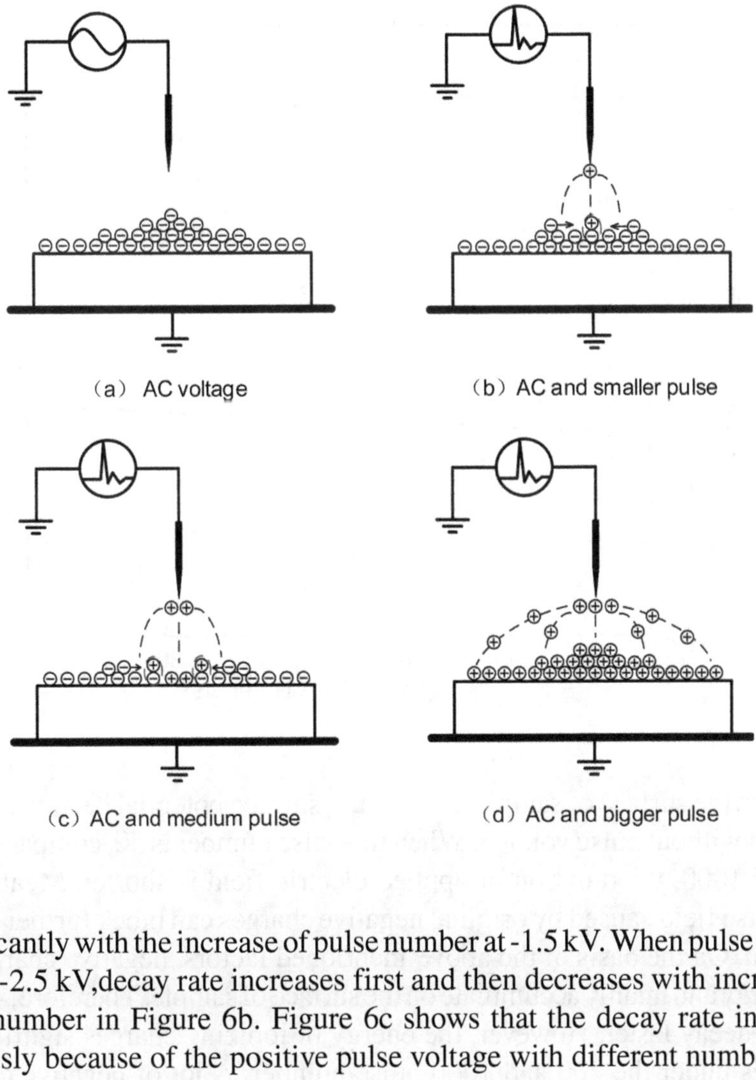

significantly with the increase of pulse number at -1.5 kV. When pulse voltage rise to -2.5 kV,decay rate increases first and then decreases with increase of pulse number in Figure 6b. Figure 6c shows that the decay rate increases obviously because of the positive pulse voltage with different number.

For the -1.5 kV pulse amplitude, under small applied electric field, surface potential decay faster with the increase in pulse number. This phenomenon shows that original negative charges which obtain more energy caused by electric field force with the increase of pulse number escape from the traps more easily. The charge decays more rapidly, as shown in Figure 6a. When pulse voltage increases to -2.5 kV in Figure 6b, charges produced by ionization accumulate enough energy with the increase of applied electric field. Thus, charges can overcome the reverse field caused by original surface charges

Figure 5. Relationship between the initial surface potential and the pulse number, after an AC voltage and a 500 Hz pulse voltage applied with different amplitudes

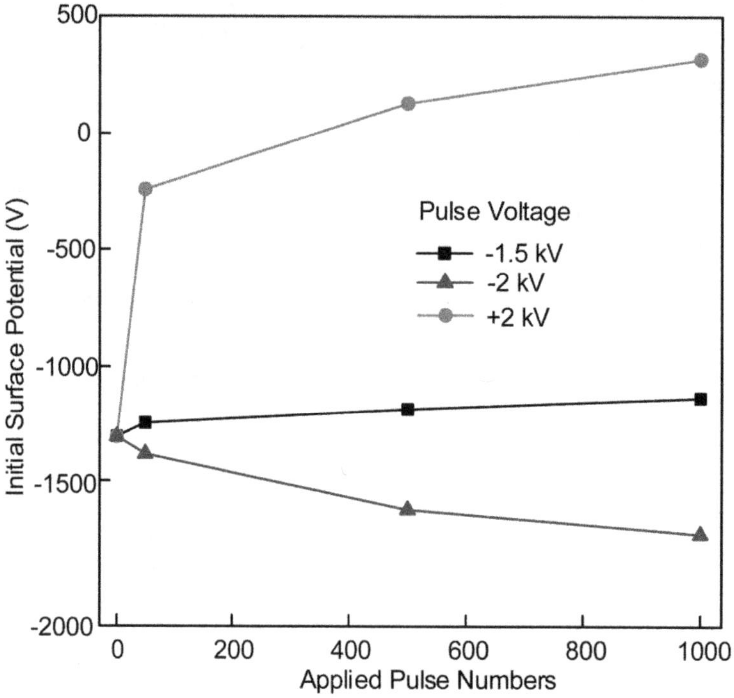

and reach the surface of sample. As a result, surface potential becomes bigger than that without pulse voltage. When the pulse number is 50, compared with 500 and 1000, the duration of applied electric field is shorter. Meanwhile, the reverse field caused by original negative charges can block further charge injection. On the basis of the above-mentioned factors, negative charges get less energy and mainly accumulate on the surface of sample. Therefore, surface charges decay faster. However, the energy of ionizing charges significantly increases under the 500 and 1000 pulse number. A lot of negative charges can overcome the reverse field and inject into the sample. Initial potential increases and decay slower. For the +2 kV pulse voltage in Figure 6c, the curves of different pules numbers are similar to those of different pulse amplitudes in Figure 3b. The initial potential is negative when the charging time is short and then it becomes positive with increasing the positive pulse number. Furthermore the surface potential decays faster at the beginning under 1000 pulse number than others. More positive pulses make more positive charges injected to the sample. The process of charge recombining increases the decay rate. From the Figures, it can be concluded that with the increase in pulse number, the influence on charge decay process will be improved.

Trap Energy Distributions

Charges can be transported by contact, corona or surface states, which are usually related to traps. The energy of the traps was related to time by considering so called demarcation energy E_m, indicating the border between emptied and occupied traps. The relationship between time and energy gap was defined as (Watson et al., 1990).

$$E_c - E_m = kT \ln(vt) \tag{1}$$

$$t\frac{dV(t)}{dt} \infty N(E_m(t))\frac{dE_m}{dt} = N(E_m(t))kT\frac{N_t}{N_c}\tau_0 \tag{2}$$

where E_c was the conduction band energy level, υ was the attempt to escape frequency, tdv/dt was proportional to the trap energy density $N(E_m)$ at energy level $E_m(t)$, N_t and N_c were trap and conduction states densities, respectively, and τ_0 was the carrier lifetime in the conduction states.

Figures 7a and 7b show trap energy distributions, after AC voltage and different pulse voltage were applied. In the Figure 7a, every curve has a peak, which can be attributed to only one changing peak of the charges in the corresponding trap depth. Compared with the 0 kV, the value of tdv/dt with -1.5 kV is similar but the value of trap depth is smaller, which reflects a faster decay process. It indicates that the smaller negative pulse voltage makes charges obtain energy to overcome the energy barrier and charges detrap more easily. As a result, the decay rate increases. However, the trap depth at the peak of -2.5 kV curve is bigger than that of 0 kV. It shows that big pulse voltages make a lot of charges trap into deeper traps, thereby reducing the decay rate. The decay rate under -2.5 kV is smaller than others.

Figure 7 b shows the trap energy distributions, after AC voltage and positive pulse voltage. The curves with pulse voltage +1 kV,+2 kV and +2.5 kV reach the first peak. Then they decrease to zero which is followed by more peaks. This is because that the process of charge dissipation divides into two stages. In the first stage, the surface potential decreases from positive value to zero. After that, it becomes negative value and then increases. Or it increases from

Figure 6. Relationship between the surface potential and the decay time, after the application of a 12 kV AC voltage and a 500 Hz pulse voltage with the different pulse numbers

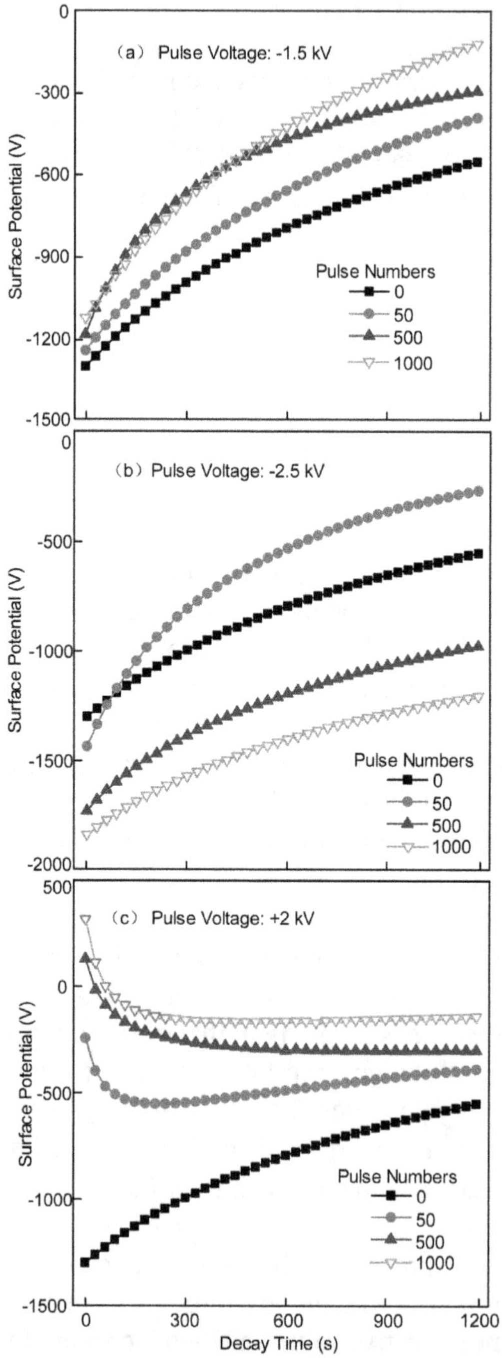

a small initial negative potential such as the curve of +1 kV in Figure 3b. In these two cases, there will be a maximum of negative value at which *t*d*v*/d*t* is 0 in the surface potential decay curve. The reason for this phenomenon is that charges of different polarity neutralize and residual negative charges move towards the center. Negative potential at the center of sample will reach the maximum when the electrical potential difference disappears. Moreover, in this stage, surface potential difference makes charges escape from the trap and neutralize with different polar charges, resulting in increase of the surface potential change rate. From the trap energy distribution standpoint, the charges get more energy to detrap and the trap energy at the peak of positive pulse *t*d*v*/d*t* curves is much smaller than that of AC curve. In the second stage, charges decay monotonously. But *t*d*v*/d*t* curves of 0 kV and +3 kV does not decrease to 0 in the middle of curves. The result indicates that negative charges are recombined completely under +3 kV pulse voltage. The surface of sample is covered by positive charges. So, it becomes a decay curve of positive charge.

Figure 8 shows *t*d*v*/d*t* curves with different pulse numbers. Figure 8a shows curves of pulse voltage with -1.5 kV. In the case of 1000, the trap energy is smaller than the cases of 0 and 50. It explains that the increasing of pulse number under -1.5 kV can increase the energy of charge and accelerate the decay of surface charge. In the Figure 8b, the curve of 50 under -2.5 kV has smaller trap energy at the peak than

that of curve 0. But the trap energy at the peak of curve under 1000 pulse number is much bigger than that of curve 0. It illustrates that 50 bigger pulses is too few to make lots of charges be injected into the sample and many charges trap in shallow traps. Therefore, decay speed is still fast. However, in the case of a large number of charging times, the decay process slows down, because more charges are captured by deep traps with increasing of the pulse application time. The trap energy of the first peak decreases with the pulse number increasing under positive voltage in the Figure 8c. It is because that positive charges injected to the sample and recombining with original negative charges increase the surface potential decay rate. In addition, more pulses can produce more positive charges and make surface potential change faster.

If the AC and negative pulse voltages are applied, the absolute value of the initial surface potential decreases at first and then increases with the rise of pulse voltage amplitude. The speed of surface charge dissipation has the same tendency. If the AC and positive pulse voltages are applied, the initial surface potential of the sample center changes from negative value to positive value with the increase in pulse amplitude. Furthermore, when the positive

Figure 7. Trap distributions with different negative and positive pulse amplitude and after 12 kV AC voltage application

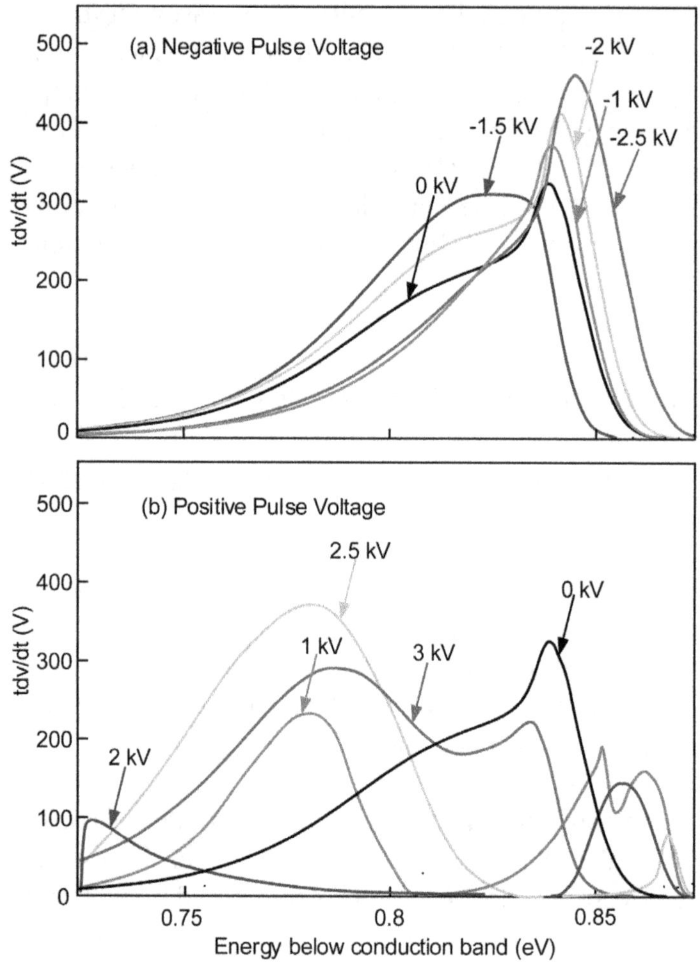

pulse voltages are small, charge dissipation is a two-stage process. In the first stage, surface potential gradually changes from positive to negative until negative surface potential reach the maximum. In the second stage, negative charges decay monotonically. Considering the pulse number effect, it is obvious that under the smaller negative pulse voltage, the initial surface potential decreases and charge decay rate increases with the rise of pulse number. But the conclusion is complete opposite under the bigger negative pulse voltage. In the case of positive pulse, with the increase of pulse number, the initial surface potential changes from negative to positive and the decay

Figure 8. Trap distributions with different negative and positive pulse numbers and after 12 kV AC voltage application

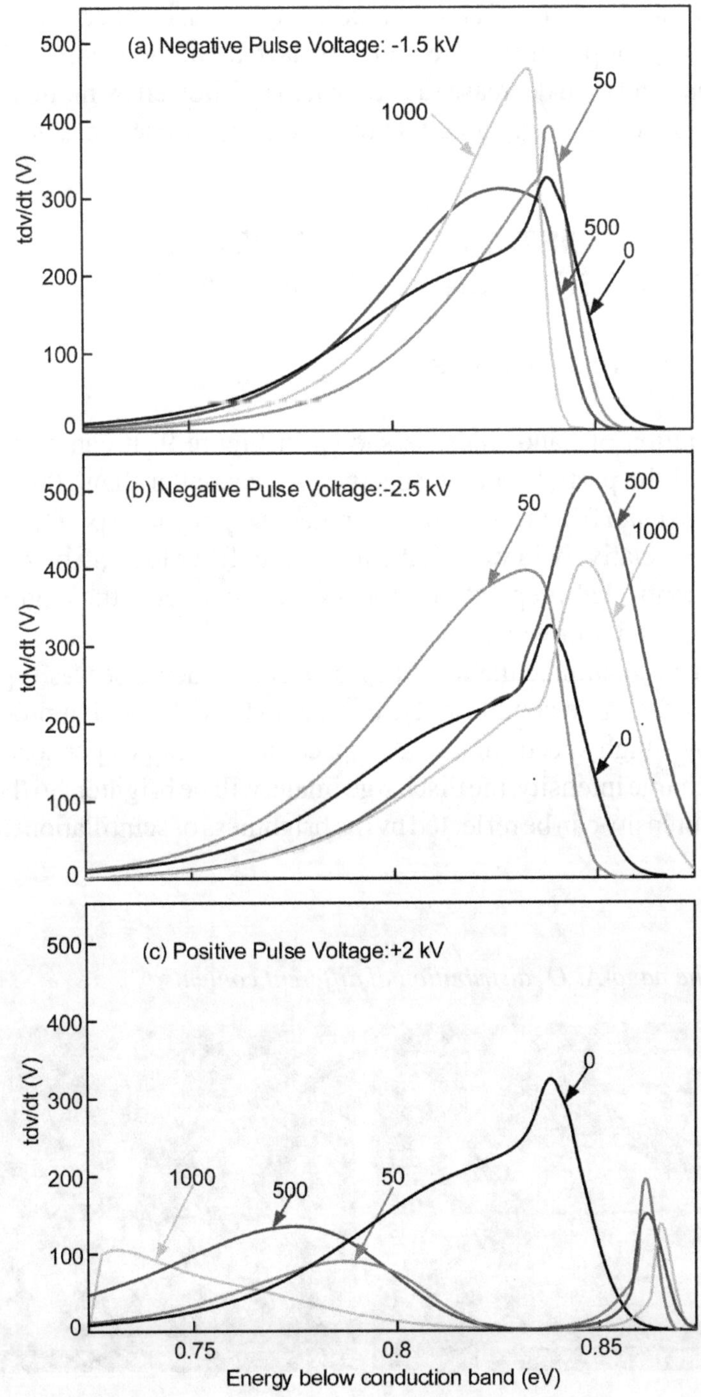

rate increases. The trap energy distribution present different trends for the different pulse voltage. The curves of trap energy for negative and positive pulse voltages reflect the different charge decay speed. Furthermore, the curves of trap energy for positive pulse voltages have a unique feature. They reach the first peak and then decrease to zero which is followed by more peaks. This phenomenon could be explained by the two-stage process of charge decay.

TRACKING RESISTANCE OF EPOXY/Al_2O_3 NANOCOMPOSITES UNDER DC VOLTAGE

Tracking Failure Process

The distribution of nano fillers is shown in Figure 9. It can be found that almost all of the particles in the specimens are smaller than 100 nm, which indicates the nano fillers are well distributed in epoxy matrix. The content of nano filler is relatively low (with the maximum filler loading of 7 wt %) and with well controlled preparation process for specimens, the agglomeration phenomenon is not obvious.

In order to guarantee the accuracy of results, each test was repeated 20 times under the same condition. Discharge images of tracking process were recorded by a high speed camera at a speed of 100 frames per second. With higher discharge intensity, the discharge image will be brighter. So the relative discharge intensity can be reflected by the brightness of scintillation discharge.

Figure 9. The nano-Al_2O_3 distribution of different contents

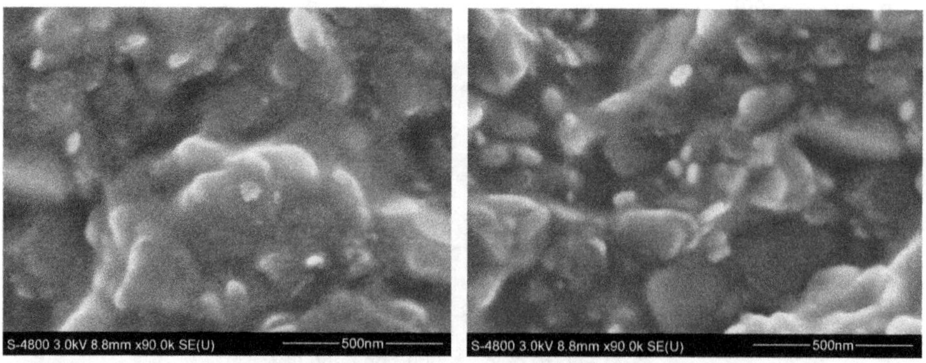

(a) 1 wt%

b) 7 wt%

When the filler content is lower than 5 wt%, as the discharge intensity shows a consistent tendency with increasing of the content, only three kinds of images are taken to reveal the effect of the content on discharge intensity. As shown in Figure 10, the relative discharge intensity from low to high is represented by the different color strip from blue to red. As time goes on, the color strip changes from blue to red, indicating the increase relative discharge intensity over time. From Figures 10a, 10b and 10c, it can be found that with increasing the nanoparticle content from 0 to 5 wt%, the tracking process lasts longer and the relative discharge intensity becomes lower, while with increasing the content from 5 to 7 wt%, the time duration of tracking process becomes shorter and the discharge intensity becomes higher. The results show that the adding of nano-Al_2O_3 particles can restrict the discharge and prolong the time during the tracking process. The enhancing effect of tracking resistance is the best when adding in a moderate amount of nano-Al_2O_3 particles. There are interfacial zones between the nano-particles and epoxy matrix. The characterization of interfaces can be explained by the multi-core model. There are three layers in the interfacial zone, including a transition layer bonded epoxy and nano-particle by chemical bonds, which can restrain the discharge which occurred on the epoxy surface. The inorganic nano-Al_2O_3 fillers have the features of good tolerance of discharge, which will help to increase the tracking resistance as well. Meanwhile, according to the theory of "dielectric discretization", the adding of nano-particles may cause the change of behavior in charge management, which would help to improve the discharge resistance. With increasing filler contents from 0 to 5 wt%, the increased nano-particles and interfacial zones will result in the degradation in the discharge intensity during the tracking process. With increasing the contents continually, the inter-particle distance decreases, resulting in the overlap of adjacent interaction zones. Due to the cluster station effects, the overlapped regions are presented as defects, which cause the discharge intensity to increase.

Figure 11 shows the relative discharge intensity of the same kind of specimen with different applied voltages. With increasing the applied voltage from 600 to 800V, the tracking process becomes shorter and the discharge intensity becomes higher. The results indicate that when the voltage becomes higher, the scintillation discharges will be more severe. Therefore, there is more energy to erode the polymer surface, which leads to the reduction of tracking process duration.

As time goes on, repeated discharges will constantly erode the specimen surface and cause the tracking failure. Figure 12 is shown to illustrate the

Figure 10. The discharge intensity of (a) 0 wt%, (b) 5 wt%, (c) 7 wt% sample under 600 V

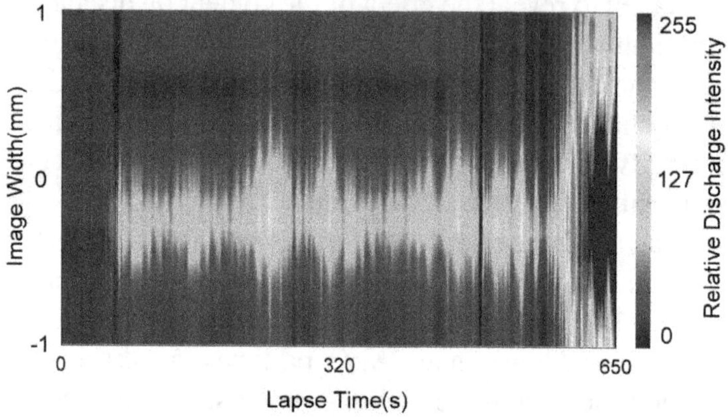

(a) 0 wt%

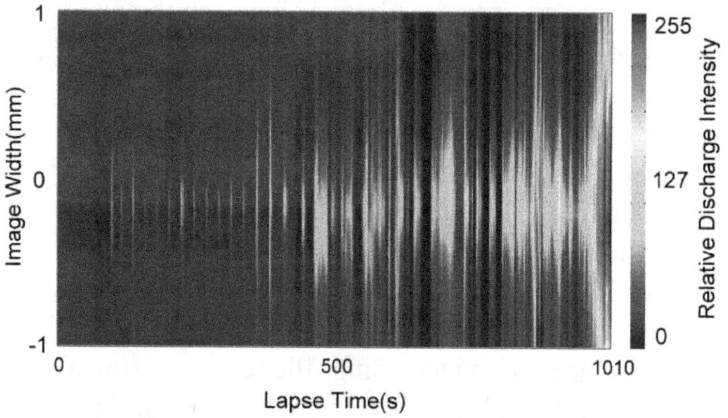

(b) 5 wt%

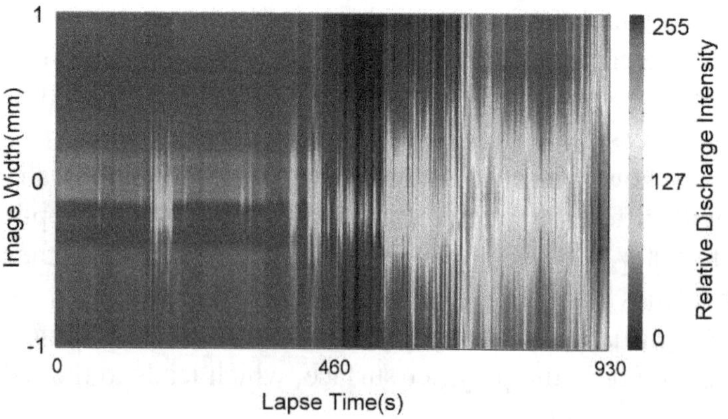

(c) 7 wt%

Figure 11. The discharge intensity of the 1 wt% specimen under different applied voltages

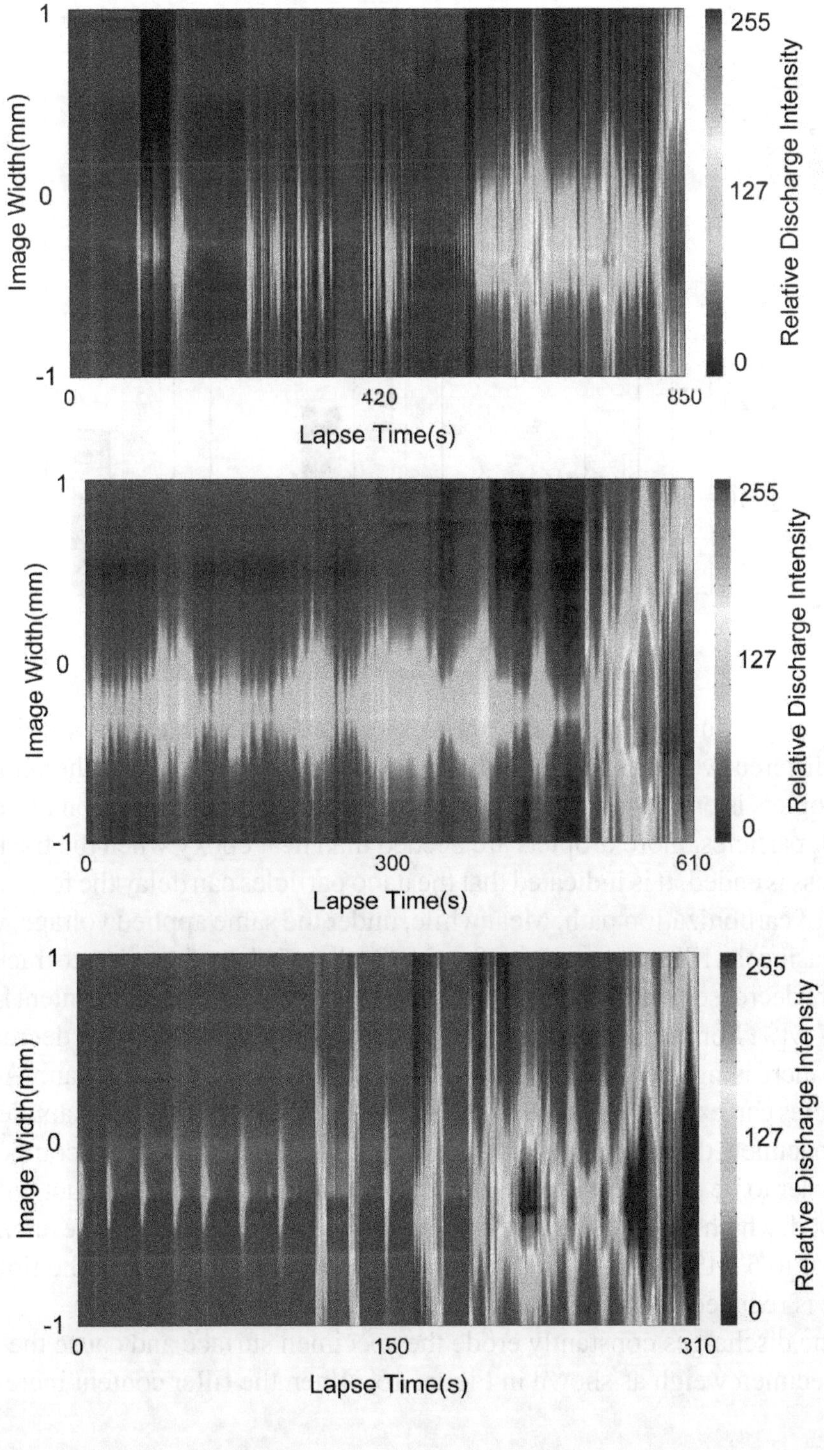

Figure 12. The duration of tracking process of different specimens under different voltages

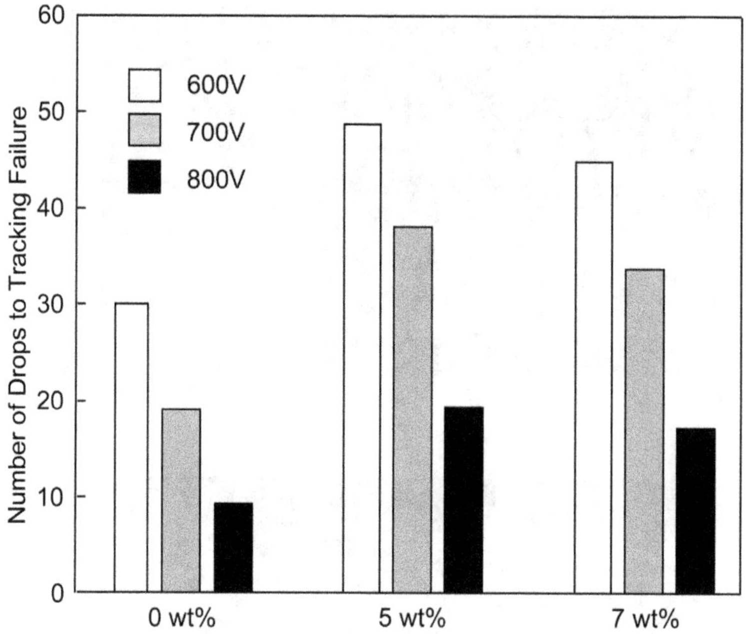

time length during the tracking process with different kinds of specimens and different voltages. Since the interval of the droplet is 20 s, the number of droplets is employed to reveal the time length. With the addition of nano-Al_2O_3 particles, more droplets are needed than neat epoxy when the tracking process is ended. It is indicated that the nano particles can delay the formation time of carbonization path. Meanwhile, under the same applied voltage, with increasing the filler content from 0 to 5 wt%, the number of droplets to tracking failure decreases, while the number increases with increasing the content from 5 to 7 wt%. For the same kind of specimen, the number of droplets decreases with increasing the applied voltage. The results show that the nano-Al_2O_3 particles can extent the tracking process and the higher the voltage applied to the specimen, the lower the resistance to tracking failure. The nano-composite is harder to be eroded and the time of generating carbonized product will be delayed, which prolongs the tracking process. The reduced discharge intensity from 0 to 5 wt% leads to the extending of the tracking process. The time to track is reduced as the discharge increases from 5 to 7 wt%.

The discharges constantly erode the specimen surface and cause the loss of specimen weigh as shown in Figure 13. When the filler content increases

Figure 13. The weight loss of different specimens under different voltages

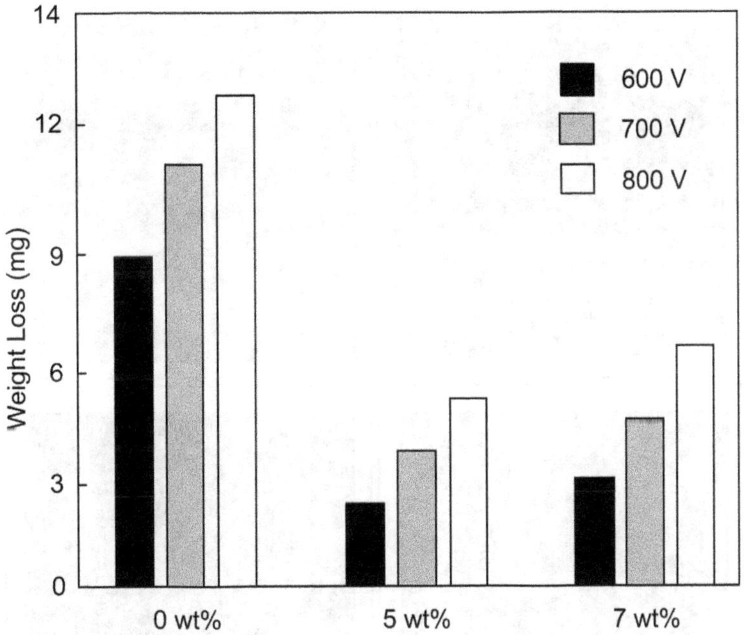

from 0 to 5 wt%, the weight loss decreases, but it increases when the content increases from 5 to 7 wt%. With the increase of applied voltage, the weight loss increases as well. The decrease of weight loss indicates that the precipitation of carbon was restrained in the composites. The results indicate that the nano-composites are harder to be eroded than the pure epoxy.

With discharges occur continually on the specimen surface, the energy is accumulated and finally breaks the molecular chain of epoxy. The carbonized particles are separated out and form the conductive path between the electrodes which causes the tracking failure as shown in Figure 14. To define the damaged area, images of tracking pattern were preprocessed to grayscale images. Although each test was repeated for 20 times, tracking patterns under the same condition were not exactly the same. Figure 14 shows the typical patterns in one test. For the 0, 1, 3, 5 wt% specimens, tracking patterns show a consistent trend, tracking pattern in pure specimen is not included in the chart. As shown in the Figure, the dark-color points in 5 wt% specimen surface are the least.

The tracking pattern area can quantifiably reflect the severity of tracking failure. Figure 15 shows the area data with different filler contents under different applied voltages. Under the same applied voltage, when the filler

Figure 14. The tracking pattern of different specimens under 600V

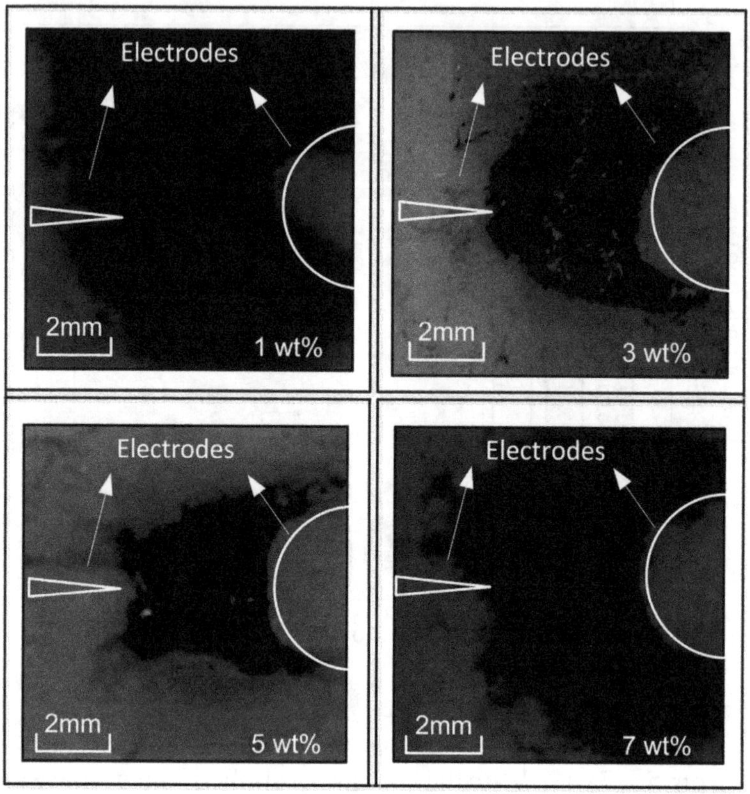

content increases from 0 to 5 wt%, the area of tracking pattern decreases, but it increases with increasing the content from 5 to 7 wt%. The tendency indicates that nano particles can restrict the formation of tracking pattern. 5 wt% is proved to be the best filler content for the restrain effect. The phenomenon is perhaps caused by the increasing chemical bonds and the restricted discharge. The chemical bonds are tightly bonded to the nano particles and epoxy by coupling agents. They are helpful to protect the epoxy against thermal and chemical erosion caused by discharges, which results in the decrease of pattern area.

Nonlinear Analysis

It is considered that the heat from scintillation discharges between the electrodes will cause the deposition of carbon on specimen surface. With continuous deposition of carbon, the electric field distribution changes constantly. The

Figure 15. The pattern area with different specimens under different voltages

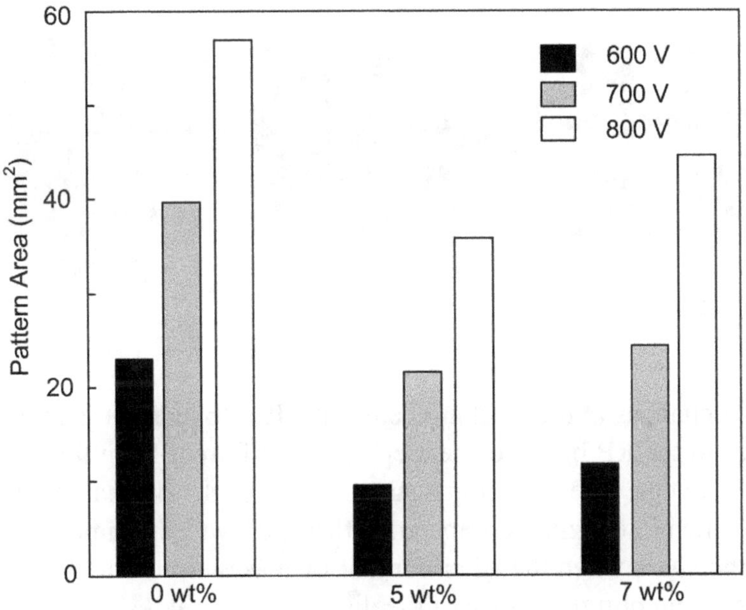

changing electric field will cause the variation of discharge intensity. Therefore, the effects of nano particles on the restriction of scintillation discharges can be gained by employing recurrence plot (RP) method on discharge currents. Meanwhile,the deposition of carbon is considered to be a random process. With the analysis of fractal dimension (FD) on the discharge image and the tracking pattern, the resistance of nano composites to tracking failure can be estimated.

The RP images of discharge currents at the prior to tracking failure stage with different kinds of specimens were obtained. As shown in Figure 16, the RP images of specimens with different contents show different geometric features. The non-linear characteristics of discharge currents can be obtained by analyzing the geometric features. The RP images consist of recurrence points and white spaces. The time series current data are reconstructed as vectors in phase space. The recurrence point means the distance of two neighboring vectors is less than or equal to the threshold value. The similarity of the vectors can be indicated by the points density. The high density means the embedding vectors are very similar to each other and change a little. The low density means the reverse. The white space segments stand for inflation process due to the high amplitude transients, which indicates that there are high

Figure 16. The RP images of the discharge currents with different specimens under 600 V

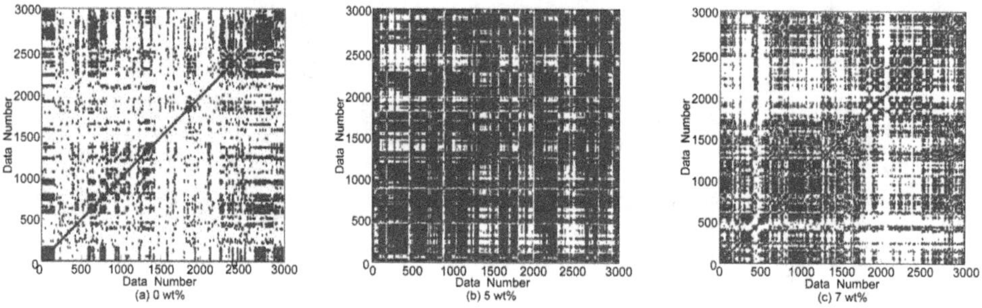

amplitude changes in the discharge currents. It is found that the topological structures in the RP image of neat epoxy are different from those of nano-Al_2O_3 specimens. The changing scale and rate of the discharge currents in pure specimens are highest compared with 5 and 7 wt% specimens. The point density increases when the filler content increases from 0 to 5 wt%, which indicates the amplitude change of the discharges is small and the discharges are more stable. However, it decreases with the increasing content from 5 to 7 wt%, indicating periodicity becomes lower and the discharges are more intensive. The tendency suggests that the discharge intensity is the lowest and the discharge currents are the most stable at 5 wt%

Figure 17 shows the RP images of discharge current at the prior to tracking failure stage with 1 wt% specimen under different dc voltages. The point density decreases with the increase of applied voltage, which shows the discharge current has a higher amplitude range and the discharge is severer under higher voltages.

Figure 17. The RP images of the discharge currents under different voltages with 1wt% specimens

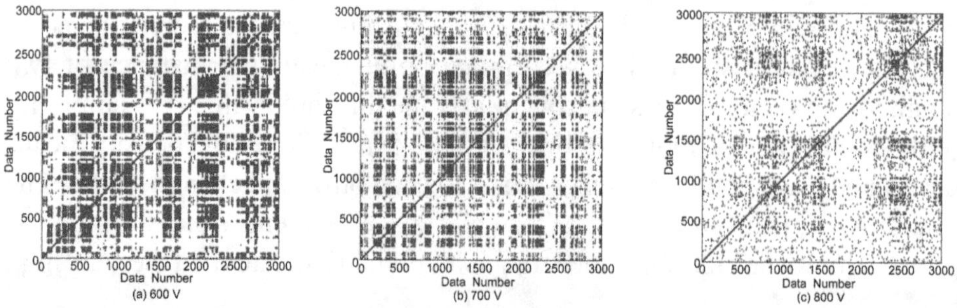

Two kinds of RP indicators are calculated to compare the effects of nano particles on tracking failure process. The indicators include recurrent rate (RR) and determinism (DET). RR indicates the distance between the vectors. The bigger value of RR means the more periodicity of the discharge current. DET can show the proportion of recurrence dots that are in parallel with diagonal lines in all the recurrence dots. The value of DET is in the range of zero to one. The closer that the value is to one, the more determinism the current is.

Figures 18 and 19 show the RP indicators with different kinds of specimens. Under the same applied voltage, with increasing the filler content from 0 to 5 wt%, the RP indicators increase, while with increasing the content from 5 to 7 wt%, the indicators decrease. For the same kind of specimen, the number of drops decreases with increasing the applied voltage. The results can also illustrate that the adding of nano Al_2O_3 particles can restrict the discharge. By adding a moderate amount of nano-Al_2O_3 particles, the enhancing effect of tracking resistance is the best.

The fractal dimension (FD) approach can identify the structures of scintillation discharge distribution and tracking patterns. The higher value of FD means the more complex structure. By analyzing the FD of these structures, effects of nano-Al_2O_3 particles on tracking failure can be illustrated. The distribution of discharge areas and the discharge traces pattern are obtained by an images processing program. A rectangle is employed to contain the discharge pattern, and then the rectangle is separated into a grid of sub-square boxes with the size of r. The minimum number of boxes covering the extracted discharge pattern is N(r). Then the value of FD is:

$$D = \lim_{r \to 0} \frac{\ln(N(r))}{-\ln(r)} \tag{3}$$

Due to the minimum size of digital image is one pixel, the box size r must be larger than the size of it. The relationship between the box number and the box size is obtained and the curve (-ln (r_k), ln $N(r_k)$) can be conformed to a double logarithmic coordinate system. The FD is obtained from the slope of regression line. The big value of FD reflects the irregular distribution of both discharge distribution and discharge pattern, while the small value represents the regular distribution (Du et al., 2011).

Figure 20 shows the FD of the discharge distribution in the whole tracking process with different kinds of specimens under different applied voltages. Under the same voltage, with increasing the filler content from 0 to 5 wt%,

Figure 18. RR and DET of the discharge currents under 600V

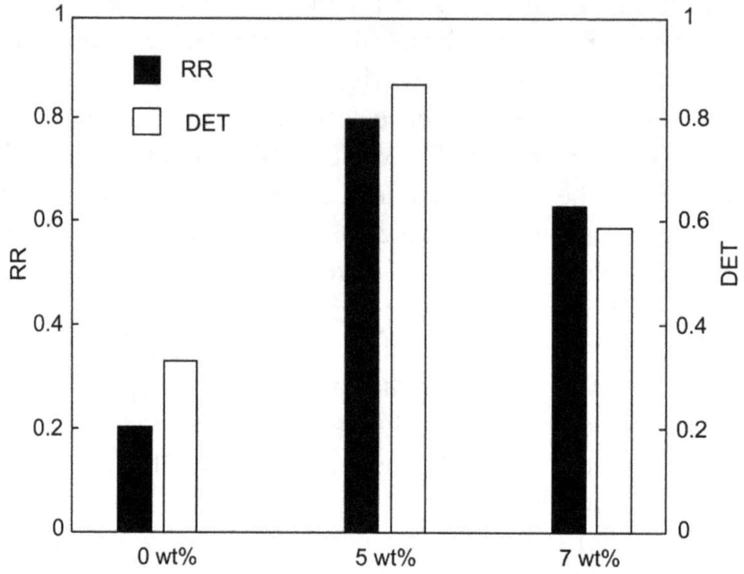

Figure 19. RR and DET of the discharge currents of 1 wt%

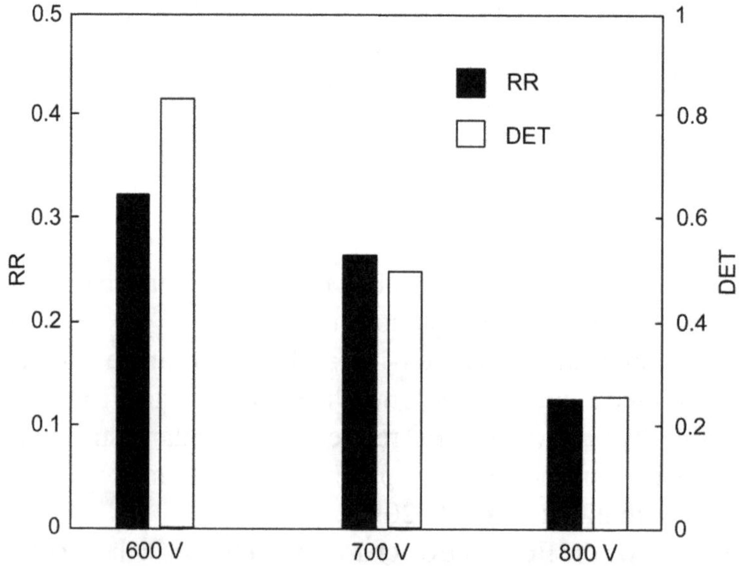

Figure 20. FD of discharge distribution of different specimens under different voltages

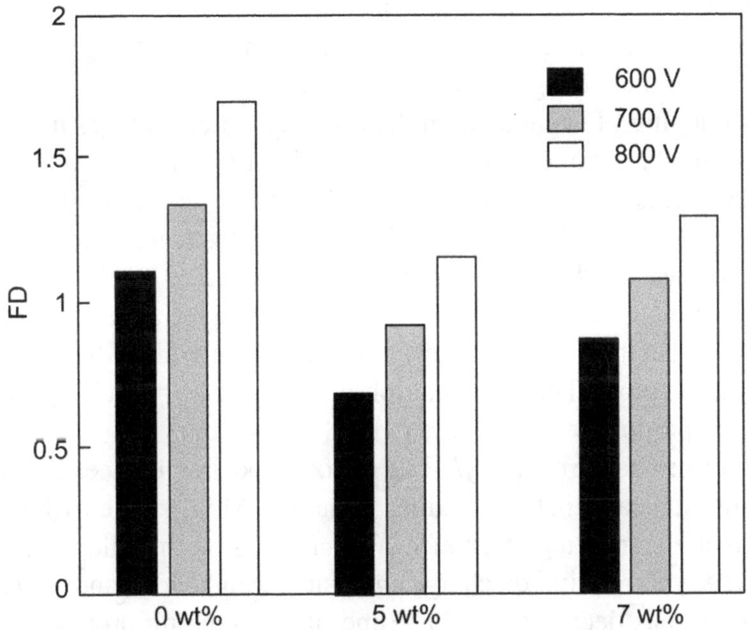

Figure 21. FD of tracking pattern of different specimens under different voltages

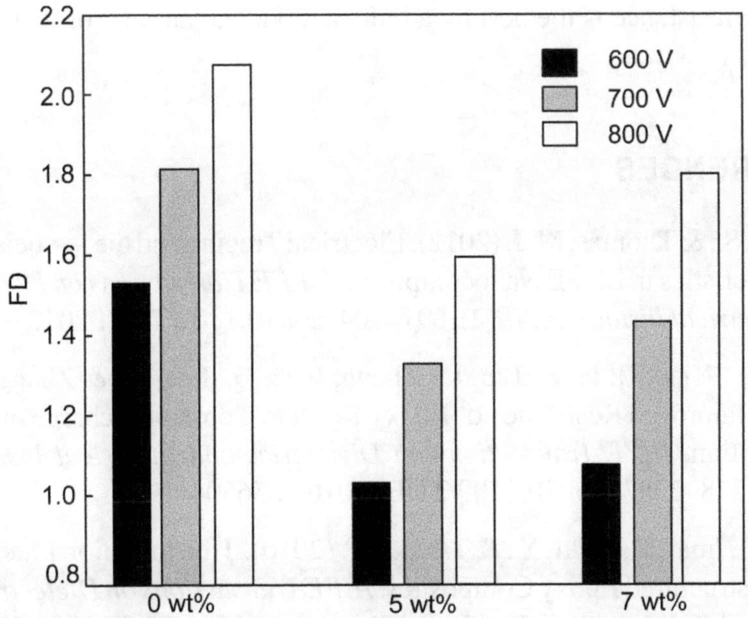

the FD decreases, while with increasing the content from 5 to 7 wt%, the FD increases. The results illustrate that the discharges are more stable when nano particles are added into epoxy. For the same kind of specimen, the FD increases with increasing the applied voltage.

By employing FD approach on the tracking pattern images in Figure 14, the FD of tracking pattern of different kinds of specimens under different applied voltages is shown in Figure 21. Under the same voltage, with increasing the content from 0 to 5 wt%, the FD of tracking pattern decreases, while it increases with increasing the content from 5 to 7 wt%. For the same kind of specimen, the FD increases with increasing the applied voltage. The results show that the tracking pattern is more regular with adding nano-Al_2O_3 particles into epoxy. The precipitation of carbonized particles is restricted by the nano particles.

Tracking resistance of epoxy/Al_2O_3 nanocomposites has been investigated. The results indicate that the adding of nano- Al_2O_3 particles has a great influence on the tracking resistance. With increasing the filler content from 0 to 5 wt%, the relative discharge intensity, weight loss and pattern area of the specimens decrease, the last time of the tracking process increases. The recurrence plot method and fractal dimension approach show that with increasing the content from 0 to 5 wt%, discharges are more stable and the tracking pattern is more regular. The reverse tendency is shown when the filler content continually increases from 5 to 7 wt%. The enhancing effect of tracking resistance is the best by adding a moderate amount of nano- Al_2O_3 particles (5 wt%).

REFERENCES

Alapati, S., & Thomas, M. J. (2012). Electrical Treeing and the Associated PD Characteristics in LDPE Nanocomposites. *IEEE Transactions on Dielectrics and Electrical Insulation, 19*(2), 697–704. doi:10.1109/TDEI.2012.6180265

An, Z. L., Xiao, H. H. F., Liu, L., Zheng, F. H. Q., Lei, Q., & Zhang, Y. W. (2016). Improved Resistance of Epoxy Resin to Corona Discharge by Direct Fluorination. *IEEE Transactions on Dielectrics and Electrical Insulation, 23*(4), 2278–2287. doi:10.1109/TDEI.2016.7556504

Chen, Y., Imai, T., Ohki, Y., & Tanaka, T. (2010). Tree Initiation Phenomena in Nanostructured Epoxy Composites. *IEEE Transactions on Dielectrics and Electrical Insulation, 17*(5), 1509–1515. doi:10.1109/TDEI.2010.5595552

Du, B. X., Liu, H. J., & Liu, Y. (2011). Effects of Magnetic Field on Tracking Failure of Gamma-ray Irradiated Polymer Insulating. *IEEE Transactions on Dielectrics and Electrical Insulation*, *18*(1), 140–148. doi:10.1109/TDEI.2011.5704503

Du, B. X., Zhang, J. W., & Gao, Y. (2012). Effects of TiO_2 Particles on Surface Charge of Epoxy Nanocomposites. *IEEE Transactions on Dielectrics and Electrical Insulation*, *19*(3), 755–762. doi:10.1109/TDEI.2012.6215077

Du, B. X., Zhang, J. W., & Liu, Y. (2012). Effect of Concentration on Tracking Failure of Epoxy/TiO_2 Nanocomposites under dc Voltage. *IEEE Transactions on Dielectrics and Electrical Insulation*, *19*(5), 1750–1759. doi:10.1109/TDEI.2012.6311524

Hayase, Y., Aoyama, H., Matsui, K., Takada, T., Tanaka, Y., & Murata, Y. (2006). Space Charge Formation in LDPE/MgO Nanocomposite Film under Ultra-high DC Electric Stress. IEE Japan Trans. *Fundamentals Materials (FM)*, *126*(11), 1084–1089. doi:10.1541/ieejfms.126.1084

He, L., & Gorur, R. S. (2016). Source Strength Impact Analysis on Polymer Insulator Flashover under Contaminated Conditions and a Comparison with Porcelain. *IEEE Transactions on Dielectrics and Electrical Insulation*, *23*(4), 2189–2195. doi:10.1109/TDEI.2016.7556494

Iwata, S. (2016). Influence of Humidity Treatment on Electrical Tree Propagation in Epoxy Resin. *IEEE Transactions on Dielectrics and Electrical Insulation*, *23*(5), 2556–2561. doi:10.1109/TDEI.2016.7736812

Jäverberg, N., Edin, H., Nordell, P., Nawaz, S., Hillborg, H., Azhdar, B., & Gedde, U. W. (2012). Dielectric Properties of Alumina-filled Poly (ethylene-co-butyl acrylate) Nanocomposites Part II- Wet Studies. *IEEE Transactions on Dielectrics and Electrical Insulation*, *19*(2), 391–399. doi:10.1109/TDEI.2012.6180230

Kimura, K., Itaya, S., & Hikita, M. (2002). *Partial Discharge Behavior on PI Film with Micro Gap under Steep-front Step Voltage*. Boston, MA: IEEE Int'l. Sympos. Electr. Insul.

Kumara, S., Alam, S., Hoque, I. R., Serdyuk, Y. V., & Gubanski, S. M. (2012). DC Flashover Characteristics of a Polymeric Insulator in Presence of Surface Charges. *IEEE Transactions on Dielectrics and Electrical Insulation*, *19*(3), 1084–1090. doi:10.1109/TDEI.2012.6215116

Lewis, T. J. (2004). Interfaces are the Dominant Feature of Dielectrics at the Nanometeric Level. *IEEE Transactions on Dielectrics and Electrical Insulation, 11*(5), 739–753. doi:10.1109/TDEI.2004.1349779

Li, P., Li, J. Z., Qin, Y. F., Gou, C., Li, Y., Zhao, J. P., & Zhang, Q. G. (2015). Wormholes Effect in Flashover Process of Alumina Filled Epoxy Resin Surfaces in SF_6. *IEEE Transactions on Dielectrics and Electrical Insulation, 22*(4), 2142–2147. doi:10.1109/TDEI.2015.004809

Li, Z., Okamoto, K., Ohki, Y., & Tanaka, T. (2010). Effects of Nano-filler Addition on Partial Discharge Resistance and Dielectric Breakdown Strength of Micro-Al_2O_3/Epoxy Composite. *IEEE Transactions on Dielectrics and Electrical Insulation, 17*(3), 653–661. doi:10.1109/TDEI.2010.5492235

Molinie, P. (2005). Measuring and Modeling Transient Insulator Response to Charging: The Contribution of Surface Potential Studies. *IEEE Transactions on Dielectrics and Electrical Insulation, 12*(5), 939–950. doi:10.1109/TDEI.2005.1522188

Molinié, P. (2005). Measuring and Modeling Transient Insulator Response to Charging: The Contribution of Surface Potential Studies. *IEEE Transactions on Dielectrics and Electrical Insulation, 12*(5), 939–950. doi:10.1109/TDEI.2005.1522188

Moraveji, M. K., & Razvarz, S. (2012). Experimental Investigation of Aluminum Oxide Nanofluid on Heat Pipe Thermal Performance. Intl. *Communications in Heat and Mass Transfer, 39*(9), 1444–1449. doi:10.1016/j.icheatmasstransfer.2012.07.024

Nath, R., & Perlman, M. M. (1990). Charge Storage in Pure and Titanium Dioxide Doped Polyarylate, and Conduction in Pure Polyarylate. *Journal of Electrostatics, 24*(3), 283–293. doi:10.1016/0304-3886(90)90015-N

Nguyen, V. T., Vaughan, A. S., Lewin, P. L., & Krivda, A. (2015). The Effect of Resin Stoichiometry and Nanoparticle Addition on Epoxy/silica Nanodielectrics. *IEEE Transactions on Dielectrics and Electrical Insulation, 22*(2), 895–905. doi:10.1109/TDEI.2015.7076790

Ning, X., Feng, H., Zhang, H. L., Liu, P., Xiang, Z., & Peng, Z. R. (2015). Dielectric Properties of Multi-layer Epoxy Resin Impregnated Crepe Paper Composites. *IEEE Transactions on Dielectrics and Electrical Insulation, 22*(1), 161–168. doi:10.1109/TDEI.2014.004526

Perlman, M. M., & Haridoss, S. (1987). Bulk Modification of Charge Trapping and Conductivity in Linear Low-density Polyethylene. *IEEE Transactions on Dielectrics and Electrical Insulation*, *22*(1), 9–12. doi:10.1109/TEI.1987.298956

Sarathi, R., Harsha, V., Vasa, N. J., Griffiths, H., & Haddad, A. (2016). Water Droplet Initiated Discharges on Epoxy Nanocomposites under DC Voltages. *IEEE Transactions on Dielectrics and Electrical Insulation*, *22*(3), 1743–1752. doi:10.1109/TDEI.2016.005387

Tagami, N., Okada, M., Hirai, N., Ohki, Y., Tanaka, T., Imai, T., & Ochi, M. et al. (2008). Dielectric Properties of Epoxy/clay Nanocomposites - Effects of Curing Agent and Clay Dispersion Method. *IEEE Transactions on Dielectrics and Electrical Insulation*, *15*(1), 24–32. doi:10.1109/T-DEI.2008.4446733

Takada, T., Hayase, Y., Tanaka, Y., & Okamoto, T. (2008). Space Charge Trapping in Electrical Potential Well Caused by Permanent and Induced Dipoles for LDPE/MgO Nanocomposite. *IEEE Transactions on Dielectrics and Electrical Insulation*, *15*(1), 152–160. doi:10.1109/T-DEI.2008.4446746

Tanaka, T., Ohki, Y., Ochi, M., Harada, M., & Imai, T. (2008). Enhanced Partial Discharge Resistance of Epoxy/Clay Nanocomposite Prepared by Newly Developed Organic Modification and Solubilization Methods. *IEEE Transactions on Dielectrics and Electrical Insulation*, *15*(1), 81–89. doi:10.1109/T-DEI.2008.4446739

Watson, P. K. (1990). The Energy Distribution of Localized States in Polystyrene, Based on Isothermal Discharge Measurements. *Journal of Physics. D, Applied Physics*, *23*(12), 1479–1484. doi:10.1088/0022-3727/23/12/002

Chapter 8
High Thermal Conductivity Polymer Insulation

ABSTRACT

In this chapter, a range of silicone rubber (SiR), polypropylene (PP) and polyethylene (PE) based composites filled with micro or nano sized boron nitride (BN) particles at different loadings were manufactured to investigate effects of thermal conductivity. The study of SiR attempts to clarify whether the addition of boron nitride (BN) particles can improve the resistance to tracking and erosion of SiR by increasing its thermal conductivity. For PP, in addition to measuring thermal conductivity of various samples, thermal dissipation was also discussed to analyze the relationship between them. Meanwhile, in order to evaluate tracking failure (a kind of surface dielectric breakdown) resistance of the manufactured samples, time to failure, erosion depth and weight loss of the test samples were measured through tracking test. As for PE, Thermal conductivity and relative permittivity were measured to characterize the basic properties of various samples. Obtained results show that several properties of the filled specimens are obviously improved.

INTRODUCTION

High voltage direct current (HVDC) technology is increasingly applied in power transmission and distribution lines with many advantages (Cavallini et al., 2015). Glass and porcelain insulators are increasingly replaced with the polymeric insulators around the world because of their lightweight, easy

DOI: 10.4018/978-1-5225-2309-3.ch008

handling and transport, better contamination performance, vandal resistance and reduced cost (Kikuchi et al., 1999; Schneider et al., 1989; Liang et al., 1999). However, reports have suggested that insulators under dc transmission lines are required to deal with significant challenges (Abbasi et al., 2014; Ghunem et al., 2014; Cherney et al., 2015). It has been reported that with the same field conditions the accumulation of pollutants under dc voltage is 1.2 to 1.5 times of that under ac voltage (Cheng et al., 1981).

Given the rapid development in power electronics technology, the trend of electric apparatus is towards high power density, high switching-frequency and miniaturization. This leads to the problem of overheat during operation. However, most insulating materials used in those devices are almost thermal insulating so that the heat generated by overload operation or intensive discharges will accumulate, which results in a rise in the operating temperature of electric apparatus (Ortiz et al., 2009). A small difference in operating temperature has a significant impact on the reliable operation of electric device (Sim et al., 2005). Once operating temperature over the critical temperature, the lifespan of an electric device will be reduced significantly and even the instrument might be damaged (Hong et al., 2012; Takezawa et al., 2001; Lee et al., 2008). In addition, a large number of insulation failures in the power system are related to thermal breakdown (Nagao et al., 1990; Choi et al., 2015). Therefore, the generated heat should be dissipated as quickly as possible to keep operating temperature at a reasonable level.

Recently, several studies have been carried out to study the thermal conductivity of different composites filled with inorganic particles. Weidenfeller et al. (2004) investigated the thermal conductivity of PP composites with various fillers in different contents. It showed that compared with neat PP, the thermal conductivity of PP/talc composites increased from 0.27 to 2.5 W/(m•K) with 30 vol% filler (Weidenfeller et al., 2004). Krupa et al. (2004) pointed out that the thermal conductivity measurements of the composites increased with increasing the graphite content nonlinearly. In addition, the thermal conductivity of filled HDPE was higher than that of filled LDPE because of high crystallization ratio (Krupa et al., 2004).

In this chapter, specimens were prepared by blending micro or nano sized h-BN particles into base matrix at different loadings. Thermal conductivity and thermal dissipation ability of various samples were measured at room temperature to analyze the relationship between them. Influence of the increased thermal conductivity on surface dielectric breakdown was revealed by discussing time to failure, erosion depth and weight loss. Moreover, DC bulk dielectric breakdown characteristic of various samples was investigated using Weibull distribution. The obtained results indicate addition of BN filler can improve the thermal conductivity of composites effectively.

EFFECTS OF THERMAL CONDUCTIVITY ON DC RESISTANCE TO EROSION OF SILICONE RUBBER/BN NANOCOMPOSITES

Thermogravimetric Analysis (TGA)

In this section, nanocomposites specimens were prepared by dispersing nano-sized BN particles into RTV SiR (pure crosslinked polydimethylsiloxane) with various weight ratios of 1, 2.5, 3, 5 and 7%. Specimens of undoped SiR were also prepared for the purpose of comparison. The thermogravimetric analysis (TGA) curves of specimens with different contents of BN are shown in Figure 1. It can been seen that the temperature at 20% weight loss for 0, 1 and 7 wt% SiR/BN composites increases with the increase of BN, which is 456, 460 and 463 °C, respectively. With the addition of BN particles, the thermal stability of SiR nanocomposite becomes better. This is probably due to the increase in physical and chemical cross-linking points created by the interactions between nanoparticles and silicone rubber (Kong et al., 2011). Figure 2 shows the distribution of nano-BN particles in 1% BN filled silicone rubber. The distribution is satisfactory as the particles are isolated and scattered uniformly in the matrix.

Effect on Erosion Resistance

To predict the relationship between the thermal conductivity and the content of BN particles, the theoretical models of geometric mean was used.

$$\lambda = \lambda_f^{v_f} \bullet \lambda_m^{(1-v_f)} \tag{1}$$

where λ, λ_m and λ_f are the thermal conductivities of nanocomposite, silicone rubber matrix and BN particles, respectively; v_f is the volume fraction of the BN particles. The measured thermal conductivity of nano-BN particles used in the tests is 150 W/m•K. The calculated thermal conductivity of the

Figure 1. TG curves of specimens with different contents of BN

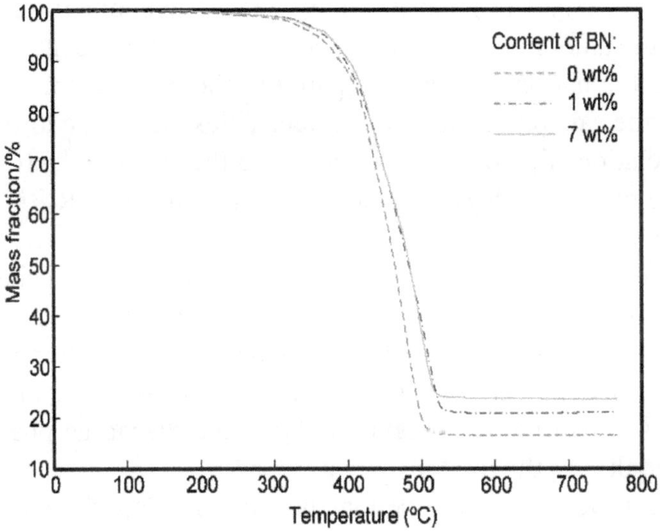

Figure 2. Scanning electron micrographs of 1% nano-BN filled silicone rubber at 25000x magnification

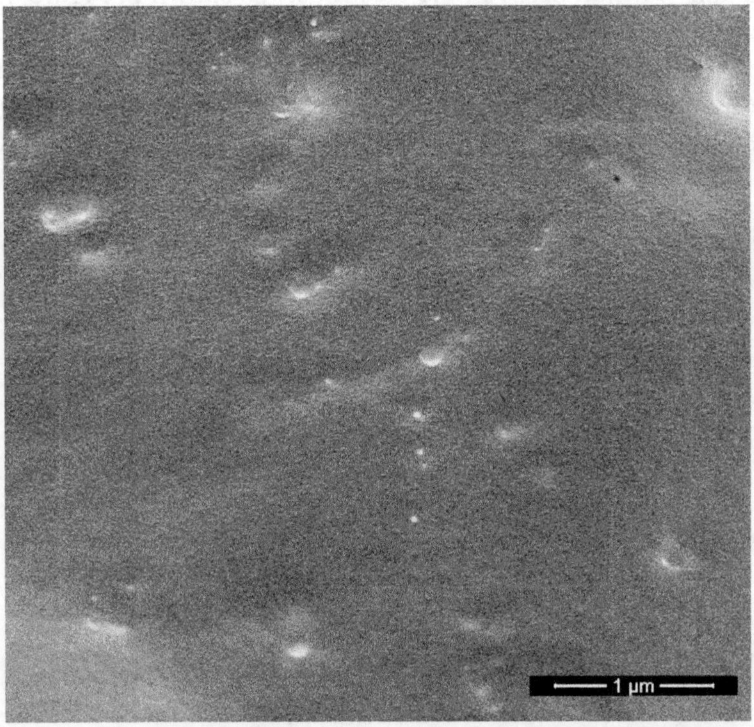

samples with weight percents of 0, 1, 2.5, 3%, 5 and 7% are respectively 0.165, 0.168, 0.176, 0.177, 0.186 and 0.194 W/m•K. The geometric mean model is based on the amount of fillers and does not take into account the size and geometry of the nanoparticles. Compared to the experimental data shown in Figure 3, the calculated thermal conductivities are a little lower. Figure 3 shows the relationship between the measured thermal conductivity and the content. The measured thermal conductivities of all the SiR/BN specimens are higher than that of the unfilled specimen.

During the inclined plane test, the two electrodes were bridged with a filament of fluid contaminant, which resulted in conduction and Joule heating of the leakage current on the surface. Dry bands formed in the conductive filament and an arc began when the voltage applied to the dry band was over the breakdown voltage. Figure 4 shows the discharge phenomena on the specimen during the test.

As the test went on, the phenomena observed on the surface of the specimen kept varying at different stages of degradation. At the initial stage, dry band arcing formed at random along the conductive filament. The heat was transferred from the arc to the SiR surface, which could cause thermal degradation of specimens, and a continuous, black and solid hydrophilic line was gradually formed between the electrodes, as shown in Figure 4. It was observed that the carbon tracking became wider as the test continued, and the position of the discharges developed down towards the bottom electrode. This was because the hydrophobicity along the filament was lost due to continuous discharges, and the water filament was becoming wider and developed into

Figure 3. Relationship between the thermal conductivity and the content of BN

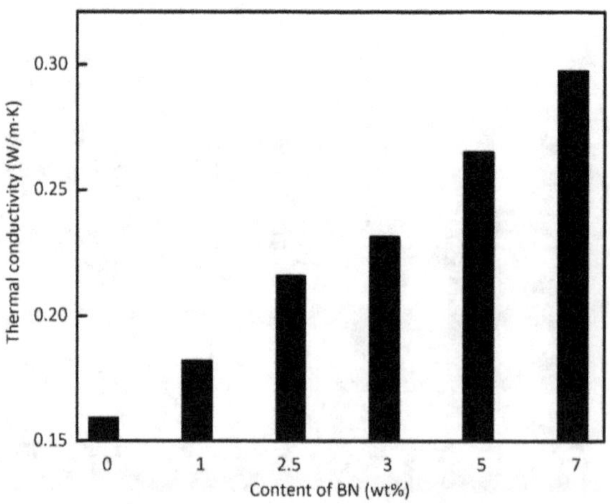

Figure 4. Photographs of specimen during the erosion test

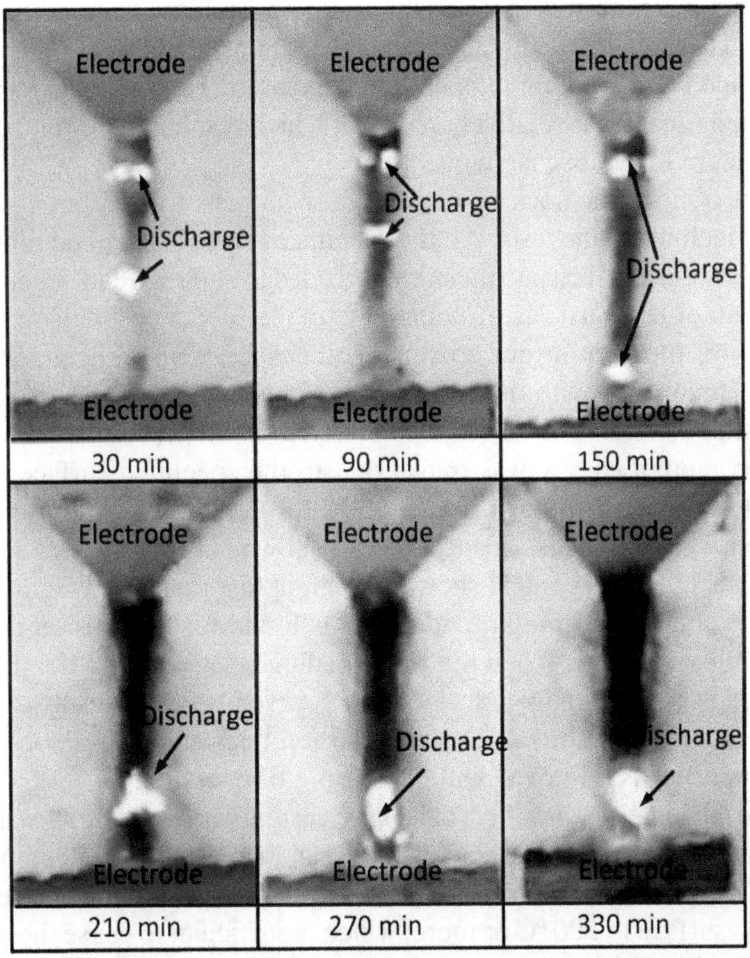

a water film. Dry bands were more likely to form at the bottom of the water film, so the dry band arcing mainly concentrated near the bottom electrode, which was the main discharge process of the test. Deep erosive material decomposition also took place near the bottom electrode.

The total discharge quantities of specimens during the test for 6 hours were calculated from the measured leakage current. Figure 5 shows the relationship between the total discharge quantity and the content of BN. It is observed that with the increase of BN content, the total discharge quantity decreases. The total discharge quantity of the undoped sample is about 278.5 C. With further addition of BN fillers, the total discharge quantity shows a decreasing trend and reaches about 163.6 C at a loading of 7 wt%. Because the addition of BN particles improves the thermal conductivity, it can be inferred that the generated heat on the specimens with high content of BN was conducted out quickly and the maximum temperature decreased. Thus, the evaporation of fluid contaminant was relatively reduced, which resulted in fewer dry bands and therefore lower discharge quantity.

Figure 6 shows the tracking patterns of specimens with different contents after the inclined plane tests. All the specimens were discharged for 6 hours during the tests and the specimens were eroded to some extend depending on the content of BN particles. Compared with the specimens with contents of 5 and 7 wt%, the tracking and erosion of other specimens were more serious. It can be found that with increasing BN content, the resistance to tracking and erosion increases.

When enough energy was transferred to the specimen surface, erosive material decomposition took place, which resulted in weight loss from the specimen. The weight loss of specimens is shown in Figure 7 (a). Specimens filled with higher BN content show less weight loss than the specimens with BN lower content. A similar tendency was found for the erosion depth, as shown in Figure 7 (b). When the filler loading was 5 and 7 wt%, the depth of erosion was remarkably reduced. It is observed from the results shown in Figures 6 and 7 that the tracking and erosion resistance of silicone rubber nanocomposites is improved with increasing filler content.

While the Joule heating effect of leakage current can result in a slow increase in specimen temperature, dry band arcing can increase the local temperature rapidly. The heat generated by discharges could bring the temperature on the specimen surface to 500 ℃ or more, which is sufficient to cause the silicone rubber backbone to break into low chain polymers and induce tracking and erosion (Ansorge et al., 2015). The temperature rise and heat dissipation have strong effect on the erosion. Adding nano-BN particles improves the thermal conductivity of silicone rubber nanocomposites, which may inhibit the heat accumulation around the discharge area and promote heat dissipation thus improving the tracking and erosion resistance.

Effect on Thermal Dissipation

The thermal information on the specimens surface were recorded with an infrared thermal imager during the inclined plane tests. It was observed that dry-band arcing mainly occurred along the filament, which increased the local

Figure 5. Relationship between the total discharge quantity and the content of BN

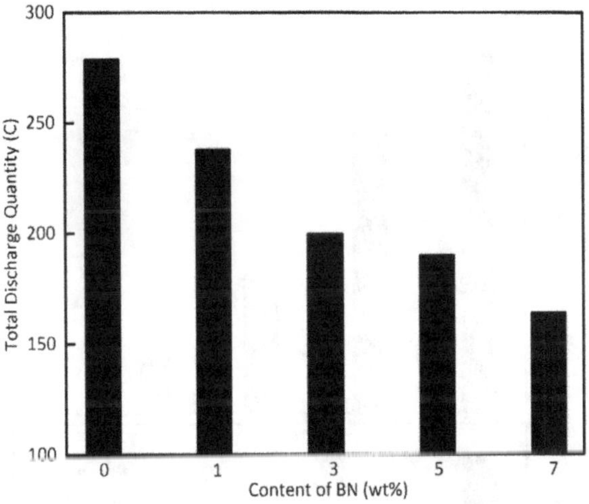

Figure 6. Tracking patterns of specimens with different contents after the inclined plane test

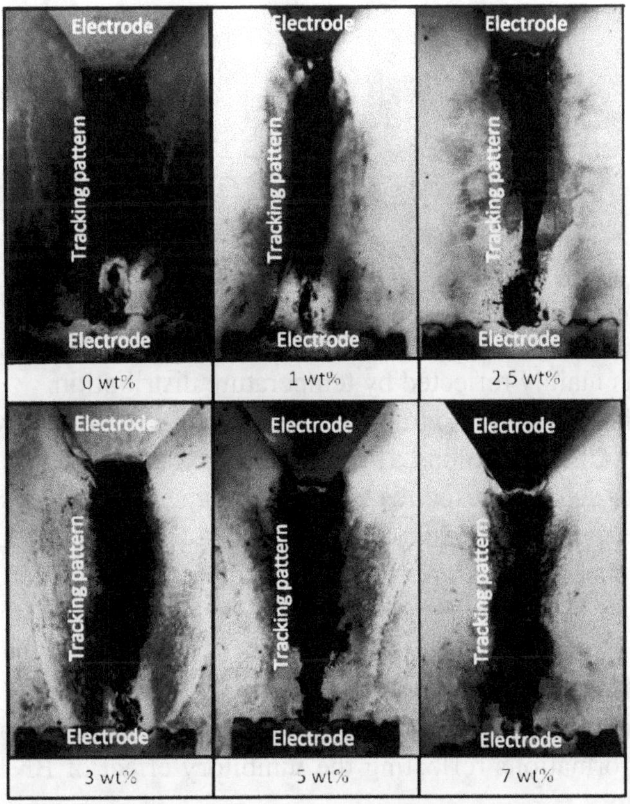

Figure 7. Relationship between the content of BN and (a) weight loss, (b) erosion depth

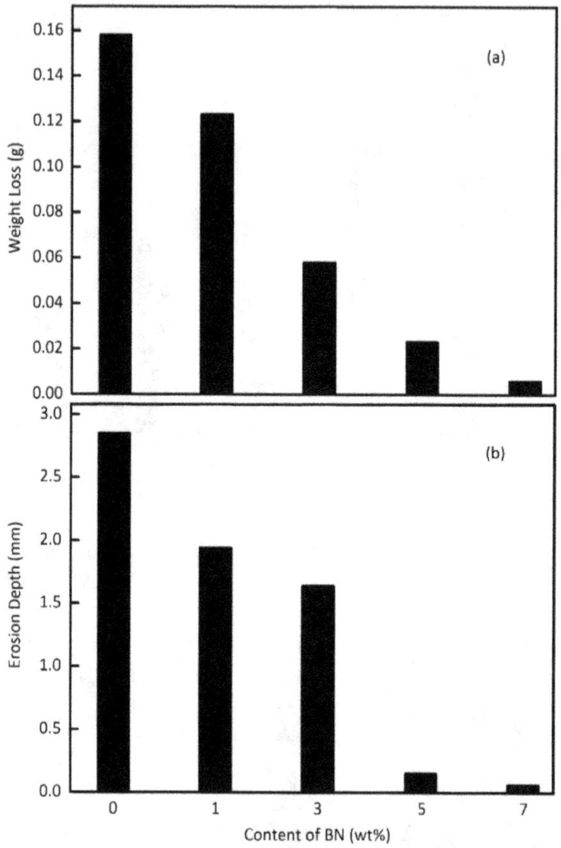

temperature significantly. And the temperature in the surrounding area also rose because of heat diffusion. The effect of thermal conductivity on tracking and erosion is mainly reflected by temperature distribution.

Figure 8 shows the thermographs of SiR/BN nanocomposites with different contents. On the thermographs, a brighter color indicates a higher temperature while a darker color corresponds to a lower temperature. In this section, the heat source region is defined, which is close to the discharge area, and has obviously higher temperature than other regions. It was found that for the same test time, the temperature in the heat source region decreased with the increasing content of BN, while the temperature around the heat source region increased with the BN content.

The temperature distribution during the inclined plane test contains some important informations reflecting the inhibitory effect of BN particles on

Figure 8. Infrared images during the test

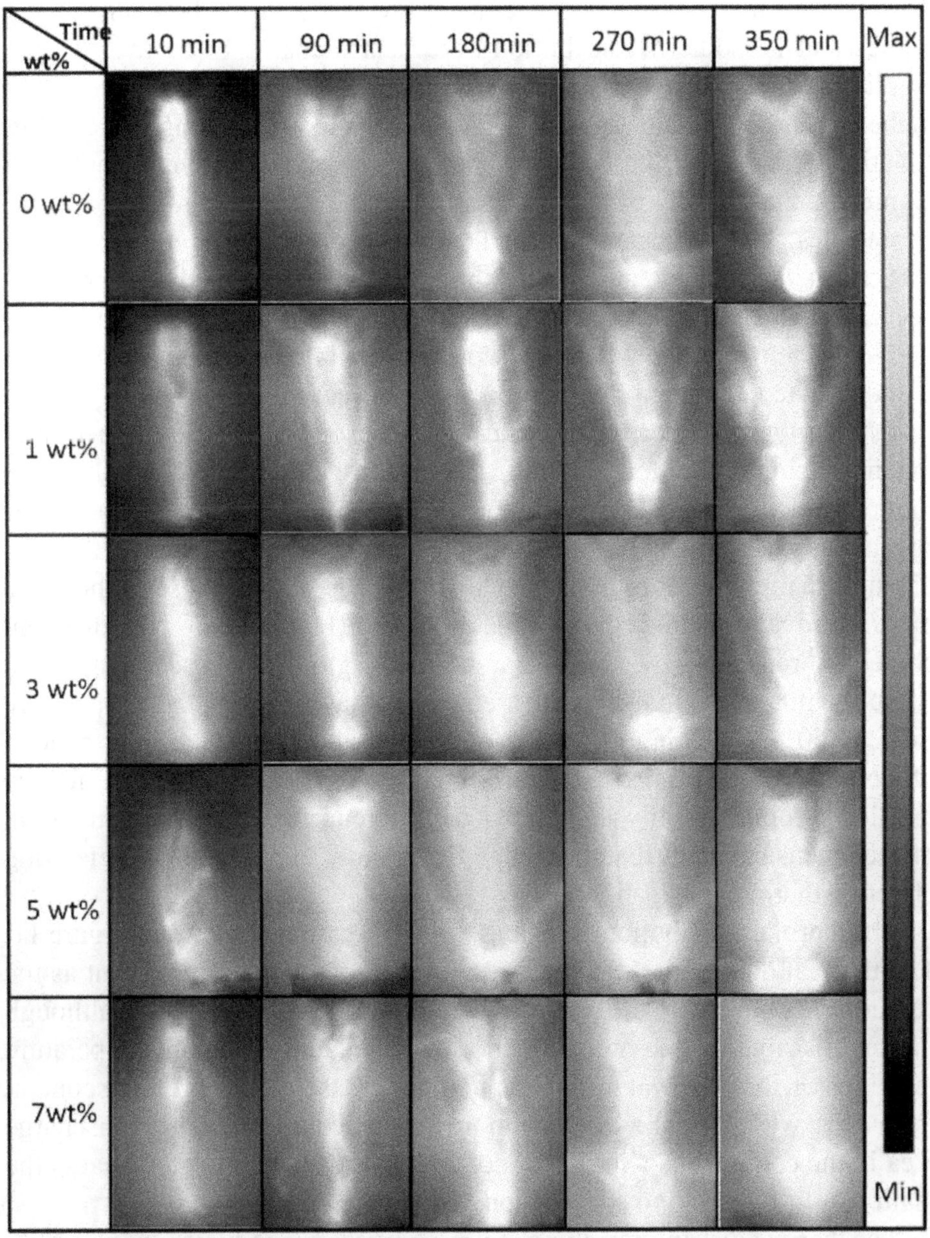

**For a more accurate representation of this figure, please see the electronic version.*

tracking and erosion. The isotherm distribution and the maximum temperature after applying the voltage for 10 min are shown in Figure 9. The background temperature on the isotherm is below 45 °C. With the increasing content of nano-BN from 0 to 7 wt%, the maximum temperature on the specimens decreases from 113 to 81 °C. The temperature of the unfilled specimen is the highest and the isotherm distribution is the most intensive. Under the same test conditions, the intensive distribution indicates that more heat is accumulated in the discharge area and cannot be dissipated to the surrounding area quickly. Relatively sparse distribution is observed in specimens with high content. It is inferred that the BN particles can improve the thermal dissipation in the composites. This indicates that the heat on specimens with higher content was conducted out more quickly and the heat in specimens with lower content accumulated at the discharge area. The heat dissipation correlates with the maximum temperature. When the heat accumulated in the discharge area was dissipated, the maximum temperature decreased.

In order to characterize the quantity of thermal dissipation, the area in the dissipation region for which the temperature exceeds 60 °C is shown in Figure 10. During the test, it was found that the lowest temperature on the specimens was about 40 °C, and the highest temperature was above 70 °C. To distinguish the difference in thermal diffusion, 60 °C was chosen. The value of S_0/S is used to evaluate the thermal dissipation in specimens with different content. S_0 represents the area of a predefined region on the specimen, which is 40×45 mm, and S indicates the area within that for which the temperature exceeds 60 °C. The highest value of S/S_0 is about 0.9 when the content of BN is 7 wt% at a lapse time of 5 h. At the same time, the value for the unfilled specimen is only about 0.6. The difference in the dissipation region of specimens indicates that the BN particles show a remarkable accelerating effect on thermal dissipation.

The thermal distribution along the Y-axis direction is shown in Figure 11. The thermal distribution lines with the maximum temperature point as the center were obtained in the way shown in Figure 11. It is found that although thermal distribution lines were obtained with the same maximum temperature in the center, the temperature at other points increases with increasing content from 0 to 7 wt%. It can be seen that more heat is accumulated in the discharge area in the composites with lower content, and the accumulated heat is the main reason for tracking and erosion.

The thermal distribution along the Y-axis direction is shown in Figure 12. The thermal distribution lines with the maximum temperature point as the center were obtained in the way shown in Figure 12. It is found that although thermal distribution lines were obtained with the same maximum temperature in the center, the temperature at other points increases with increasing content from 0 to 7 wt%. It can be seen that more heat is accumulated in the discharge

Figure 9. Relationship between the isotherm distribution and the content at 10min after the voltage applied

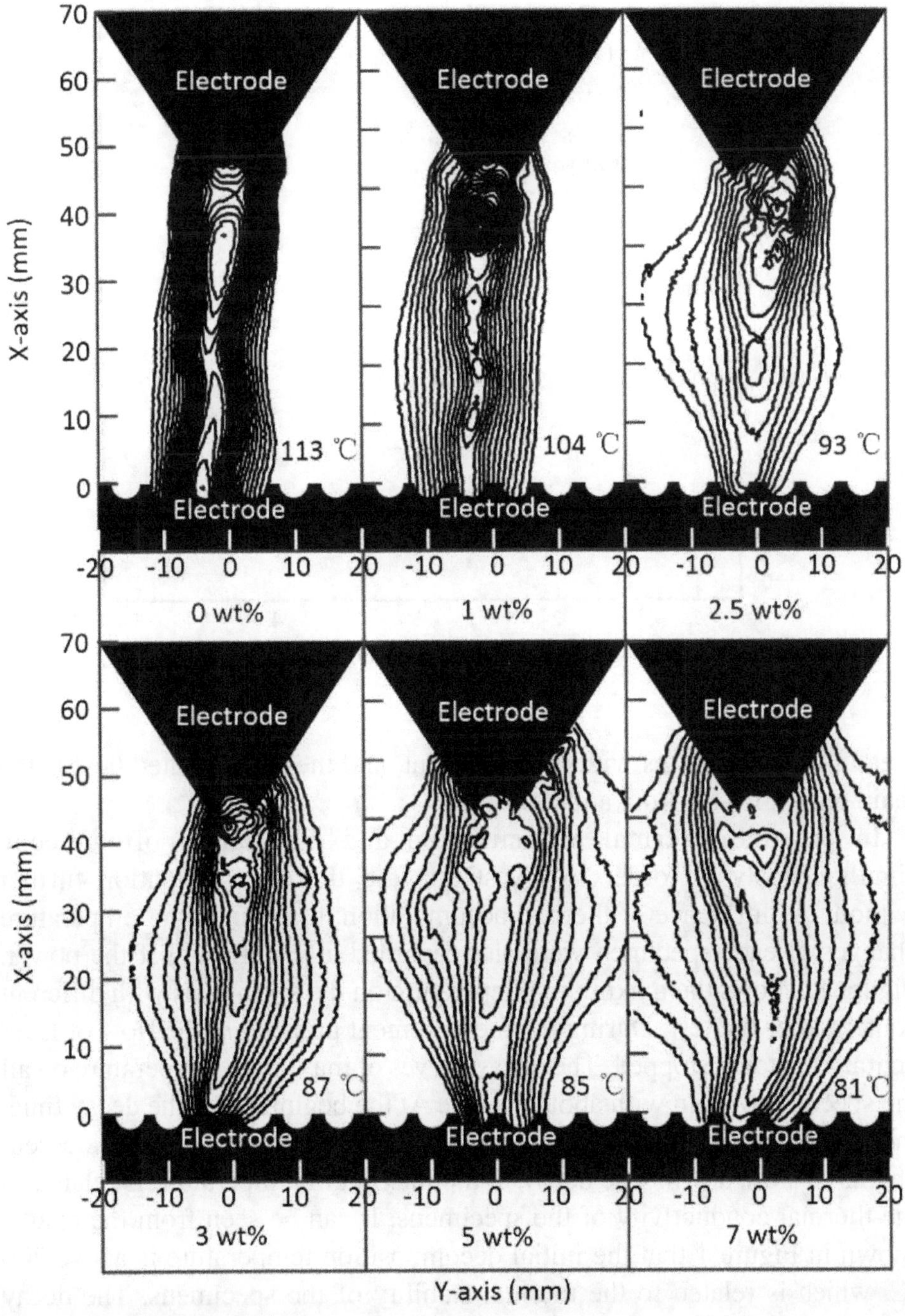

Figure 10. Relationship between the dissipation region that temperature exceeds 60 °C and the lapse time in the test process

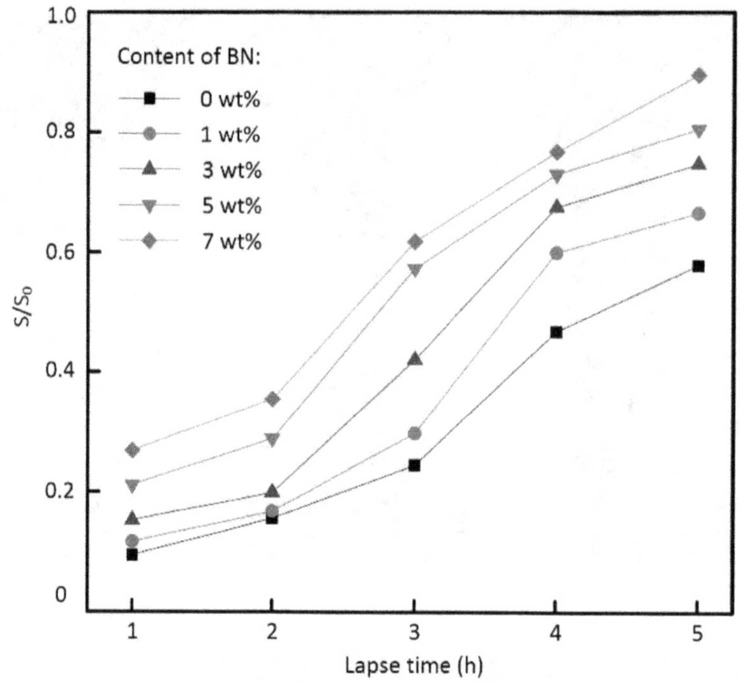

area in the composites with lower content, and the accumulated heat is the main reason for tracking and erosion.

In practice, thermal accumulation and dissipation often occur simultaneously. In order to investigate the thermal dissipation further without the influence of thermal accumulation, the maximum temperature changes of each specimen were also recorded after turning off the power. Figure 12 shows the maximum temperature on the specimens with different content after the test. During the measurement procedure, the flow of fluid contaminant was stopped. The decay curves of maximum temperature on all the specimens begin with about 100 °C. At the beginning of the decay time, the maximum temperature decreases rapidly, and then the rate of descent begins to slow down. The decay in the maximum temperature is related to the thermal conductivity of the specimens. It can be seen from the results shown in Figure 1 that the initial decomposition temperature is above 200 °C, which is related to the thermal stability of the specimens. The decay curves and TGA curves respectively show different thermal properties. The

Figure 11. The thermal distribution along the Y-axis direction with different lapse time in the test process

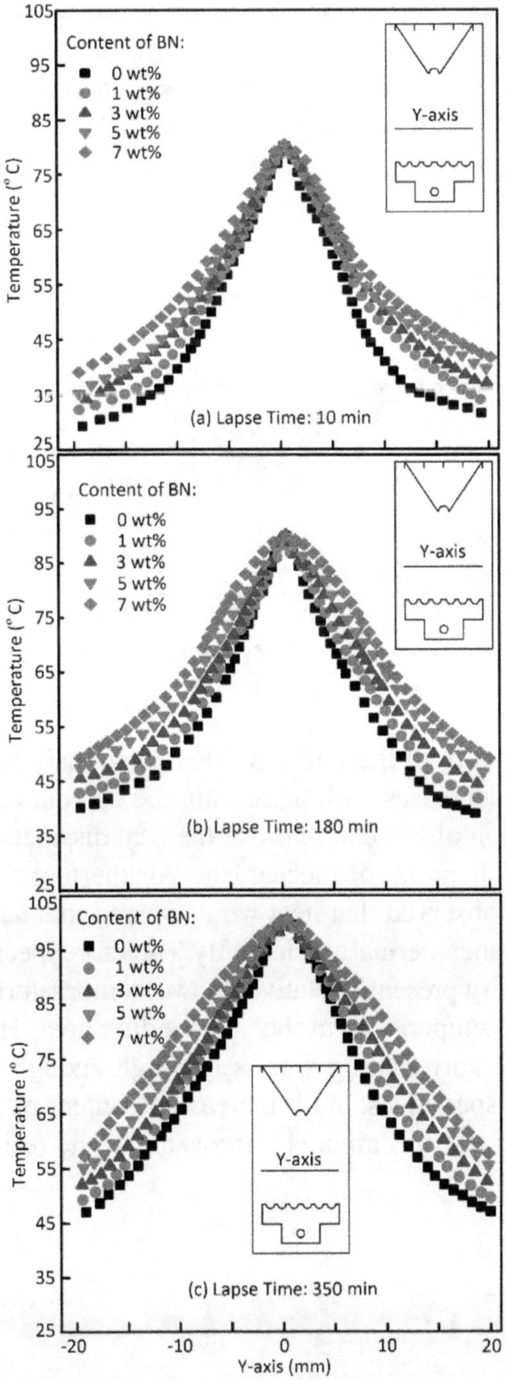

Figure 12. Relationship between the maximum temperature and the decay time after the discharges for 6 hours

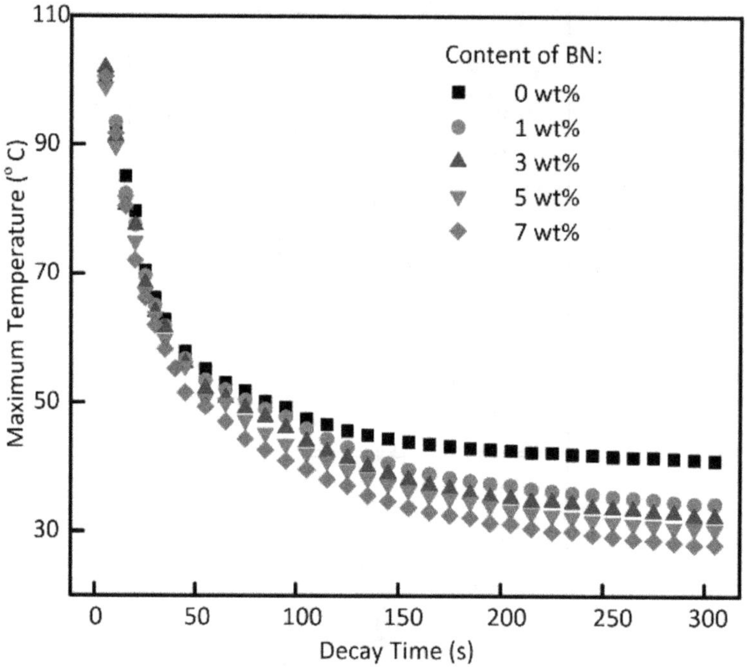

maximum temperature of the unfilled SiR is obviously higher than that of filled SiR, and it decreases with increasing the content of BN. The results indicate the addition of BN can improve the heat dissipation of specimens.

A schematic illustration of mechanisms for thermal diffusion is shown in Figure 13. It is observed that heat would spread to a larger region of the specimen with a higher thermal conductivity. Therefore specimens with higher thermal conductivity present a relatively lower temperature in the discharge area and a higher temperature in the surrounding area. However, the high temperature in the surrounding area is not high enough to cause thermal decomposition of specimens. With increasing content of BN particles, the heat dissipation in the specimen is improved and the resistance to erosion increases.

Figure 13. Schematic illustration of mechanisms for thermal diffusion

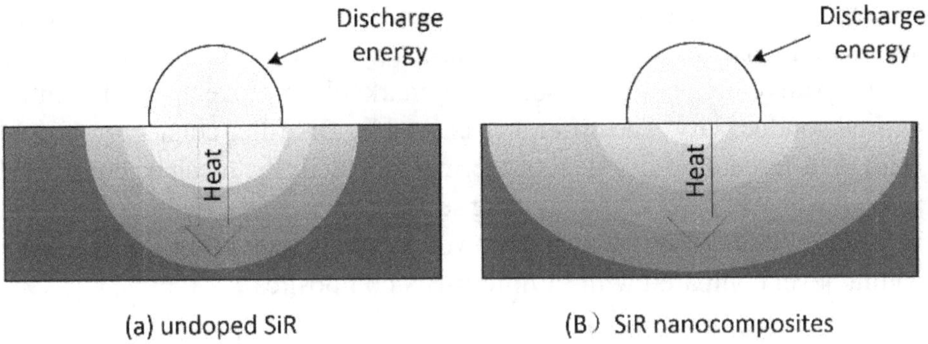

EFFECTS OF THERMAL CONDUCTIVITY ON SURFACE DIELECTRIC BREAKDOWN OF MICRO OR NANO SIZED BN FILLED POLYPROPYLENE COMPOSITES

Morphological Analysis

In this section, Microcomposites were prepared by mixing micro flake-like BN particles (median diameter: 10 μm, thickness: 1 μm) into PP matrix with different contents (3, 6 and 9 wt%). Meanwhile, nano-BN spherical-like particles (median diameter: 50 nm) in different weight ratios of 3, 6 and 9 had been introduced with the PP matrix to manufacture PP/BN nanocomposites.

In order to investigate structure morphology of the doped samples and BN fillers (marked by red arrow) distribution in the matrix, SEM images of the fractured surface of the compounds are illustrated in Figure 14. Obviously, the cross section of the pure sample is very smooth, while the cross-section of PP/micro-BN samples becomes rougher with increasing filler concentration. It indicates the interfacial interaction between micro-BN filler and PP matrix is very strong. It also shows that most nano-BN particles are uniformly dispersed in base matrix with various loading levels, which means the interfacial interaction between nano BN filler and PP matrix is good. Under the same conditions, in comparison with micro sized BN particles, the nano sized BN particles are extremely dispersed in the PP matrix. Every nano particle is distributed like an isolated point so that nano particles cannot touch each other even at 9 wt% filler loading, which may not favor the formation of thermal conductive paths in the matrix. Whereas, the flaky micro-sized particles in PP/micro-BN composites create many short lines. Those lines gradually connect each other with increasing loading level (marked by the white circle). Then some thermal paths emerge, which results in a remarkable improvement of the entire thermal conductivity. The difference caused by BN filler connection would result in different improvement of thermal conductivity. It can be confirmed by the result revealed in the following section, "Thermal Conductivity," that the nano-BN filled samples show lower thermal conductivity at the same loading level compared with PP/micro-BN composites.

Thermal Conductivity

Thermal conductivities of various PP/BN composites with different loadings are summarized in Table 1. Data presented in Table 1 are statistical means of 5 samples for each kind of compound. Obviously, thermal conductivity of the doped sample is related to filler size and filler loading.

Owing to the high thermal conductivity of BN filler, addition of BN particles can enhance the entire thermal conductivity of PP compound notably. Compared with neat sample, whose thermal conductivity is 0.23 W/(m•K), all the doped PP composites obtain higher thermal conductivity. Especially, for samples filled with micro sized BN fillers, the max thermal conductivity can be improved up to 0.46 W/(m•K) at concentration of 9 wt%, which is two times as that of pure sample. Moreover, the thermal conductivity of micro-BN filled PP composite is higher than that of PP/nano-BN composite at any loading levels.

As shown in Figure 15, the measured thermal conductivity of the doped sample is also demonstrated as a percentage of increase compared with the neat resin. It is clearly revealed that the higher the loading level is, the higher the thermal conductivity in the BN filled sample (Kochetov et al., 2010). Along with increasing the loading level, some thermal conductive pathways emerge so that thermal conductivity of the composite is improved more significantly (Huang et al., 2011). In addition, compared with the sample coated micro BN particles, the nano sized BN doped one represents a lower thermal conductivity at any given loading levels. It may be due to the low aspect ratio of nano sized BN filler which goes against to form thermal conductive paths in the matrix (Kemaloglu et al., 2010).

Figure 14. SEM pictures of the fracture surface of the compounds (a)nano-BN filler; (b)micro-BN filler

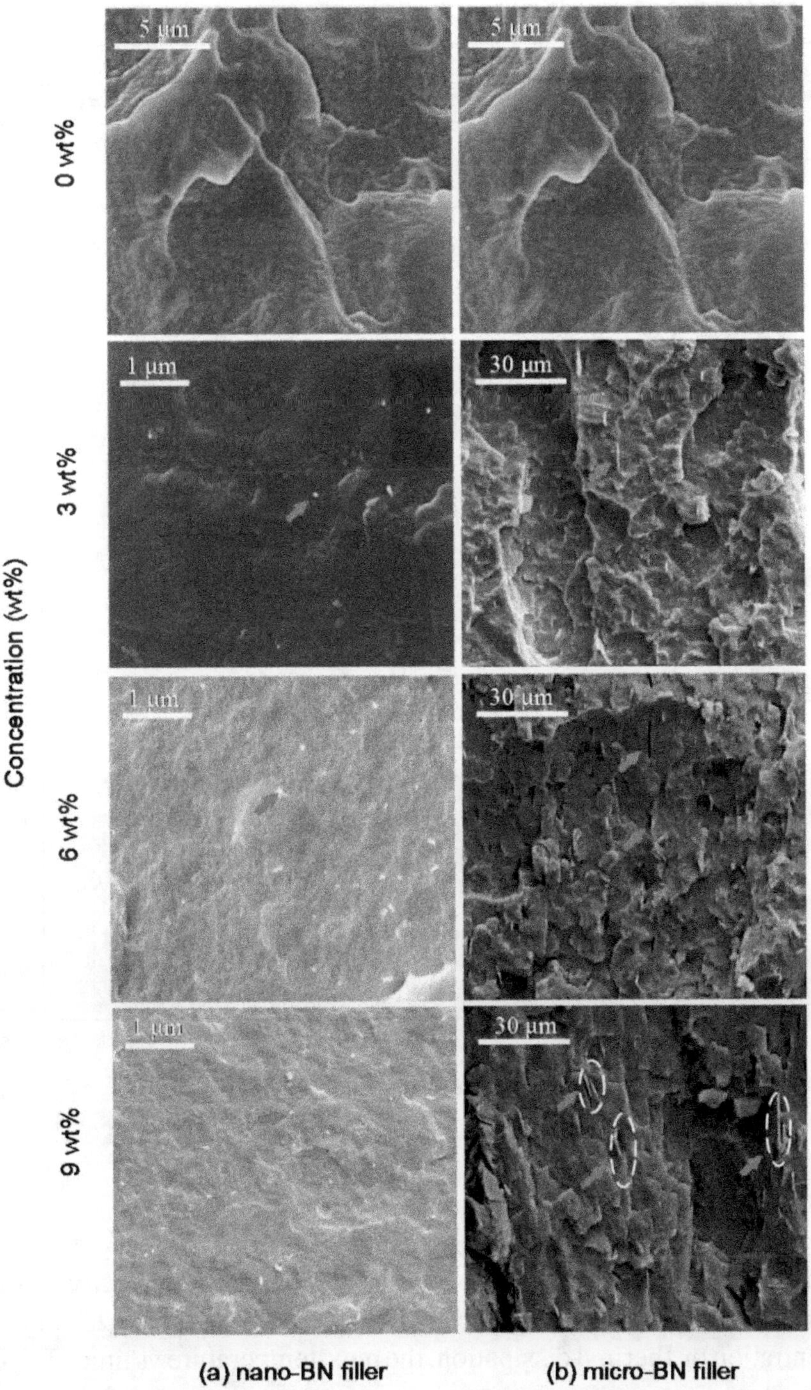

Table 1. Thermal conductivity (λ (W/(m•K)) of the tested samples

ω (wt%)	λ (micro-BN)	λ (nano-BN)
0	0.23	0.23
3	0.30	0.28
6	0.38	0.35
9	0.46	0.40

Figure 15. Percentage of the increased thermal conductivity of the doped sample compared with the one of the neat matrix

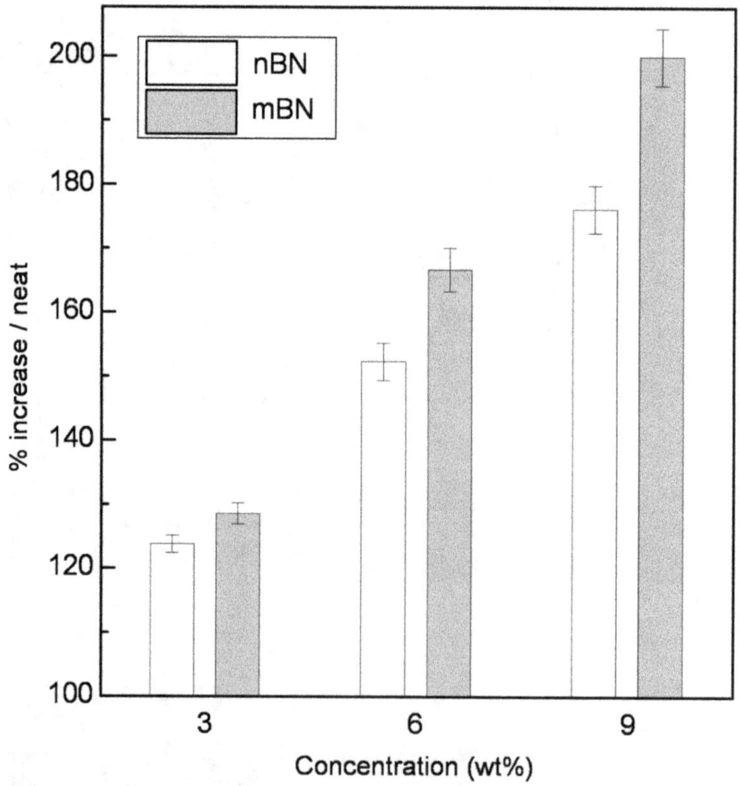

Thermal Dissipation

Thermal dissipation is closely related to the thermal conductivity of the polymeric materials. In order to clarify effects of filler size and filler concentration on thermal dissipation, the max-temperature vs time (T_{max}-t) on the sample surface was investigated at room temperature. Figure 16 illustrates the schematic experimental diagram of the thermal dissipation test setup. A range of plate-like samples (20 mm × 20 mm × 2 mm) were tested in this part.

In this section, thermal dissipation ability is reflected by cooling rate. The higher cooling rate implies higher thermal dissipation ability. Here, the slope of the tangent at every point on the T_{max}-t decay curve is noted as the cooling rate.

The T_{max}-t decay curves of various samples with different filler concentrations are plotted in Figure 17. The initial temperature is set to 120 °C for all curves, while the end is 24 °C. Every point on the curve is a statistical mean from five samples for each kind of composite. It shows that the max temperature decays exponentially in the first 100 s and then slowly decreases until the end. Actually, heat transfers rapidly in the initial period because of the large temperature contrast between direct-heated region and its surrounding. Then the temperature contrast becomes small along with the prolonged time, hence the max temperature on the sample surface slowly decreases and practically stabilizes to room temperature. In addition, the max temperature of the doped samples decays apparently faster than that of pure sample, independently of the size of BN filler, even more rapidly so with increasing filler loading.

Figure 16. Schematic diagram of the thermal dissipation test system

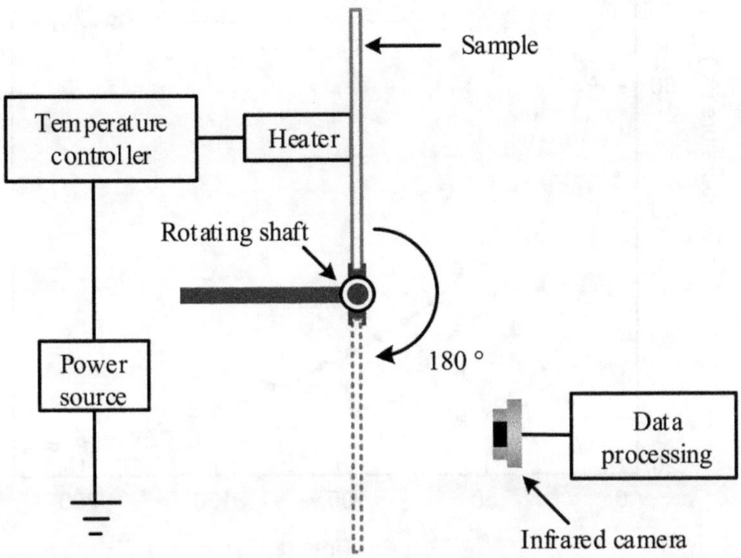

Figure 17. Surface max temperature-time decay curves of the composites with different loadings. (a) coated nano-BN. (b) coated micro-BN

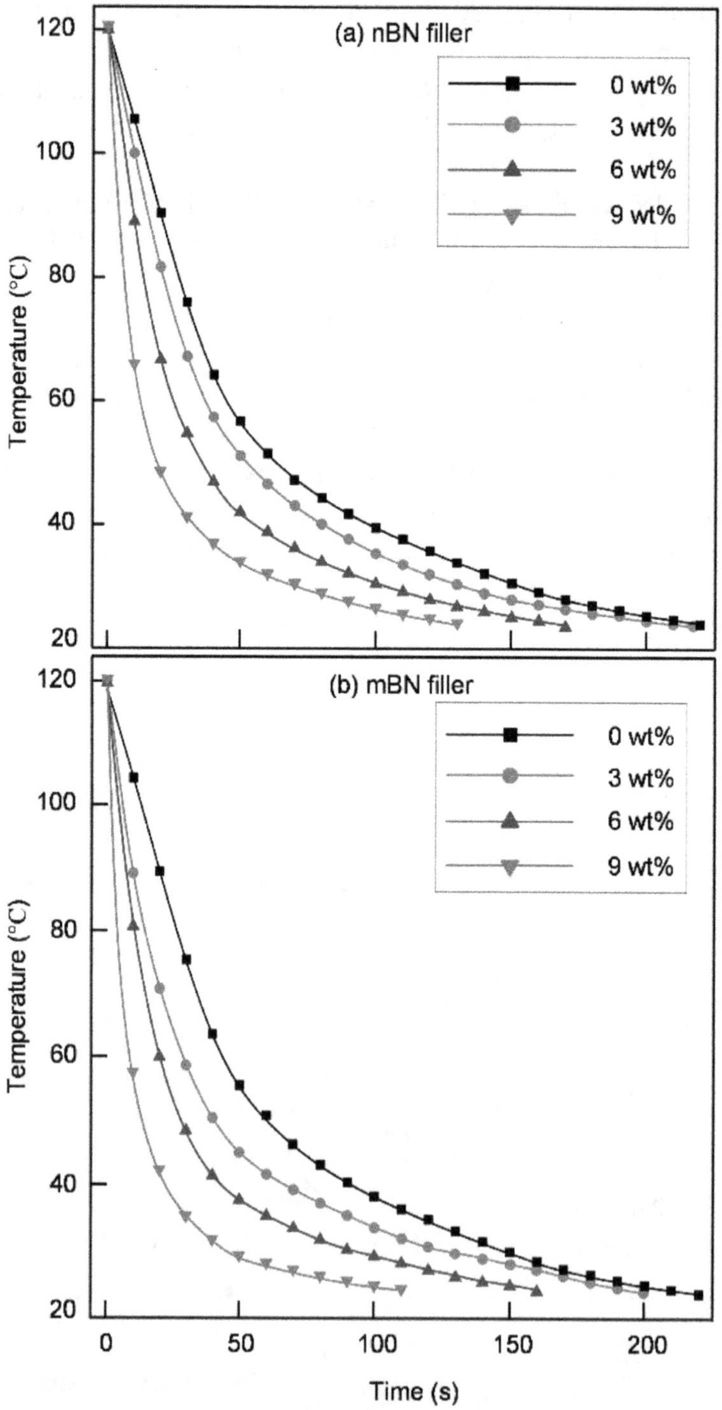

In order to better illustrate effects of filler size and filler concentration on cooling rate, average cooling rates of various samples are illustrated in Figure 18. Average cooling rate is calculated from the equation (2) as follows:

$$ACR = \frac{T_0 - T_{room}}{t_{decay}} \tag{2}$$

where *ACR* is average cooling rate, T_0 is initial temperature, T_{room} is room temperature and t_{decay} is decay time.

It can be seen that *ACR* of the doped sample is higher than that of neat PP and shows an increase trend with increasing filler loading level. This result can be supported by the conclusion in "Thermal Conductivity" that thermal conductivity is improved in the BN filled sample. The higher filler loading is, the higher thermal conductivity obtains, which significantly influences average cooling rate or thermal dissipation ability. Moreover, PP micro-composite shows a higher *ACR* at the same filler concentration due to its higher thermal conductivity. Therefore, the thermal dissipation ability of the manufactured sample is positive correlated with its thermal conductivity.

Surface Dielectric Breakdown

It is evident that thermal accumulation during partial discharge is the major reason to cause the tracking failure. When thermal accumulation raises to a level at which a large number of molecular chains are destroyed on the sample surface, the sample surface starts to melt and even burn. This will lead to destruction in form of tracking on the composite surface and surface dielectric breakdown finally. In the present study, tracking resistance of various samples was evaluated by measuring time to failure, weight loss and erosion depth.

Figure 19 depicts variations of time to failure for all the manufactured specimens with respect to filler concentration at different pulse frequencies. It is found that time to failure represents an increasing tendency with increasing the loading level of BN fillers, irrespective of both BN filler size and pulse frequency. As the obtained conclusion in "Thermal Dissipation", introduction of BN filler into polypropylene matrix will enhance the entire thermal dissipation ability due to the increased thermal conductivity, which has a very significant inhibitory effect on thermal accumulation. Furthermore, the incorporation of BN filler can reinforce the thermal resistance of composites (Zhou et al., 2007). All of those factors can effectively help to delay the

Figure 18. Average cooling rate of the PP/BN composites in different contents

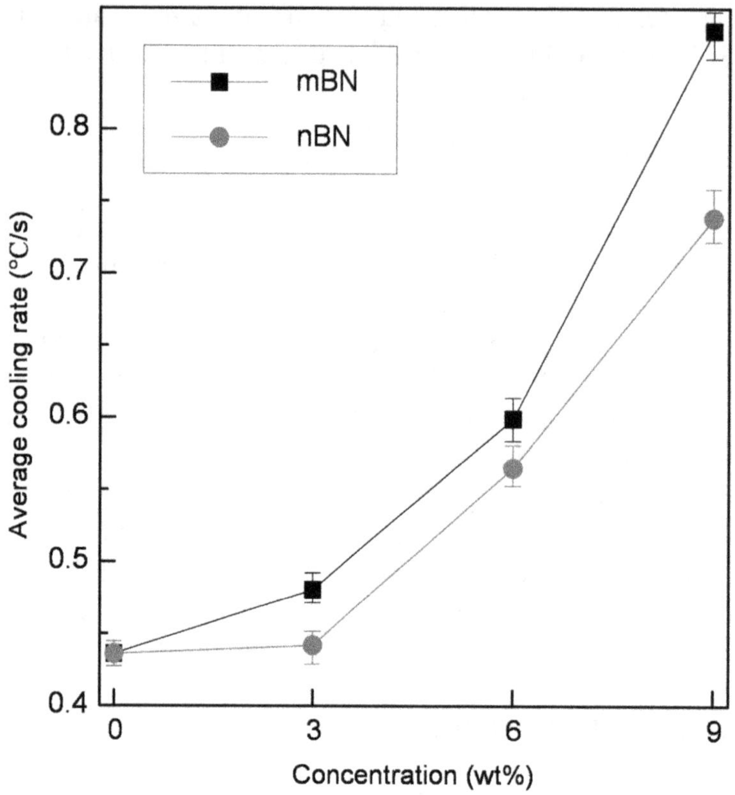

occurrence of dielectric failure. At the same filler concentration, the difference in pulse frequency also results in variations in time to failure. Generally, time to failure shows a decreasing trend with increasing pulse frequency (from 150 to 250 Hz). It can be explained that thermal accumulation would increase with the increase of pulse frequency which makes molecular chains be broken up more rapidly. Therefore, time to failure decreases with respect to pulse frequency.

In order to make a comparison of time to failure between two kinds of compounds, various samples were electrically stressed under the same pulse frequency. It's found that the obtained conclusion is same irrespective of pulse frequency. Thus, the selected results from 200 Hz are reported here. As illustrated in Figure 20, although time to failure shows a similar rise tendency with increasing filler concentration independently of the type of sample, PP/micro-BN composites exhibit a longer time to failure at high loading level, especially filler loading up to 9 wt%. Incorporation of micro-BN particles

Figure 19. Variations of time to failure with respect to filler loading at different pulse frequencies

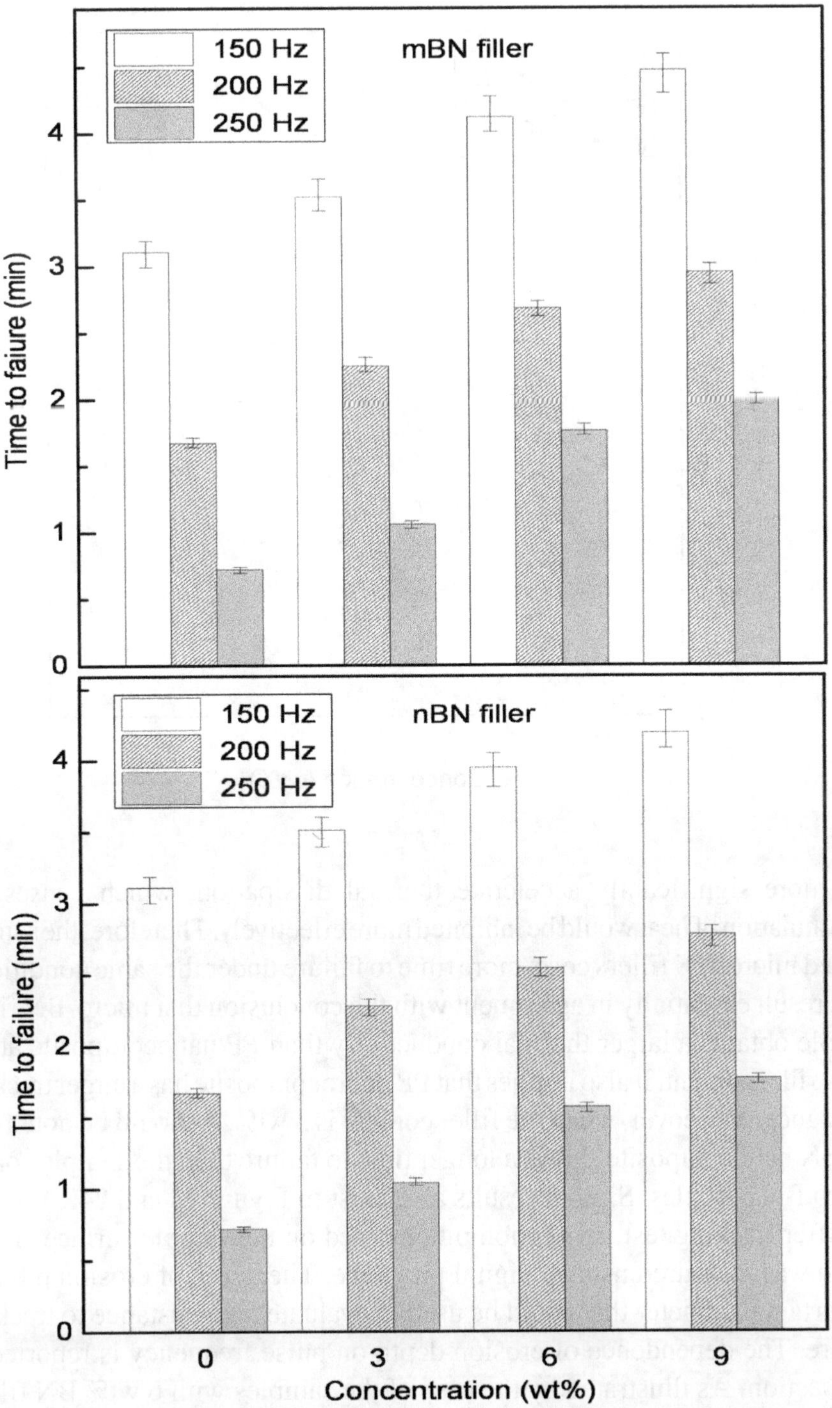

Figure 20. Comparison of time to failure between micro-BN filled PP and PP/BN nanocomposites with applying repetitive pulse at 200 Hz

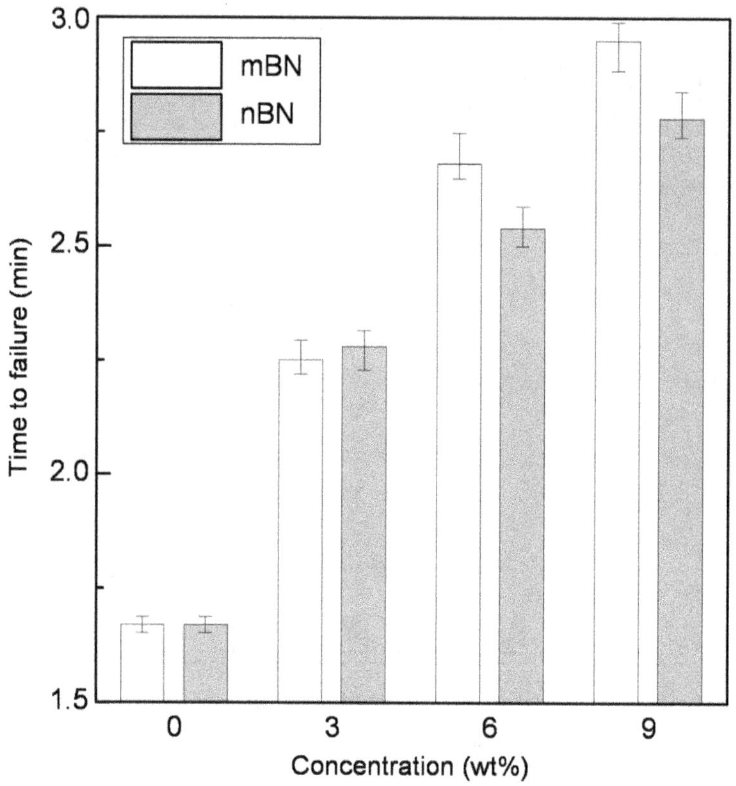

can more significantly accelerate thermal dissipation, which causes the accumulation of heat would be inhibited more effectively. Therefore, the sample coated micro-BN fillers costs more time to failure under the same conditions. This result essentially in agreement with the conclusion that micro-BN filled sample obtains a larger thermal conductivity than PP nanocomposite at the same filler content. It also implies that PP microcomposite has a larger tracking resistance. Moreover, when the filler content is 3 wt%, it should be noted that PP/BN nanocomposite shows a longer time to failure than the sample coated micro BN particles. Similar results also exist in Figure 23 and 24.

After tracking test, an erosion pit emerged on the sample surface and its depth was measured using a digital pit gauge. The depth of erosion pit is an important parameter that could be used to evaluate the resistance to tracking failure. The dependence of erosion depth on pulse frequency is reported in this section. As illustrated by the case of the samples with 6 wt% BN fillers in Figure 21, the depth of erosion increases with increasing pulse frequency from 150 to 250 Hz. In fact, more heat could be accompanied by increasing pulse frequency. Hence, more molecular chains would be broken up which promotes forming a deeper erosion pit. It is worth noting that this conclusion is independent of the size of BN filler.

The dependence of erosion depth on both time to failure and filler loading is similar under repetitive pulse at different frequencies. Hence, the selected results from 200 Hz are illustrated in Figure 22 to limit the length of paper. It has known that a number of BN particles filled in PP matrix can accelerate thermal dissipation so that the thermal accumulation would be inhibited. Meanwhile, inorganic BN particles do not take part in the process of erosion because of the excellent high temperature resistance. Considering all these factors, the development of erosion could be restrained effectively in the BN doped sample. In other words, the increased thermal conductivity of the

Figure 21. Erosion depth as a function of pulse frequency for the samples with 6 wt% BN fillers

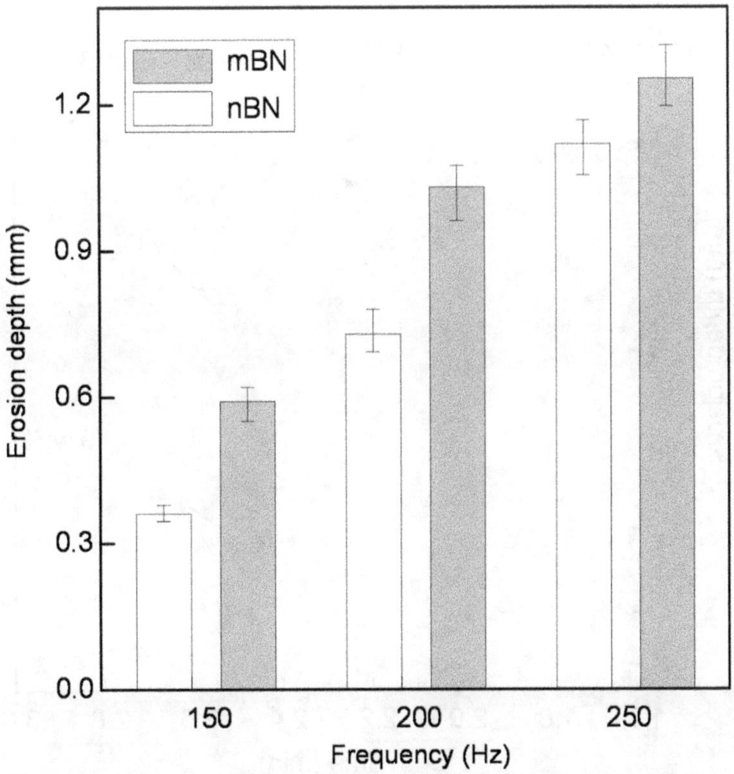

doped sample has a positive effect on the resistance to erosion. Generally, under the same conditions, erosion depth of the sample that after discharge exposure decreases evidently with increasing filler concentration. When the filler loading is 9 wt%, the erosion depth of BN filled sample is reduced dramatically. This result is also independent of filler size.

In order to better illustrate the effects of filler size on erosion depth and weight loss of the test samples with different loadings, all the samples were stressed by repetitive pulse discharges for 1 minute. Then the erosion depth and the weight loss of various samples were measured. The selected results from 200 Hz are reported here to limit the length of paper since the changing trends of erosion depth and weight loss is similar from 150 to 250 Hz.

The comparison on erosion depth respects to filler concentration between two categories of composites is shown in Figure 23. As can be seen, erosion depth decreases with increasing filler mass fraction. At the same time, the

Figure 22. Dependence of erosion depth on both time to failure and filler loading under repetitive pulse at 200 Hz

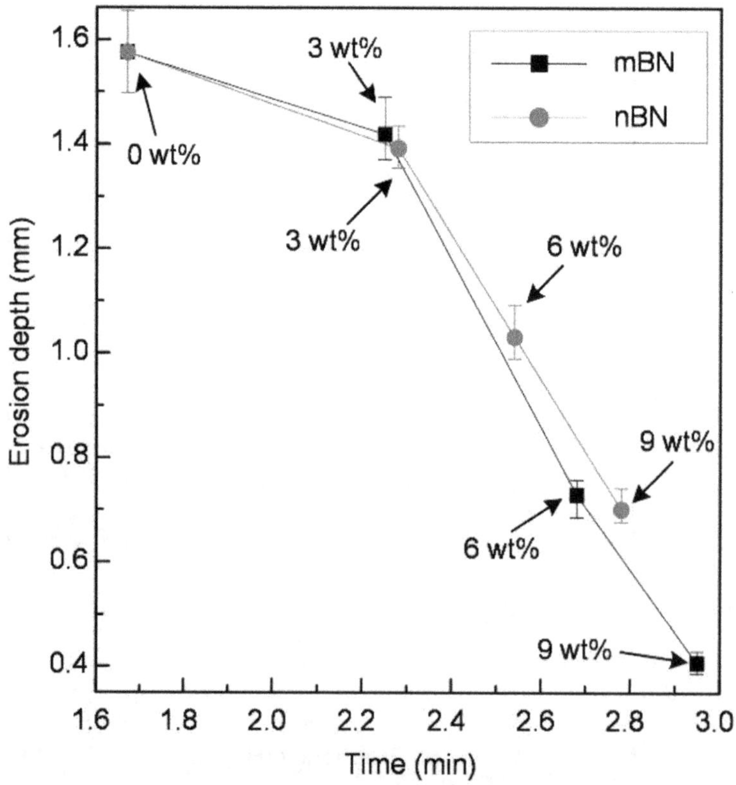

erosion depth of micro-BN filled sample is lower than that of nano-BN filled virgin in relatively high loading level (> 3 wt%). Unexpectedly, when the concentration is 3 wt%, the nano-BN filled sample shows a lower erosion depth instead. This exceptional result will be discussed later.

Weight loss as a measurement of various test samples filled with different loadings is depicted in Figure 24. Similarly, weight loss decreases with increasing filler content and micro-BN filled sample shows a lower weight loss at the same loading level except 3 wt%, where the sample filled with nano-BN particles exhibits a lower weight loss. This result is well correlated with both the time to failure and the erosion depth investigated.

In fact, the time to failure, the erosion depth and the weight loss of nano-BN filled PP composites are also affected by nano-effects, especially bonding strength between nano-sized filler and matrix (Kozako et al., 2004). Both the bonding strength and the increased thermal conductivity can help to increase

Figure 23. Comparison of erosion depth with different loadings between the PP coated micro BN particles and PP/nano-BN composites after applying repetitive pulse at 200 Hz for 1 min

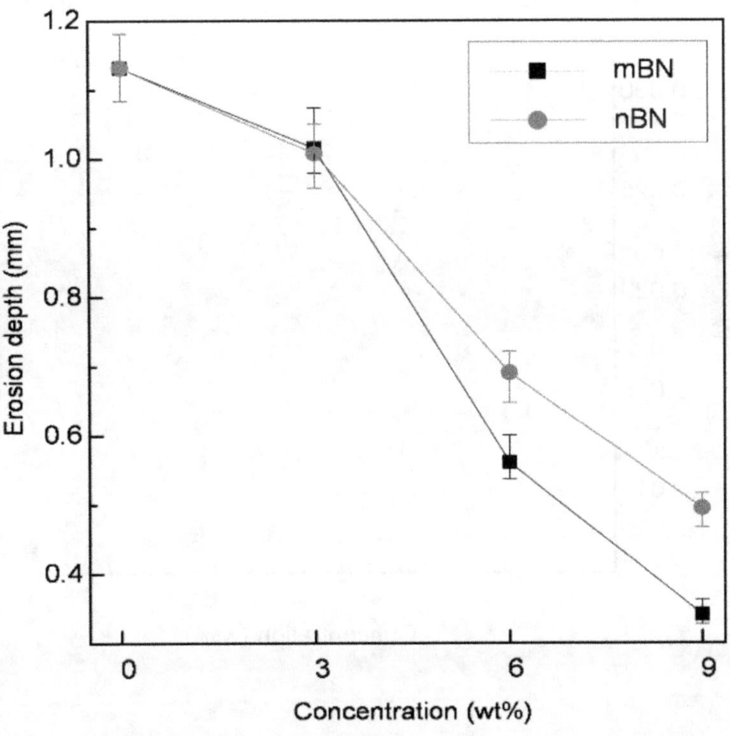

the resistance to tracking, prolong the time to failure and reduce the degree of erosion in the PP matrix coated nano BN fillers. Therefore, when the filler content is 3 wt%, compared with PP/micro- BN sample, nano BN filled sample shows a longer time to failure, a smaller erosion depth and weight loss. However, along with increasing filler loading into higher loading level, the increased thermal conductivity would play the dominant role in affecting improvement of the resistance. Therefore, micro BN filled sample shows higher tracking resistance at relatively high loading level.

Considering all above-mentioned result, the main conclusion can be drawn that addition of BN filler into pure PP contributes to improvement of tracking resistance. Compared with nano BN filled samples, PP/micro-BN samples obtain higher tracking resistance at relatively high loading level.

Figure 24. Relation between BN filler content and weight loss after applying repetitive pulse at 200 Hz for 1 min

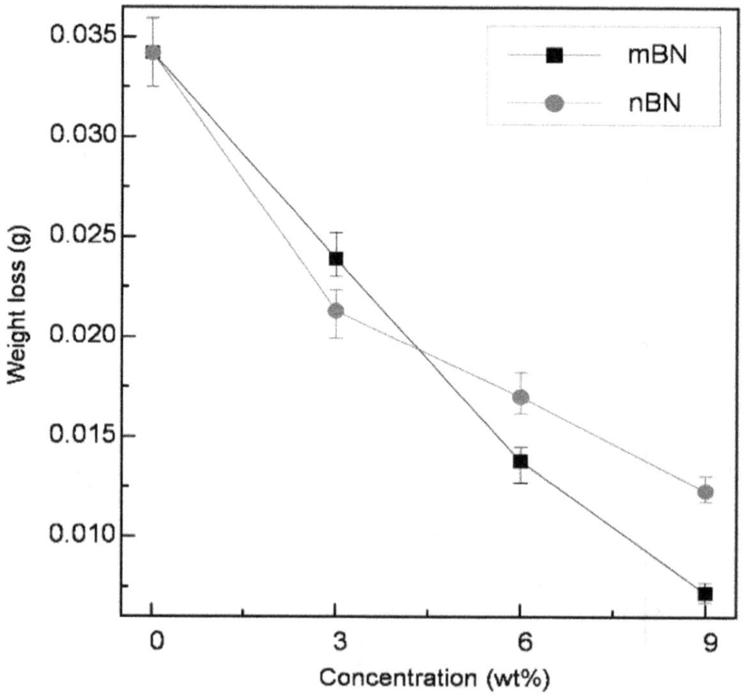

Bulk Dielectric Breakdown

In order to evaluate bulk dielectric breakdown characteristic, DC breakdown strength of the manufactured sample was measured in this section. It was performed using a classical sphere-to-sphere electrode system in compliance with IEC 60243 standards. For the manufactured sample, its bulk dielectric breakdown characteristic has to be validated, especially for DC high voltage applications. In this section, DC breakdown strength of various samples is measured through test and the obtained data is analyzed using 2-parameter Weibull distribution method (Tuncer et al., 2006). DC breakdown strength at the Weibull probability of 63.2% is noted as the characteristic breakdown strength ($\hat{\alpha}$). In addition, the shape parameter ($\hat{\beta}$) of plots is also discussed to evaluate robustness of the manufactured samples.

Figure 25 and Figure 26 depict the Weibull plots for the measured DC breakdown strength data of various manufactured samples. The characteristic breakdown strength ($\hat{\alpha}$) and the shape parameter ($\hat{\beta}$) of Weibull distribution for all tested samples are summarized in Table 2.

As illustrated in Figure 25, the nano-BN filled composite has a higher DC breakdown strength than neat PP. With increasing filler loading, DC breakdown strength increases. The maximum breakdown strength of PP/BN nanocomposites is 336.2 kV/mm with 9 wt% filler, increased by about 16.5% when compared with that of pure PP. It is hypothesized that the strong bonding strength between nano sized filler and matrix can help to improve the DC breakdown strength. On the other hand, nano-sized filler shows a higher breakdown strength than the base matrix and discharges channels tend to end up with nano-sized fillers (Fuse et al., 2008; Li et al., 2010). Therefore, addition of nano-sized BN filler into pure PP can improve the DC breakdown strength of PP nanocomposites. It is also worth noting that with increasing loading level, the shape parameter of PP/BN nanocomposites increases firstly and then decreases, even lower than that of the neat matrix in filler content of 9 wt%. It is assumed that some small voids may emerge in the matrix at high loading level, which makes the doped sample presents a lower shape parameter than neat PP.

For the sample coated micro BN particles, its DC breakdown strength is lower than that of neat PP and it shows a substantial negative correlation to filler loading. As demonstrated in Figure 26, more BN particles are embedded into the PP matrix, lower DC breakdown strength is. At the same time, the shape parameter decreases significantly in high filler loading, which implies

Figure 25. Weibull plots of DC dielectric breakdown strength for the PP/nano-BN samples (Confidence Interval: 90%)

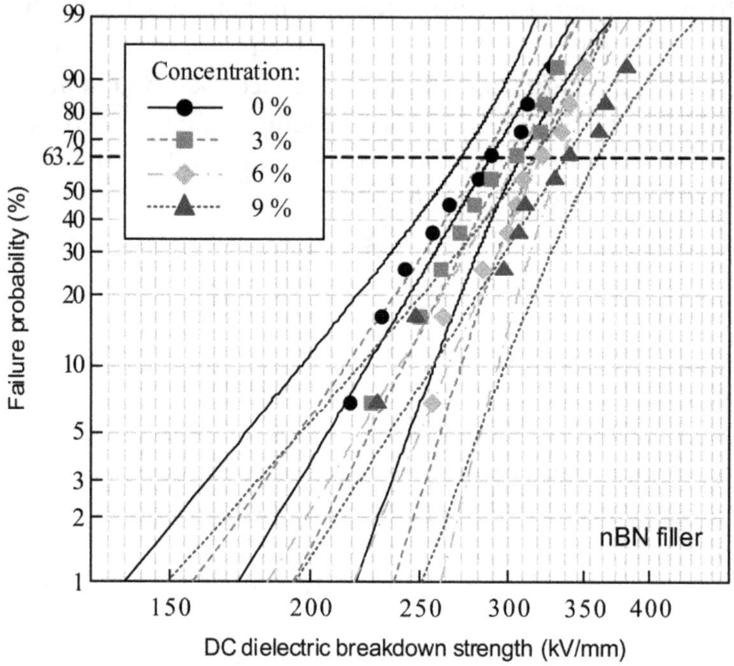

inferior robustness. In fact, after dispersing micro sized BN fillers into neat PP, some defects are also formed in the matrix because of the interfacial interactions and voids. And when increases filler loading, more voids tend to be formed in the compound (Wang et al., 2011). Hence, it is evident that addition of micro BN filler results in a reduction in DC breakdown strength.

Through the comparison between two kinds of manufactured samples, the main conclusions can be drawn that introduction of nano BN filler can evidently improve the entire DC breakdown strength. Whereas, all the doped samples show worse robustness at high filler loading level.

Figure 26. Weibull plots of DC dielectric breakdown strength for the PP/micro-BN samples (Confidence Interval: 90%)

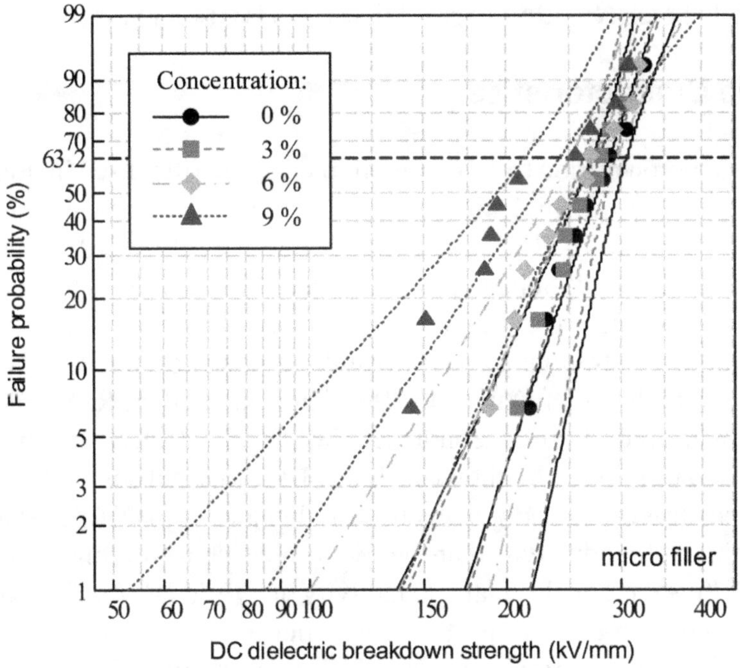

Table 2. Values of the scales parameter estimators $\hat{\alpha}$ and of the shape estimator parameter $\hat{\beta}$

Sample	$\hat{\alpha}$ (kV/mm)	$\hat{\beta}$
Pure PP	288.6	8.962
PP + 3 wt% nano-BN	299.8	10.45
PP + 6 wt% nano-BN	319.5	12.11
PP + 9 wt% nano-BN	336.2	8.335
PP + 3 wt% micro-BN	278.7	9.953
PP + 6 wt% micro-BN	273.3	6.693
PP + 9 wt% micro-BN	242.7	4.407

THERMAL CONDUCTIVITY AND ARCING RESISTANCE OF MICRO OR HYBRID BN FILLED POLYETHYLENE UNDER PULSE STRENGTH

Sample Characteristics

Hexagonal boron nitride (h-BN) was adopted as the filler used in this section because of its novel high thermal conductivity and excellent insulating properties (Mo et al., 2015; Zhou et al., 2007; Li et al., 2010; Zhi et al., 2009; Lin et al., 2012; Song et al., 2012), and PE was used as the base resin. In order to maximize the thermal conductivity of PE/BN composite at the same filler concentration, the powder mixing was chosen to mix BN fillers and PE powder together (Yu et al., 2002; Zhu et al., 2015; Ren et al., 2016).

The measured thermal conductivities of all PE/BN specimens are summarized in Table 3. As shown in Table 3, the thermal conductivity of PE/BN composite is expressed as a function of filler loading. BN filled sample has a higher thermal conductivity than the neat sample whose thermal conductivity is 0.31 W/(m•K). The thermal conductivity of hybrid BN reinforced sample almost reaches up to 1 W/(m•K) at the filler loading of 40 wt%, which is twice higher than that of the pure sample. Representative SEM images of the fractured surface of samples at different loadings are illustrated in Figure 27. It was confirmed that the cross-section of the pure sample is very smooth. Meanwhile, the cross-section of BN filled sample becomes rougher and rougher with increasing filler loading. The rough fractured surfaces of BN filled samples as shown in SEM images indicate strong interfacial interactions between BN particles and PE matrix.

Actually, adding BN particles into PE matrix can form some thermal conductive pathways. With increasing filler loading, much more thermal conductive pathways will take shape in the base resin. As shown in Figure 27, when filler content exceeds 20 wt%, so many thermal paths take shape and cross each other to make up thermal conductive networks. This leads to a much more dramatic improvement in thermal conductivity, as presented in Table 3.

Moreover, PE/hybrid-BN composite shows a higher thermal conductivity at the same filler loading. It is well known that the quantity of nanoparticles is larger than that of micro-particles at the same mass fraction. And nanoparticle has a smaller grain diameter as well as a higher aspect ratio. Both of them result in more numerous and denser BN particles could be embedded into

Figure 27. Representative SEM images of the fractured surface of samples

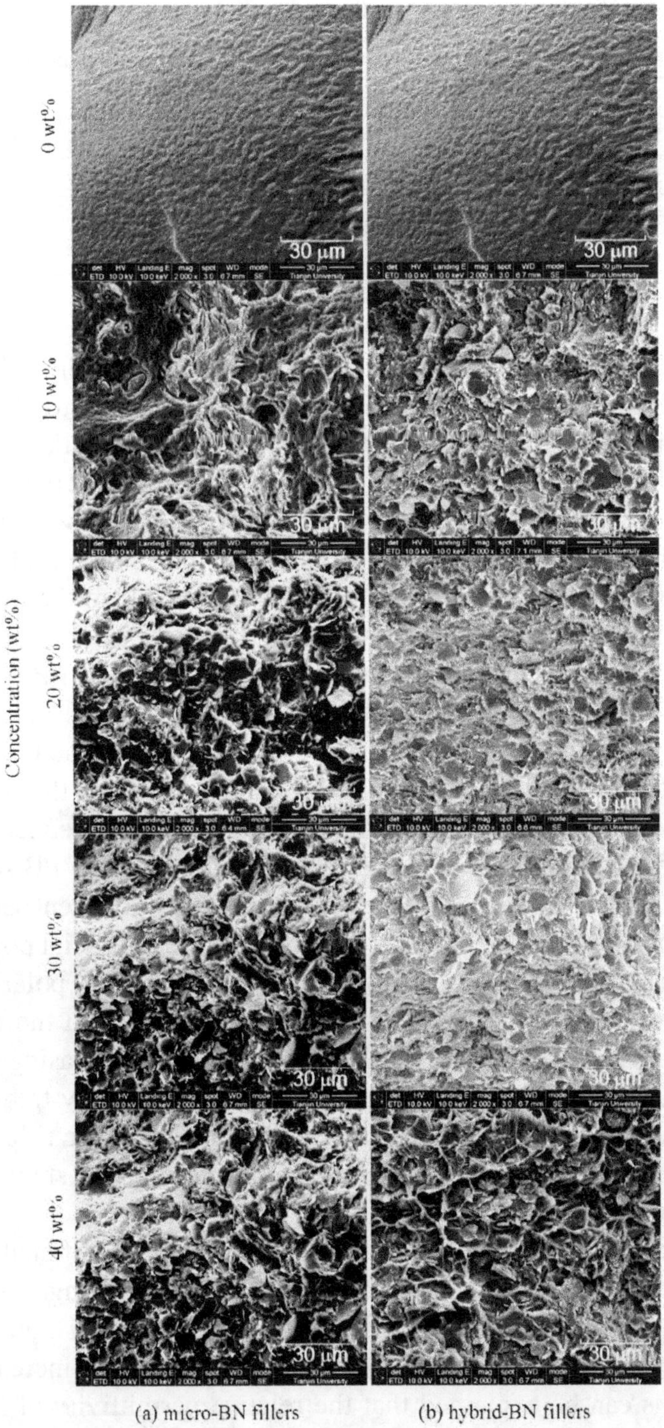

Table 3. Thermal conductivity (λ (W/(m•K)) of PE/BN composites in different loading levels ((wt%))

ω (wt%)	λ (micro-BN)	λ (hybrid-BN)
0	0.31	0.31
10	0.50	0.56
20	0.56	0.68
30	0.66	0.79
40	0.85	0.97

PE/hybrid-BN at the same filler loading. Figure 28 is the schematic of PE/hybrid-BN composite at high filler loading (>20%) to illustrate the microstructure. As shown in Figure 28, main conductive paths in PE matrix are formed by micro-sized BN particles. Nano-sized BN particles play roles in connecting micro-sized BN particles with each other, which results in additional thermal conductive paths take shape. It indicates that adding hybrid BN particles into the polymer matrix is more favorable to the formation of thermal conductive pathways. Therefore, hybrid BN fillers reinforced PE composite has a higher thermal conductivity than PE/micro-BN composite at the same filler concentration.

Relative permittivity is a function of frequency for all the manufactured samples, which is shown in Figure 29. It can be observed that the relative permittivity of PE resin decreases slightly with increasing frequency. In fact, as a kind of non-polar polymer, the relative permittivity of PE should be a constant with respect to frequency. However, it is slightly different between very high and very low frequency due to the existence of interfacial polarizations are mainly caused by the impurities in PE. More interfacial polarizations in the matrix would result in higher relative permittivity and the number of those interfacial polarizations will slightly reduce with increasing frequency. Therefore, the relative permittivity is relatively high at low frequency and it decreases gradually with increasing frequency. For all the filled samples, the influence of frequency on the relative permittivity also shows a similar trend to that of the neat sample.

On the other hand, filler concentration also has a significant influence on the relative permittivity. Over the whole frequency range, the relative permittivity of micro-BN particles filled composite is clearly higher than that of the pure sample and it becomes higher and higher with increasing filler loading. This can be explained that the relative permittivity of BN (about

Figure 28. Schematic of PE/hybrid-BN composite to illustrate its microstructure

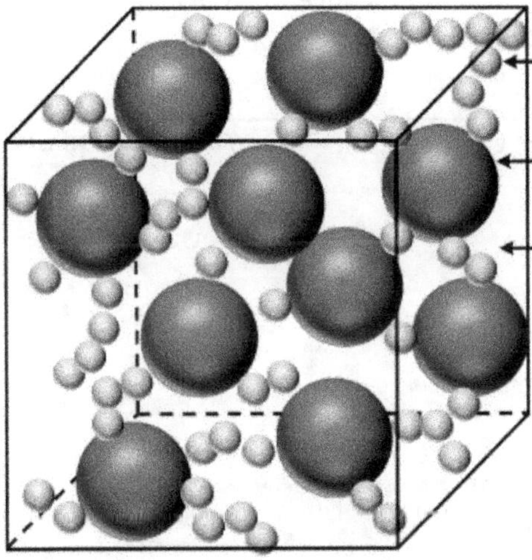

4.0) filler is higher than that of the base resin. Hence, addition of micro-sized BN filler can increase the relative permittivity of the whole composite. With the increase of micro-filler loading in PE matrix, the influence of BN filler permittivity on the relative permittivity of composites becomes more and more largely. However, for the specimen coated hybrid-BN fillers, the relation between its relative permittivity and filler concentration is different and complex. As shown in Figure 29 (b), when loading level is 10 wt%, the relative permittivity of PE/hybrid-BN composite is about 2.30 at 1000 Hz and even lower than that of the pure sample. However, when the concentration of BN filler reaches 20 wt%, the relative permittivity at 1000 Hz for PE/hybrid-BN composite becomes higher than that of the pure sample and it increases with increasing filler concentration. Actually, when filler loading is below the threshold value, the influence of filler permittivity on the whole composite permittivity is relatively small. After embedding a few nanoparticles into the matrix, there forms an immobile and compact nanolayer in the interfaces between polymer chains and nanoparticles due to the strong physical and chemical bonding of the molecular chains with nanoparticles. Therefore, the mobility of the molecular chains in the matrix is hindered, which results in a reduction of the relative permittivity for the whole composite. In the same time, the immobility of polymer chains is a function of filler concentration. Embedding more nano-sized particles into the polymer matrix causes more polymer chains to be immobile, and hence the relative permittivity of the whole composites would further decrease with increasing nano-sized filler

Figure 29. The variations of relative permittivity with respect to frequency for all the manufactured samples at different filler concentrations: (a) micro-BN fillers reinforced samples (b) hybrid-BN fillers reinforced samples

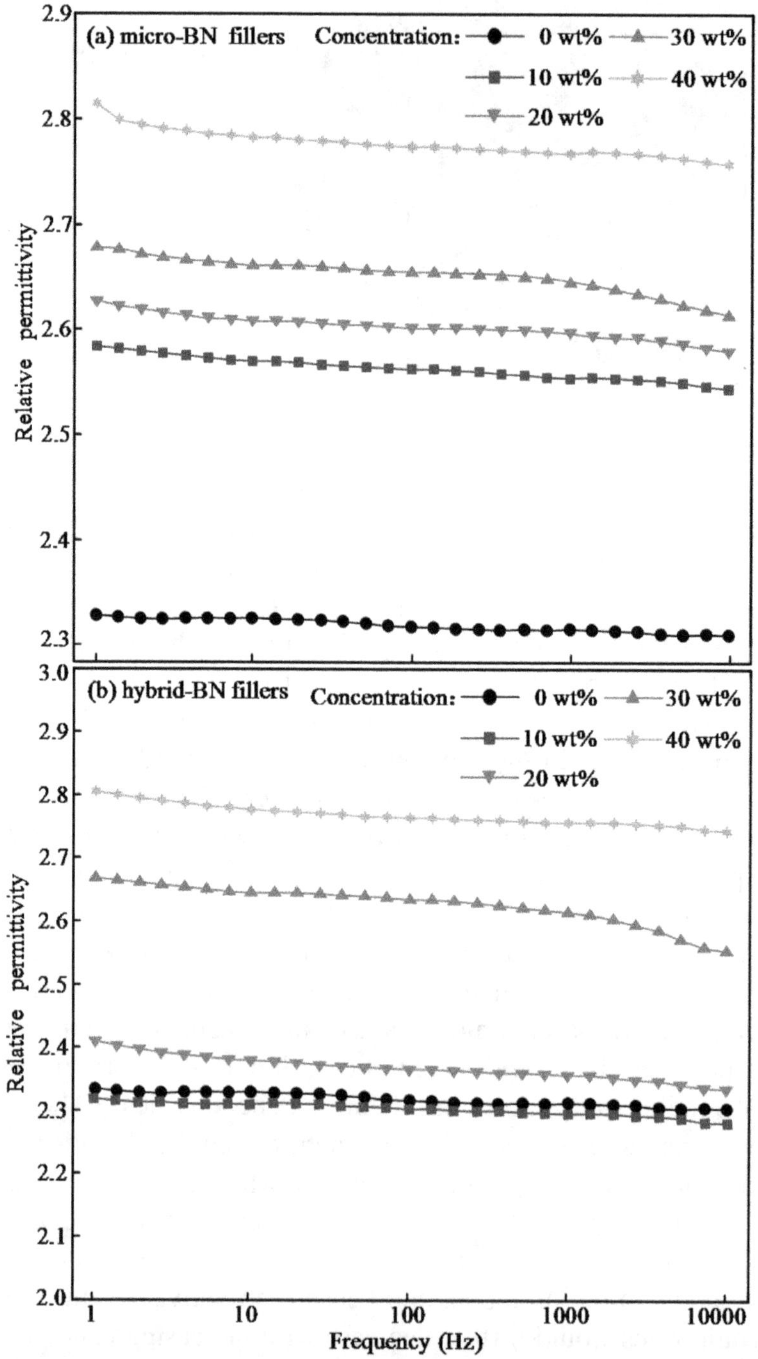

concentration. Therefore, it is hypothesized that the relative permittivity of the whole composite is lower than that of the base resin when the filler concentration is below the threshold value. However, once filler loading exceeds the threshold value, influence of

BN filler permittivity would come into playing the dominant role, which results in an improvement in the relative permittivity of composites. Moreover, this improvement in relative permittivity is positively correlated with filler concentration. Therefore, relative permittivity increases with increasing filler concentration.

Figure 29 also shows that the relative permittivity of BN filled composites increases from 2.31 at 1000 Hz for PE to about 2.76 at 1000 Hz at the filler concentration of 40 wt%. Therefore, the relative permittivity of BN filled composites could maintain a relatively low value when the filler concentration is appropriate.

Thermal Accumulation

In order to show the effects of BN filler loading and size on thermal accumulation, the thermographs of the test samples after applying pulse discharge at 200 Hz for 100s were recorded. It was observed that the sample surface was damaged initially around the spot with the maximum temperature during the test. Therefore, the maximum temperature of the sample surface is a very representative parameter that should be analyzed. After data processing, the maximum temperature of the sample surface with respect to filler loading is represented in Figure 30. It can be seen that the maximum temperature decreases with increasing filler content under the same discharge conditions. When filler loading is 10 wt%, the maximum temperature of PE/hybrid-BN composite is 288.4 °C, while only 157.6 °C at 40 wt%.

As mentioned in "Sample Characteristics," embedding more BN fillers into PE matrix could result in higher thermal conductivity. The increase of thermal conductivity can improve the ability of thermal dissipation in the discharge area, and hence the heat generated from discharge will be conducted out quickly. Then the effect of thermal accumulation will be reduced. Moreover, as a kind of inorganic material with remarkable chemical and thermal stabilities, BN does not participate in the process of surface dielectric breakdown. Considering all these factors, the thermal properties of the whole PE composite can be improved significantly with increasing filler concentration and hence the maximum temperature of the sample surface shows a descending trend.

Figure 30. Relation between the maximum temperature of the sample surface and filler concentration after applying pulse discharge at 200 Hz for 100 s

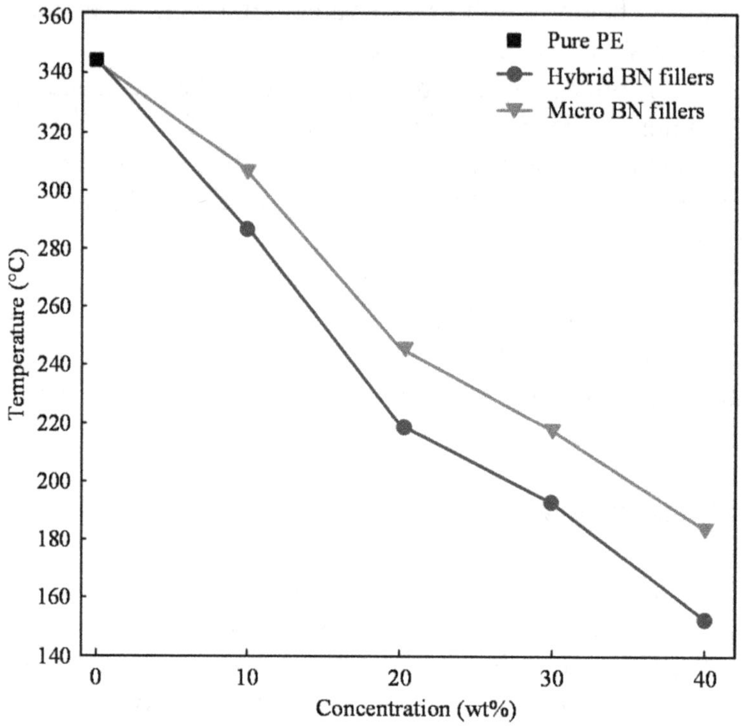

Figure 30 also shows that the maximum temperature of PE/hybrid-BN sample surface is lower than that of micro-sized BN particles reinforced sample surface at the same filler concentration. When filler loading is 20 wt%, the maximum temperature of PE/micro-BN sample surface is 247.8 °C, while only 219.5 °C for the sample coated hybrid-BN particles. It implies that addition of hybrid-BN particles is more helpful to inhibit the thermal accumulation and reduce the maximum temperature of the sample surface. As discussed in "Sample Characteristics," compared with micro-BN filled sample, much more thermo-conductive pathways can be formed in PE/hybrid-BN composite at the same filler loading. Therefore, thermal accumulation could be suppressed more effectively in the composite coated hybrid-BN particles.

A further analysis of the differences in the temperature distribution between two categories of BN filled samples is illustrated in Figure 31. The vertical Y-axis is defined as the direction along plate-to-needle electrode, and the number on the horizontal X-axis is symmetrical around point 0. As shown

Figure 31. Comparison of the isotherm distributions after applying pulse at 200 Hz for 100 s between two categories of BN filled samples

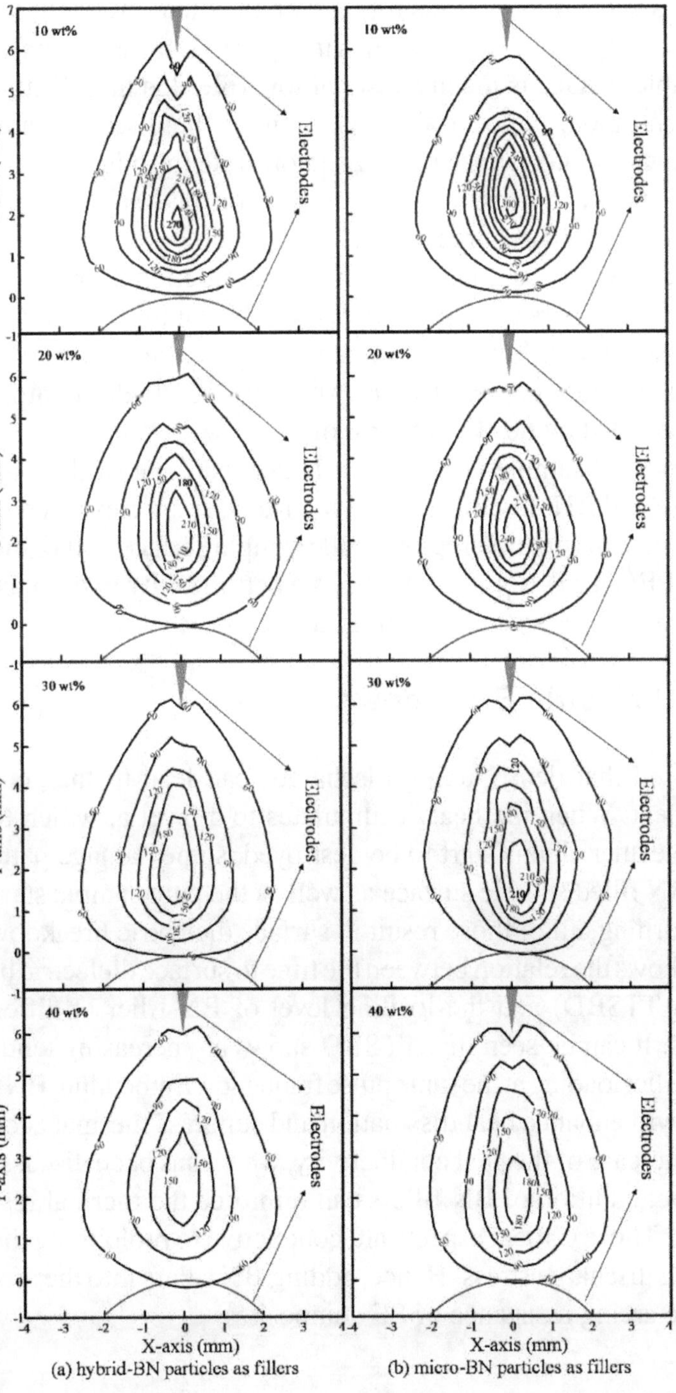

in Figure 31, the temperature at the narrow center is higher than that at any other positions. It implies that most of the heat generated from discharge concentrates at the narrow discharge center and then spreads to surrounding regions gradually. Moreover, as shown in Figure 31, the isotherm distribution on the sample surface is much denser at low filler loading. With increasing filler concentration, the central temperature in the discharge area shows a clear downward trend. Those results demonstrate introduction of BN fillers can accelerate the process of heat dissipation and suppress the thermal accumulation in the discharge area.

At the same filler loading, the isotherm distribution of the sample coated hybrid fillers is sparser, while PE/micro-BN composite shows a denser isotherm distribution. Moreover, there is a significant difference in the size of the isothermal diagram between two series of PE/BN composites. The region surrounded by 60 °C isotherm of

PE/hybrid-BN sample is obviously larger than that of the sample filled with micro sized BN. It indicates that hybrid BN particles filled sample has a larger thermal dissipation region on the sample surface. All of those results verify that PE/hybrid-BN composite has a better ability to dissipate the heat generated from discharge at the same filler content.

Surface Dielectric Breakdown

It is proposed that dense arcing discharges can lead to the generation of sufficient heat. When the heat accumulates to a level at which the sample surface molecular chains start to be destroyed, some damage patterns show up on the BN filled sample surface as well as the pure sample starts melting and even burning, all of those result in surface dielectric breakdown finally. Figure 32 shows the relation between the time to surface dielectric breakdown (shorted as TTSBD) and the loading level of BN filler at different pulse frequencies. It can be seen that TTSBD shows an increasing tendency with increasing filler loading at the same pulse frequency. Embedding BN fillers into the matrix will enhance heat dissipation and suppress thermal accumulation due to the increase of thermal conductivity, which has been discussed before. Furthermore, addition of BN fillers can reinforce the thermal resistance of composites. Those critical aspects are conductive to prolonging the duration of the arcing discharge stress. Hence, adding BN fillers into the the resin will improve the arcing resistance of PE composites.

Meanwhile, differences in pulse frequency could cause a variation in TTSBD of the manufactured sample at a constant filler loading level. Here, selected results from 40 wt% BN filled samples are reported to limit the length of paper, since the changing trend of the time to surface dielectric breakdown with respect to filler loading from 0 to 40 wt% is similar. In fact, more heat could be accompanied by increasing pulse frequency, and hence more molecular chains would be broken up. Therefore, TTSBD decreases with increasing pulse frequency from 150 to 250 Hz as illustrated in Figure 33. The data presented in Figure 33 also reveal that PE/hybrid-BN composite exhibits a longer TTSBD than micro-sized BN filled composite at the same filler concentration. As discussed in "Thermal Accumulation," adding hybrid-BN particles into the resin can inhibit thermal accumulation and accelerate thermal dissipation more significantly than just adding micro sized BN. Therefore, the composite coated hybrid-BN particles costs more time to surface dielectric breakdown, which reveals that PE/hybrid-BN composite has a larger arcing discharge resistance than PE/micro-BN composite at the same filler loading.

Figure 34 shows the relative damage degree curves that are derived from the images of the PE/hybrid-BN samples after surface dielectric breakdown at different frequencies. Here, the relative damage degree of the base resin is not discussed. Actually, when surface dielectric breakdown occurred in the pure sample, the products of decomposition were mainly H_2O and CO_2. Hence, PE did not produce a typical carbon-like track. Repetitive discharges only caused the erosion of the pure sample without obvious surface color change.

As shown in Figure 34, under the same pulse frequency, the size of the relative damage degree curve changes with filler loading dramatically. Obviously, the higher level of the filler loading is, the smaller size of the curve is. It can be seen that the relative damage degree curve shows a higher peak at low hybrid filler content. With increasing filler loading from 10 to 40 wt%, the peak on the curve decreases from 192 to 93 under 150 Hz. At the same time, the area under the relative damage degree curve also shows a gradual decrease trend with increasing filler loading. Smaller area under the relative damage degree curve implies smaller area be damaged on the sample surface. These experimental results verify that addition of BN fillers can suppress the formation of damage pattern and decrease the damage degree because of the improvement in the resistance to arcing discharge. Moreover, for any given filler loading, the relative damage degree distribution gradually becomes higher and wider with the increase of pulse frequency. It can be explained that much more heat would be generated from pulse discharge

Figure 32. Relation between the time to surface dielectric breakdown and filler concentration with the application of different pulse frequencies

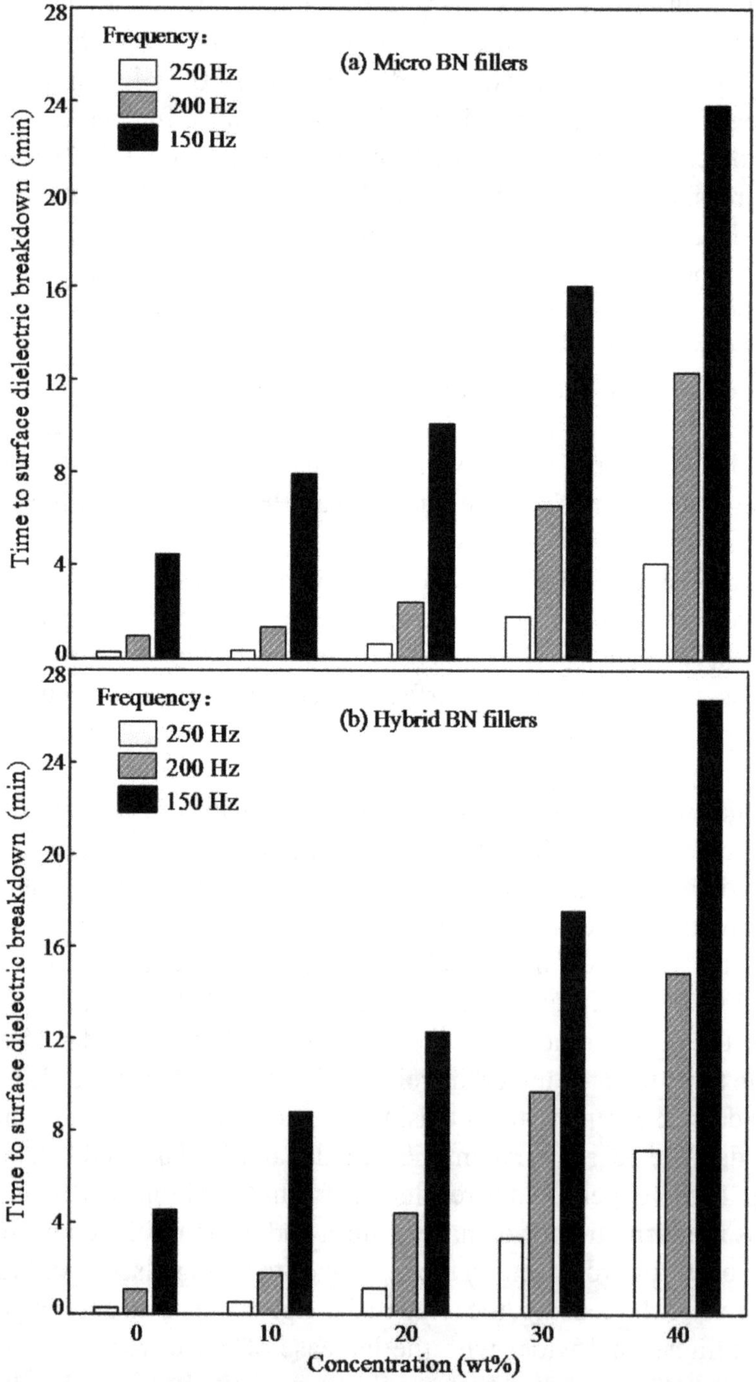

Figure 33. Comparison of the time to surface dielectric breakdown between two categories of composites at different pulse frequencies

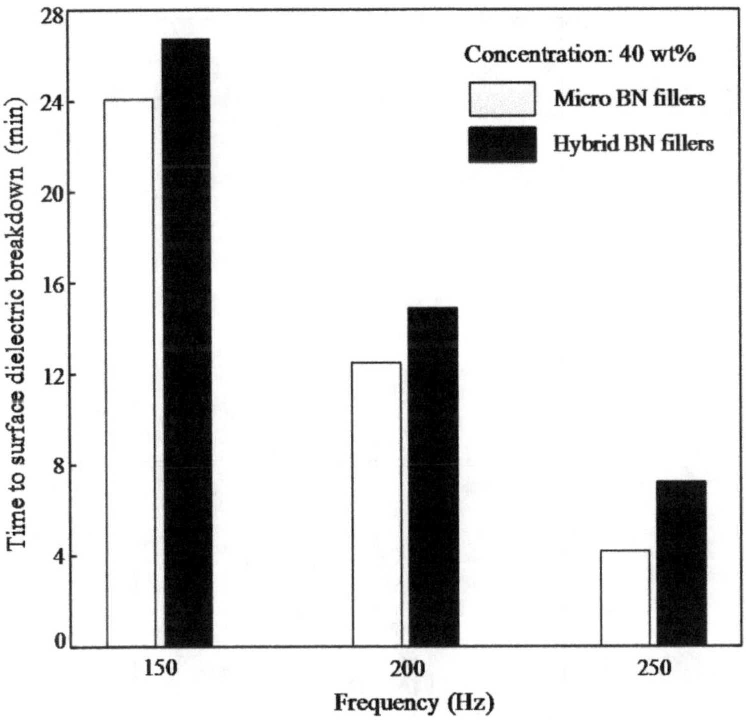

under high frequency, which results in more areas on the sample surface to be destroyed. This result indicates that the damage degree will be increased under high discharge frequency.

Consider again the case of PE/hybrid-BN samples, as shown in Figure 35, damage area of the sample surface is expressed as a function of filler concentration at different pulse frequencies. It reveals that damage area of the sample surface shows a significant decrease trend with increasing hybrid BN filler content. This tendency is consistent in 150, 200 and 250 Hz. Meanwhile, damage area becomes larger under high pulse frequency at the same mass fraction. Therefore, the damage area of the sample surface grows with increasing pulse frequency but decreases with the increase of filler concentration. Considering all the obtained results, it can be concluded that the damage degree of the damage pattern on the sample surface is affected by both pulse frequency and filler loading.

Figure 34. The relative damage degree distributions of the surface dielectric breakdown images for PE/hybrid-BN composites at different frequencies

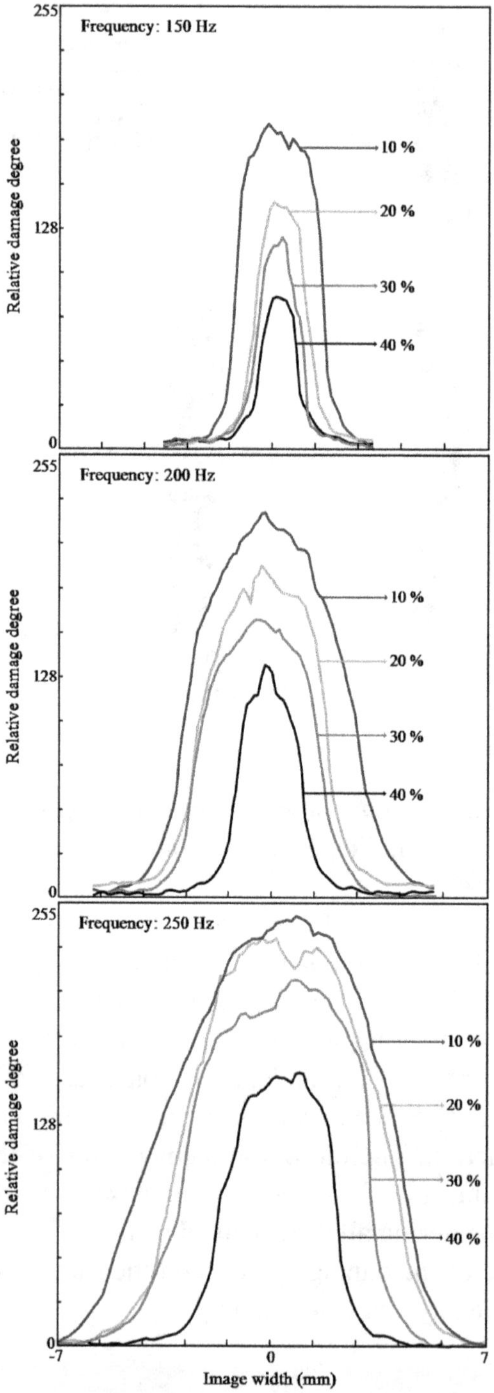

Figure 35. Relation between damage area and hybrid-BN filler concentration at different pulse frequencies

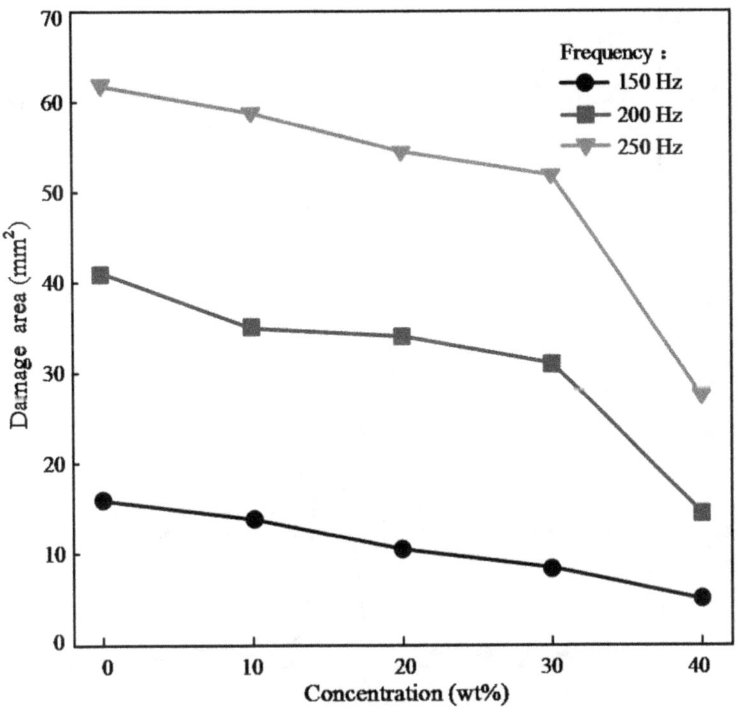

Figure 36 shows the comparison of the relative damage degree distributions between two series of the composites. At the same filler loading, the relative damage degree distribution of the sample filled with hybrid-BN is lower and narrower than that of micro-BN coated sample. As mentioned above, surface dielectric breakdown happened in composite is strongly related to the thermal accumulation during discharge.

For the sample coated hybrid-BN fillers, the heat can spread out more rapidly from discharge area to surrounding regions at the same filler loading because of its higher thermal conductivity. On the other hand, BN filler does not take part in the process of surface dielectric breakdown because of its exceptional thermal stability. These results indicate that adding hybrid-BN fillers can restrain the sample surface to be damaged much more effectively.

Based on all the discussions above, it can be concluded that adding hybrid BN fillers into the composite can reduce the relative damage degree and improve sample's arcing resistance much more effectively.

Figure 36. Comparison of the relative damage degree distributions after applying pulse at 200 Hz for 100 s between two categories of PE composites

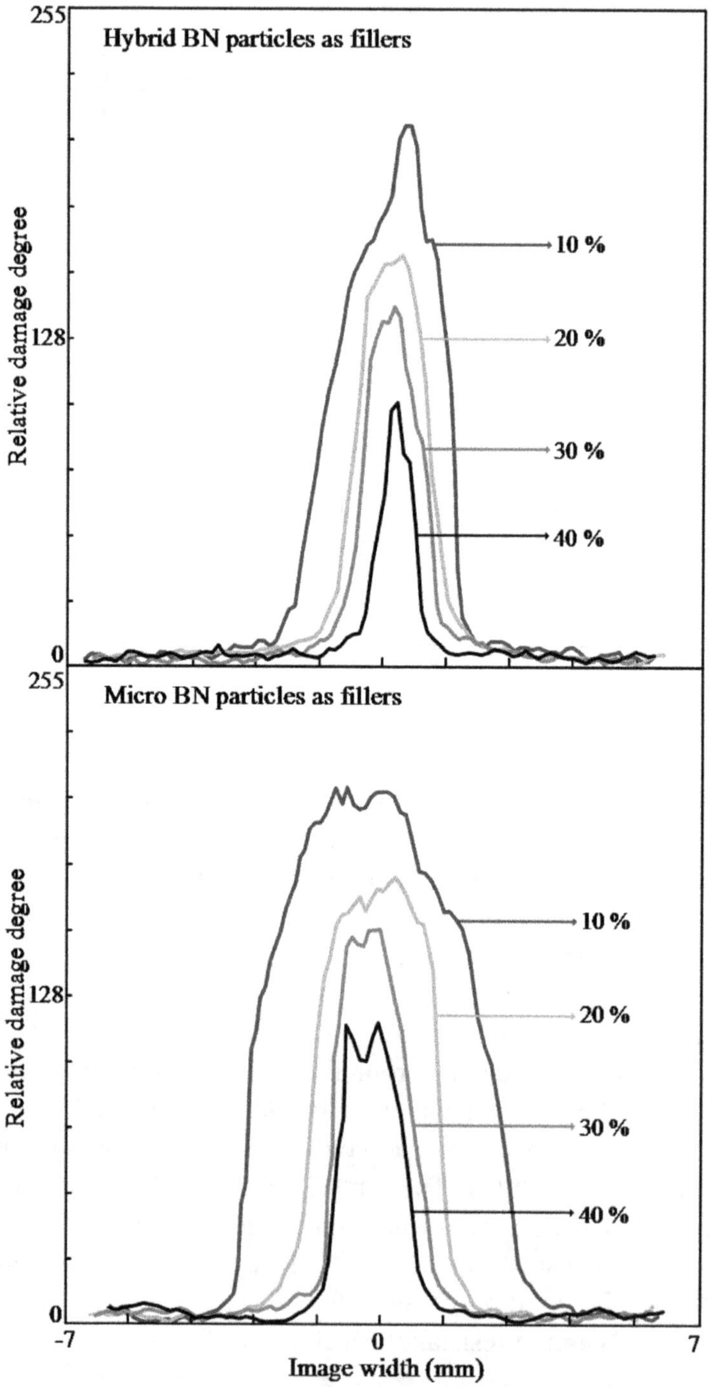

REFERENCES

Abbasi, A., Shayegani, A., & Niayesh, K. (2014). Pollution Performance of HVDC SiR Insulators at Extra Heavy Pollution Conditions. *IEEE Transactions on Dielectrics and Electrical Insulation*, *21*(2), 721–728. doi:10.1109/TDEI.2013.003990

Ansorge, S., Schmuck, F., & Papailiou, K. O. (2015). Impact of Different Fillers and Filler Treatments on The Erosion Suppression Mechanism of Silicone Rubber for Use as Outdoor Insulation Material. *IEEE Transactions on Dielectrics and Electrical Insulation*, *22*(2), 979–988. doi:10.1109/TDEI.2015.7076799

Cavallini, A., Montanari, G. C., & Morshuis, P. H. (2015). HVDC Insulation and Diagnostics. *IEEE Transactions on Dielectrics and Electrical Insulation*, *22*(1), 4–6. doi:10.1109/MEI.2015.7048130

Cheng, T. C., & Wu, C. T. (1981). *Contamination Flashover of HVDC Insulators*. EL-2016, Research Project 848-1, Final report, EPRI.

Cherney, E. A., Gorur, R. S., Krivda, A., Jayaram, S., Rowland, S., Li, S., & Ramirez, I. et al. (2015). DC Inclined-Plane Tracking and Erosion Test of Insulating Materials. *IEEE Transactions on Dielectrics and Electrical Insulation*, *22*(1), 211–217. doi:10.1109/TDEI.2014.004991

Choi, D. H., Randall, C., & Lanagan, M. (2015). Combined Electronic and Thermal Breakdown Models for Polyethylene and Polymer Laminates. *Materials Letters*, *141*, 14–19. doi:10.1016/j.matlet.2014.11.028

Fuse, N., Ohki, Y., Kozako, M., & Tanaka, T. (2008). Possible Mechanisms of Superior Resistance of Polyamide Nanocomposites to Partial Discharges and Plasma. *IEEE Transactions on Dielectrics and Electrical Insulation*, *15*(1), 161–169. doi:10.1109/T-DEI.2008.4446747

Ghunem, R. A., Jayaram, S. H., & Cherney, E. A. (2014). Investigation into the Eroding Dry-Band Arcing of Filled Silicone Rubber under DC Using Wavelet-Based Multiresolution Analysis. *IEEE Transactions on Dielectrics and Electrical Insulation*, *21*(2), 713–720. doi:10.1109/TDEI.2013.004018

Hong, J. P., Yoon, S. W., Hwang, T., Oh, J. S., Hong, S. C., Lee, Y., & Nam, J. D. (2012). High Thermal Conductivity Epoxy Composites with Bimodal Distribution of Aluminum Nitrideand Boron Nitride Fillers. *Thermochimica Acta*, *537*, 70–75. doi:10.1016/j.tca.2012.03.002

Huang, X. Y., Jiang, P. K., & Tanaka, T. A. (2011). Review of Dielectric Polymer Composites with High Thermal Conductivity. *IEEE Electr. Insul. Mag.*, *27*(4), 8–16. doi:10.1109/MEI.2011.5954064

Kemaloglu, S., Ozkoc, G., & Aytac, A. (2010). Properties of Thermally Conductive Micro and Nano Size Boron Nitride Reinforced Silicon Rubber Composites. *Thermochimica Acta*, *499*(1), 40–47. doi:10.1016/j.tca.2009.10.020

Kikuchi, T., Nishimura, S., Nagao, M., Izumi, K., Kubota, Y., & Sakata, M. (1999). Survey on the Use of Non-ceramic Composite Insulators. *IEEE Transactions on Dielectrics and Electrical Insulation*, *6*(5), 548–556. doi:10.1109/94.798113

Kochetov, R., Andritsch, T., Morshuis, P. H. F., & Smit, J. J. (2010). Effect of Filler Size on Complex Permittivity and Thermal Conductivity of Epoxy-Based Composites Filled with BN Particles. *IEEE Conf. Electr. Insul. Dielectr. Phenom.*, 1-4. doi:10.1109/CEIDP.2010.5723962

Kong, S. M., Mariatti, M., & Busfield, J. J. C. (2011). Effects of Types of Fillers and Filler Loading on the Properties of Silicone Rubber Composites. *Journal of Reinforced Plastics and Composites*, *30*(13), 1087–1096. doi:10.1177/0731684411416267

Kozako, M., Kido, R., Fuse, N., Ohki, Y., Okamoto, T., & Tanaka, T. (2004). Difference in Surface Degradation due to Partial Discharges between Polyamide Nanocomposites and Microcomposites. *IEEE Conf. Electr. Insul. Dielectr. Phenomena (CEIDP)*, 398-401. doi:10.1109/CEIDP.2004.1364271

Krupa, I., Novák, I., & Chodák, I. (2004). Electrically and Thermally Conductive Polyethylene/Graphite Composites and Their Mechanical Properties. *Synthetic Metals*, *145*(2), 245–252. doi:10.1016/j.synthmet.2004.05.007

Lee, K. Y., Kim, K. Y., Han, W., & Park, D. (2008). Thermal, Electrical Characteristics and Morphology of Poly (Ethylene-Co-Ethyl Acrylate) / CNT Nano-composites. *IEEE Transactions on Dielectrics and Electrical Insulation*, *15*(1), 205–213. doi:10.1109/T-DEI.2008.4446752

Li, T. L., & Hsu, S. L. C. (2010). Enhanced Thermal Conductivity of Polyimide Films via A Hybrid of Micro-and Nano-Sized Boron Nitride. *The Journal of Physical Chemistry B*, *114*(20), 6825–6829. doi:10.1021/jp101857w PMID:20433158

Li, Z., Okamoto, K., Ohki, Y., & Tanaka, T. (2010). Effects of Nano-filler Addition on Partial Discharge Resistance and Dielectric Breakdown Strength of Micro-Al2O3/Epoxy Composite. *IEEE Transactions on Dielectrics and Electrical Insulation*, *17*(3), 653–661. doi:10.1109/TDEI.2010.5492235

Liang, X., Wang, S., Fan, J., & Guan, Z. (1999). Development of Composite Insulators in China. *IEEE Transactions on Dielectrics and Electrical Insulation*, *6*(5), 586–594. doi:10.1109/94.798115

Lin, Z. Y., Yao, Y. G., Mcnamara, A., Moon, K. S., & Wong, C. P. (2012). Single/Few-Layer Boron Nitride-Based Nanocomposites for High Thermal Conductivity Underfills. *IEEE 62nd Electron.Compon. Technol. Conf.*, 1437-1441.

Mo, H., Huang, X., Liu, F., Yang, K., Li, S., & Jiang, P. (2015). Nanostructured Electrical Insulating Epoxy Thermosets with High Thermal Conductivity, High Thermal Stability, High Glass Transition Temperatures and Excellent Dielectric Properties. *IEEE Transactions on Dielectrics and Electrical Insulation*, *22*(2), 906–915. doi:10.1109/TDEI.2015.7076791

Nagao, M., Kimura, T., Mizuno, Y., Kosaki, M., & Ieda, M. (1990). Detection of Joule Heating Before Dielectric Breakdown in Polyethylene Films. *IEEE Transactions on Electrical Insulation*, *25*(4), 715–722. doi:10.1109/14.57095

Ortiz, C., Skorek, A. W., Lavoie, M., & Bénard, P. (2009). Parallel CFD Analysis of Conjugate Heat Transfer in A Dry-Type Transformer. *IEEE Transactions on Industry Applications*, *45*(4), 1530–1534. doi:10.1109/TIA.2009.2023561

Ren, P. G., Si, X. H., Sun, Z. F., Ren, F., Pei, L., & Hou, S.-Y. (2016). Synergistic Effect of BN and MWCNT Hybrid Fillers on Thermal Conductivity and Thermal Stability of Ultra-high-molecular-weight Polyethylene Composites with A Segregated Structure. *Journal of Polymer Research*, *23*(2), 1–11. doi:10.1007/s10965-015-0908-y

Schneider, H. M., Hall, J. F., Karady, G., & Renowden, J. (1989). Nonceramic Insulators for Transmission Lines. *IEEE Transactions on Power Delivery*, *4*(4), 2214–2221. doi:10.1109/61.35649

Sim, L. C., Ramanan, S. R., Ismail, H., Seetharamub, K. N., & Gohc, T. J. (2005). Thermal Characterization of Al_2O_3 and ZnO Reinforced Silicone Rubber as Thermal Pads for Heat Dissipation Purposes. *Thermochimica Acta*, *430*(1), 155–165. doi:10.1016/j.tca.2004.12.024

Song, W. L., Wang, P., Cao, L., Anderson, A., Meziani, M. J., Farr, A. J., & Sun, Y.-P. (2012). Polymer/Boron Nitride Nanocomposite Materials for Superior Thermal Transport Performance. *Angewandte Chemie International Edition*, *51*(26), 6498–6501. doi:10.1002/anie.201201689 PMID:22589022

Takezawa, Y., Saeki, M., Yoshida, H., & Saito, A. (2001). Thermal Deterioration Diagnosis by Optical Fiber Sensors for Mica-epoxy Insulation of HV Induction Motors. *IEEE Transactions on Dielectrics and Electrical Insulation*, *8*(1), 104–110. doi:10.1109/94.910432

Tuncer, E., James, D. R., Sauers, I., Ellis, A. R., & Pace, M. O. (2006). On Dielectric Breakdown Statistic. *Journal of Physics. D, Applied Physics*, *39*(19), 4257–4268. doi:10.1088/0022-3727/39/19/020

Wang, Z. B., Iizuka, T., Kozako, M., Ohki, Y., & Tanaka, T. (2011). Development of Epoxy/BN Composites with High Thermal Conductivity and Sufficient Dielectric Breakdown Strength Part II-Breakdown Strength. *IEEE Transactions on Dielectrics and Electrical Insulation*, *18*(6), 1973–1983. doi:10.1109/TDEI.2011.6118635

Weidenfeller, B., Höfer, M., & Schilling, F. R. (2004). Thermal Conductivity, Thermal Diffusivity, and Specific Heat Capacity of Particle Filled Polypropylene. *Compos. Part. A-Appl. S., 35*(4), 423-429.

Yu, S., Hing, P., & Hu, X. (2002). Thermal Conductivity of Polystyrene-Aluminum Nitride Composite. *Compos. Part. A-Appl. S., 33*(2), 289-292.

Zhi, C., Bando, Y., Terao, T., Tang, C., Kuwahara, H., & Golberg, D. (2009). Towards Thermoconductive, Electrically Insulating Polymeric Composites with Boron Nitride Nanotubes as Fillers. *Advanced Functional Materials*, *19*(12), 1857–1862. doi:10.1002/adfm.200801435

Zhou, W. Y., Qi, S. H., An, Q. L., Zhao, H. Z., & Liu, N. L. (2007). Thermal Conductivity of Boron Nitride Reinforced Polyethylene Composites. *Materials Research Bulletin*, *42*(10), 1863–1873. doi:10.1016/j.materresbull.2006.11.047

Zhou, W. Y., Qi, S. H., Tu, C. C., & Zhao, H. Z. (2007). Novel Heat-Conductive Composite Silicone Rubber. *Journal of Applied Polymer Science*, *104*(4), 2478–2483. doi:10.1002/app.25479

Zhu, B. L., Wang, J., Zheng, H., Ma, J., Wu, J., & Wu, R. (2015). Investigation of Thermal Conductivity and Dielectric Properties of LDPE-Matrix Composites Filled with Hybrid Filler of Hollow Glass Microspheres and Nitride Particles. *Composites. Part B, Engineering*, *69*, 496–506. doi:10.1016/j.compositesb.2014.10.035

Chapter 9
Polymer Insulation in Nuclear Power Station

ABSTRACT

Nuclear power is giving an increasing share of power resource in China since it produces much less air pollution compared with power plant consuming fossil fuel. The power system insulation in nuclear power plant takes the risk of radiation. This chapter provides some basic research on the effect of radiation on different materials. The gamma-ray irradiation inhibiting surface charge accumulation on polyethylene is shown in this chapter. The quantity of surface charge is relative with the direction and the magnitude of voltage. Another interesting study presents the effects of atmospheric pressure on tracking failure of gamma-ray irradiated polymer insulating materials. Finally this chapter puts forward a research revealing that the Magnetic could influence the field tracking failure of gamma-ray irradiated polybutylene polymers.

INTRODUCTION

Polymer insulating materials are widely used in a large variety of applications in electrical and electronic devices. Once electric field applied between electrodes reaches the breakdown strength of gas, discharge can be introduced onto polymer surface (Du et al., 2008). As sufficiently intensive discharge lasts for a considerable time, the decomposed carbon products, with some parts of the channel carbonized, are progressive and rapidly deposit on the

DOI: 10.4018/978-1-5225-2309-3.ch009

surface (Du et al., 2008). Once the carbonized products bridge the electrodes, dielectric surface breakdown that is termed tracking failure, occurs with a sudden decrease of resistance between the electrodes and a permanent loss of the insulation property (Du et al., 1995). The tracking failure has been found an important reason for polymer degradation which possibly results in severe damage to the devices (Sierota et al., 1998). Accordingly, from the viewpoint of safety, it is necessary to gain a good understanding of the behavior of tracking failure. A number of papers with respect to the behavior have been published, and the importance of such information for preventing fires, short-circuit and insulation failure has been proposed (Yoshimura et al., 1997).

Recently, polymers installed in electrical and electronic devices are required to be used in nuclear reactors and radiation facilities (Garrett et al., 2000; Chen et al., 1998; Lorenzi et al., 2009), where the materials are exposed to high energetic radioactive rays, such as gamma-ray, electrons and protons (Takada et al., 2006). For instance, outside and inside secondary shield in the containment vessel of 40-year old nuclear power plants, the maximum radiation dose rates are 0.01 Gy/ h and 1 Gy/ h. The dose rate in nuclear power plants varies widely from 10 μGy/ h to 10 kGy/ h with a potential of total exposure of 1000 kGy (Kyoto et al., 1992). The molecular structures of the polymers can be altered as a consequence of the irradiation through mechanisms like chain scission, oxidation and cross-linking (Du et al., 2009), which results in the variation of their insulation properties. It was reported that when polyethylene (PE) was exposed to 4 MeV electron beam for a total dose up to 2 MGy, its bulk breakdown strength was approximately twice comparing with PE irradiated under the same condition for a total dose of 1.5 MGy. In either case, the breakdown strength was higher than that of un-irradiated PE (Stark et al., 1955). On the contrary, the breakdown strength of polystyrene (PS) was observed decreasing from 63 kV/ mm to 51 kV/ mm as it subjected to radiation with total dose of 36 MGy (Goetzel et al., 1965). It is proposed that the dielectric strength is improved by irradiation for cross-linking type materials, but the conclusion for degradation type materials is opposite (Laghari et al., 1990).

Although radiation effect on bulk breakdown of polymers has been concerned, previous studies fell short of taking proper account of such effect on tracking failure. Du et al. investigated the resistance to tracking of gamma-irradiated polyethylene and modified polycarbonate under dc voltage, the results revealed that the tracking behavior was dependent upon the radiation dose and was varied as a function of polymer types (Du, 2001). Yoshimura *et al.* has proposed that, for the safe use of polymers in radioactive environment, further studies should be carried out so as to clarify the radiation effect on the tracking failure (Yoshimura et al., 1997). It must be brought in mind that, rather than solely under electric field, polymers in practical operation are always subjected to combined electric and magnetic fields. The magnetic field can be introduced by either the application of permanent magnets (Nakamura et al., 2003) or the excitation of current flow in electric circuit (Larbalestier, 2001). The flux density of magnetic field around power equipments varies with an extensively large range from the order of micro Tesla to tens of Tesla (Nakamura et al., 2003; Larbalestier et al., 2001). It has been observed that a magnetic field with flux density of 180 mT can significantly change surface discharge characteristics on polymer (Du et al., 2005), which is likely to alter the behavior of tracking failure hence influences the reliability of the material. Therefore, designing electrical insulation using polymers, one should take into consideration the effect of magnetic field on tracking behavior as well.

Polyethylene (PE) has been used in an extensively large variety of applications in electrical and electronic engineering. The dielectric property of PE is changed in response to gamma-ray irradiation (Laghari et al., 1990), which gives rise to the alternation of charge trapping characteristics. It's revealed that traps in gamma-ray irradiated semi-crystalline low density polyethylene (LDPE) are based on the defects induced in the process of irradiation. The formation of these defects can be in the amorphous phase or on the surface of the crystallites (Kostoski et al., 2006). Chen has pointed out that the trapping characteristic of LDPE irradiated with low dose levels (10 kGy) is remarkably different from that irradiated with high dose levels (from 50 kGy to 500 kGy), which is ascribed to the occurrence of carbonyl group (C=O) generated by radiation-induced oxidation reactions (Chen et al., 1998). Das-Gupta has reported that surface charge density on LDPE irradiated by gamma-ray with 60 kGy is reduced by a factor of 2 compared with that of the un-irradiated LDPE. It is suggested that if the materials subject to appropriate dosage of gamma-ray irradiation, their capacities of surface charge are weakened (Das-Gupta et al., 1995). PBN, PET and PBT are widely used in electrical insulation because of their high breakdown strength, high resistivity and low dielectric loss (Du et al., 2004). PBN is widely used in electric engineering, chemical, metallurgy and medical equipment. PBT is usually used for electric coupler, blind plug, switch and insulating cover. PET is one of the most widely used polymers for electrical and electronic industries. For example, the PET film is widely used as an insulating material in the electrical industry (used for making cables, capacitors, etc). Whether the dielectric property of the polymers is changed by irradiation and atmospheric pressure or not are worth investigating. In this paper, the effects of total dose of gamma-ray irradiation and atmospheric pressure on tracking failure of PBN, PET and PBT were studied. The results revealed that the tracking failure properties are greatly changed by gamma-ray irradiation and reduced atmospheric pressure. The present work is carried out with the purpose of understanding the effect of gamma-ray irradiation as well as magnetic field on tracking failure. Obtained results show that the behavior of tracking is dependent upon the radiation dose and is varied as a function of $\vec{E}\times\vec{B}$.

GAMMA-RAY IRRADIATION INHIBITING SURFACE CHARGE ACCUMULATION ON POLYETHYLENE

Analysis Method

The methods of getting surface charge density from probe response have been highly developed during the past 30 years (Ootera et al., 1988; Takuma et al., 1998; Faircloth et al., 2003; Kumada et al., 2004), among which the linear calibration method is involved (Jaitly et al., 1988). In the method, the relation between the induced voltage and the surface charge density beneath the probe was considered approximately linear, which could be expressed as $\sigma=MV_0$, where V_0 was the output of the follower, M was a calibration constant and σ was the surface charge density. The calibration constant can be given by $M=C_e/S$, where C_e is the equivalent capacitance of the induction rod against earth, S is the area of cross-section of the rod (Jaitly et al., 1988). For the present probe, C_e is 56 pF hence M is 0.018 pC/mV•mm^2. Surface charge density for the case of irradiation with 0 kGy, 100 kGy and 1000 kGy under ±2 kV, ±3 kV, ±4 kV was obtained.

Effect of Gamma-Ray Irradiation on Surface Charge Accumulation

Surface charge accumulation has been found strongly depending upon the shape of electric field. As for the field employed in the present work, corona discharge occurring around the tip of needle electrode can work as a source of charges. There are very few electrons in the corona at any potential that can get to the polymers as their range in air is much smaller than the air gap from the tip to the sample surface (Das-Gupta, 1990). It has been pointed out that positive corona can provide hydrated versions of H^+, NO^+ and NO_2^+ ions while negative discharge generates CO_3^- ions. The ions are then encouraged by the electric force to move onto the sample surface (Das-Gupta, 1990). With their arrival, they become trapped by localized surface states. Electron transfer subsequently takes place and these charges become electronic in nature (Toomer et al., 1980). The charges may transport laterally along the surface and through the bulk to the back electrode under the field generated by the applied voltage and by the charge itself (Das-Gupta, 1990), which gives rise to the surface charge distribution.

Figure 1(a) and 1(b) show typical profiles of surface charge distribution on un-irradiated PE charged at 3 kV and – 3 kV, respectively. The two profiles show similar distribution of the charges which is characterized by a sharp peak locating in the middle of surface that tends to decrease towards the edge. This is attributed to the fact that the electric field beneath the tip is intensive, a large amount of charges are deposited there and the max charge density is formed. The charges are deposited in a shallow layer of the sample surface since they're introduced by corona discharge for a short time (Fang et al., 1996). Thus, characteristics of surface charge accumulation strongly depend upon the surface property, in which the behavior of localized surface states plays an important role.

Figure 2 and 3 show the charge distribution varying as a function of the applied voltage for samples irradiated with 100 kGy. With the increase of the applied voltage, the max charge density for both positively and negatively charged samples increases. By comparing Figure 2 and 3, it is found that the negative charges are deposited with a higher peak than that of the positive charges. Figure 4 shows the dependence of total surface charge upon the applied voltage as a function of polarity. With the increase of absolute value of the applied voltage, the total surface charge increases. The number of the negative charges is larger than that of the positive charges. Based on the results obtained from Figure 2 to 4, it is considered that the mechanism of surface charging is dependent upon the polarity of charging particles.

It is known that the corona discharge is encouraged by the growth of applied voltage, which provides larger number of ions that can be trapped. In addition, the enhancement of the applied voltage also results in the increase of intensity of electric field beneath the needle tip. Consequently, both the max

Figure 1. Surface charge distribution on un-irradiated sample

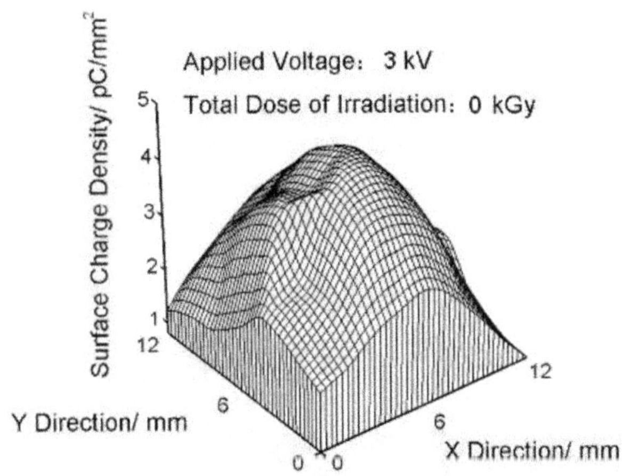

(a) Positively charged

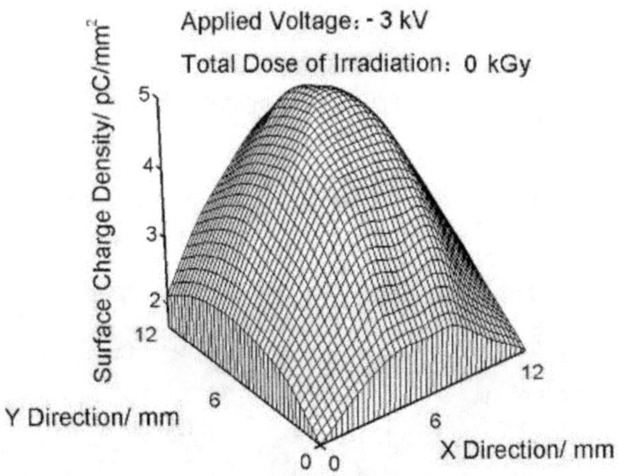

charge density and the total surface charge, as shown in Figure 2, 3 and 4, appear to increase. Because negative corona can be induced more easily than positive corona with identical applied voltage, it is assumed that the number of negative charges generated during the period of corona discharge is larger than that of positive charges. Thus, more negative charges are likely to be accumulated on the sample surface. Since positive and negative charges are captured by donor-like hole traps and accepter-like electron traps, respectively. It may be suggested by Figure 4 that the number of hole trap is smaller than that of electron trap at the surface layer, which is in consistent with the case in the bulk (Kao, 2000). Accordingly, it is likely to understand why more

Figure 2. Positively charged sample surface with the irradiation dose of 100 kGy by applying different bias voltage

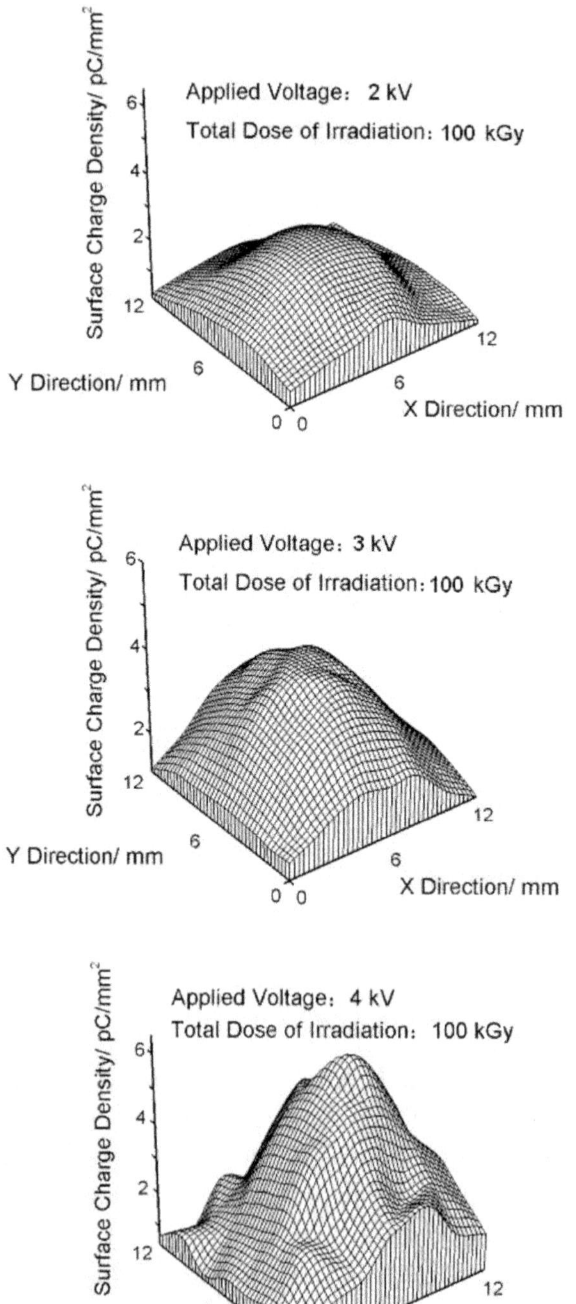

Figure 3. Negatively charged sample surface with the irradiation dose of 100 kGy by applying different bias voltage

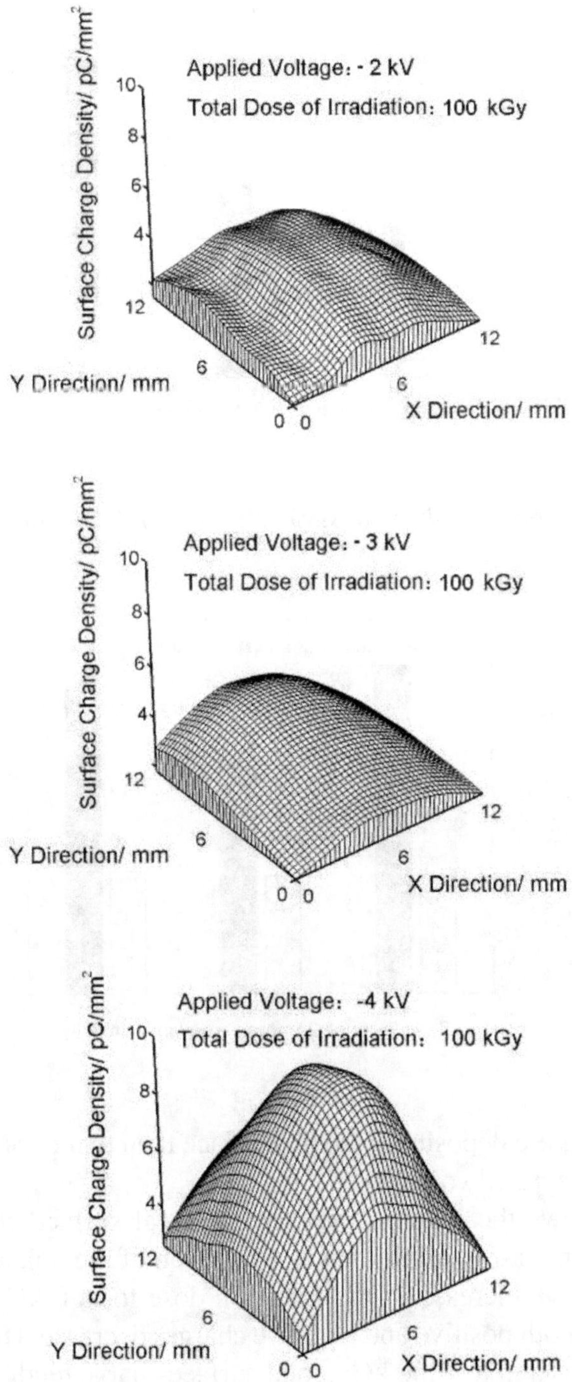

Figure 4. Relation between the total surface charge and the absolute value of applied voltage

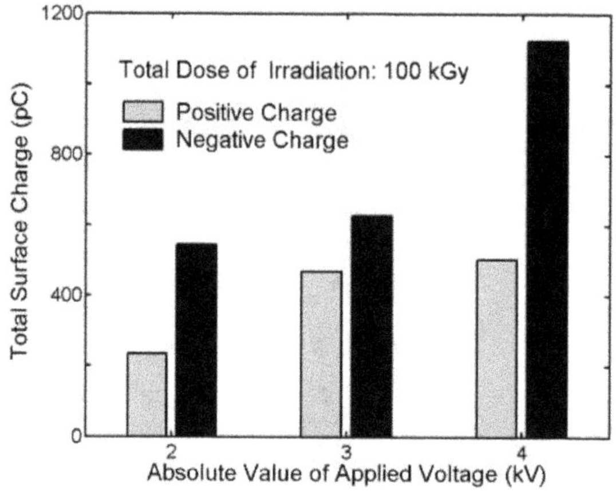

Figure 5. Relation between the total surface charge and the total dose of gamma-ray irradiation

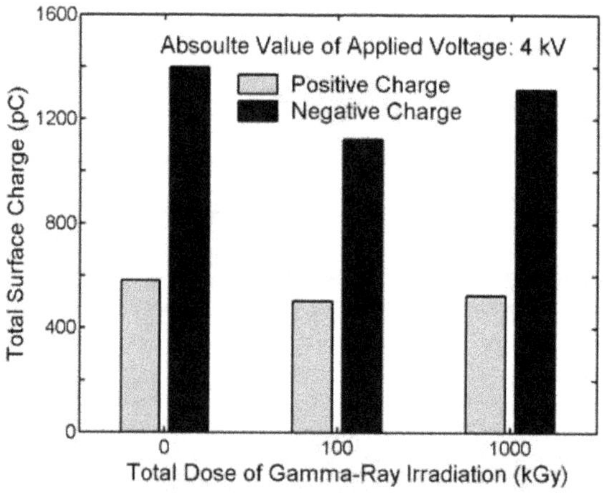

negative charges are deposited onto the surface than that of positive charges under identical applied voltage.

Figure 5 shows the relation between the total surface charge and the total dose of gamma-ray irradiation as a function of the polarity of charged particles. With the increase of the radiation dose form 0 kGy to 100 kGy, the numbers of both positive and negative charges decrease. However, as the radiation dose rise up to 1000 kGy, total surface charge tends to increase. It reveals that the capacity of surface charge is initially weakened at lower dose of irradiation, but becomes increased as the radiation dose grows. Moreover, obtained results show that the change of number of negative charges is more remarkable in comparison to that of positive charges, indicating that gamma-ray irradiation affects the surface charge accumulation varying as a function of the polarity of charging particles.

Figure 6 shows surface charge distribution on the samples irradiated with 1000 kGy charged at ±4 kV. For the positively charged, as shown in Figure 6(a), charges are accumulated with a peak of gentle slope at the surface center. For the negatively charged, as shown in Figure 6(b), the charge distribution is characterized by a lower charge peak locating near the center that is surrounded with several sub-peaks, which indicates that the charges are more mobile than those on un-irradiated samples.

Figure 7 shows the relation between the total surface charge and the total dose of gamma-ray irradiation as a function of the polarity of charged particles. With the increase of the radiation dose form 0 kGy to 100 kGy, the numbers of both positive and negative charges decrease. However, as the radiation dose rise up to 1000 kGy, total surface charge tends to increase. It reveals that the capacity of surface charge is initially weakened at lower dose of irradiation, but becomes increased as the radiation dose grows. Moreover, obtained results show that the change of number of negative charges is more remarkable in comparison to that of positive charges, indicating that gamma-ray irradiation affects the surface charge accumulation varying as a function of the polarity of charging particles.

The alternation of chemical structures induced by gamma-ray irradiation has received a lot of attentions during the past decades (Chen et al., 1998; Kao et al., 2000; Chen et al., 1999; Das-Gupta, 1992). It has been well established that the atoms that make up the polymers are bounded together by weak covalent bonds which can be disrupted easily as the energy of gamma-ray radiation is absorbed and polymer radicals are formed (Du et al., 2007). Generally, considering PE, such rupture can take place in the main chain or in the side branch. Subsequently, alternation of structure of polymers occurs by chemical radical reactions through mechanisms like chain scission, oxidation and cross-linking (Chen et al., 1999). The productions of the reaction are considered to be remarkably dependent upon the atmosphere in which the radiation is carried out. With the presence of oxygen, one oxidation production that is observed readily emerging within LDPE is carbonyl group (C=O) (Chen et al., 1998). The concentration of carbonyl groups increases as a

Figure 6. Surface charge characteristics for PE irradiated with 1000 kGy

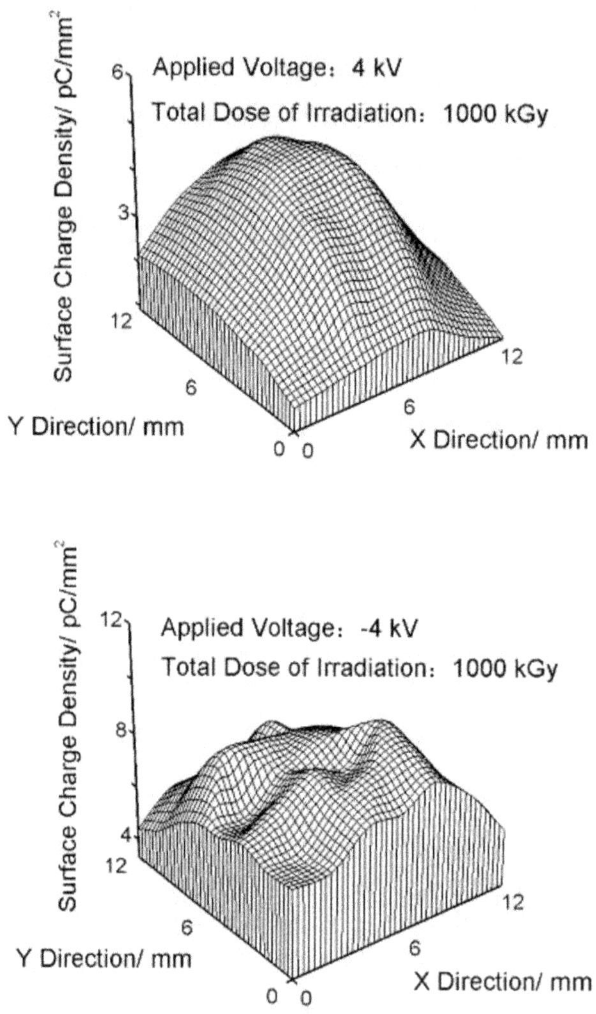

consequence of the growth of radiation dose, which introduces abundant shallow traps at the surface layer that are available to assist charge migration. Accordingly, chain scission and subsequently oxidation, namely degradation reaction of molecules, tend to extend the density of localized surface states (Chen et al., 1998). On the other hand, as the polymer radicals are formed, pairs of adjacent radicals can cross-link readily leading to the decrease of the active point at surface layer, as a result of which the density of localized surface states appears to decrease (Das-Gupta, 1992).

Figure 7. Relation between the total surface charge and the total dose of gamma-ray irradiation

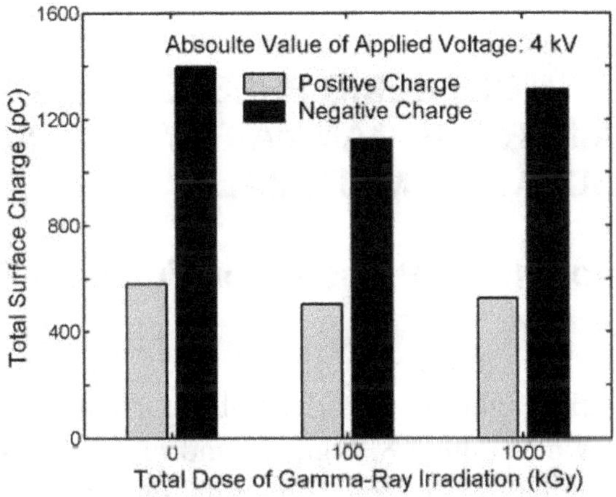

In practice, degradation and cross-linking reactions occur simultaneously, the reaction result is determined by the one dominant. With the increase of the radiation dose from 0 kGy to 100 kGy, the total surface charge reduces, which reveals that the reaction result is dominated by cross-linking. This is in well agreement with the fact presented by Das-Gupta (Das-Gupta et al., 1995). It is assumed that the characteristics of surface charge accumulation is strongly dependent upon the reaction result of cross-linking at relatively low dose levels of irradiation. As the radiation dose rise up to 1000 kGy, the total surface charge tends to increase, although the number of charge is smaller than that of the un-irradiated sample. It indicates that the degradation reaction appears to be strengthened with the radiation dose and the reaction result is still dominated by cross-linking. This is similar to the point presented by Laghari that the reaction result is dominated by cross-linking when the radiation dose is lower than ~ 800 kGy but tends to be degradation dominated when the radiation dose grows up (Laghari et al., 1990). The presence of sufficient oxygen in the radiation environment that sustains the oxidation reactions finally gives rise to the enhancement of the number of traps with shallow energy level. The obvious transportation of the charges may be attributed to the formation of such traps. Furthermore, electron appears to transport along the surface with remarkably higher mobility than that of hole, which suggests that the alteration of characteristics of electron-trap is greater than that of hole-trap. As a result, the distribution of negative charge on sample irradiated with 1000 kGy differs from the un-irradiated one more obviously than that of the positive charge.

EFFECTS OF ATMOSPHERIC PRESSURE ON TRACKING FAILURE OF GAMMA-RAY IRRADIATED POLYMER INSULATING MATERIALS

Effects of Atmospheric Pressure on Tracking Failure

The properties of PBT and PBN are shown here. The relation between the time to tracking failure and the total dose of the irradiation under 100 kPa and 1 kPa was shown in Figure 8, Figure 9 and Figure 10. With increasing the total dose of irradiation, the time to tracking failure increased with PBN and PET, but decreased with PBT for the investigated range of the irradiation. With the decrease of the atmospheric pressure, the time to tracking failure of all samples increased. With decreasing the pulse interval, the time to tracking failure of all samples decreased.

With the increase of the total dose, the time to tracking failure of PBN and PET increase, which indicates the thermal properties of tracking resistance are improved. However, the time to tracking failure of PBT decreases with the increase of the total dose, which indicates that the thermal properties of tracking resistance are worsened. Evidently, the total dose has a different effect on the polymers. A cross-linking reaction of an organic material is one main factor for improving tracking resistance and conversely a degradation reaction is conceivable as a factor for decreasing tracking resistance. The atoms that make up a polymer are bounded together by weak covalent bonds that are disrupted easily by gamma-ray radiation, and as bonds are broken, new ones are formed and the structure of the polymer is altered (Banford et al., 1999). In practice, cross-linking and degradation reaction often occur simultaneously, and the reaction result is determined by the one that is dominant (Kuriyama et al., 1978). For PBN and PET, it is supposed that the cross-linking reaction is superior to the degradation reaction (Blodgett et al., 1969) and the combination between the molecules extends the three-dimensional networks. For PBT, it is supposed that the degradation reaction is superior to the cross-linking reaction and the scission of the main chain bonds results in the formation of low-molecular-weight chain fragments.

Figure 8. Relation between the time to tracking failure and the total dose of irradiation for PBN

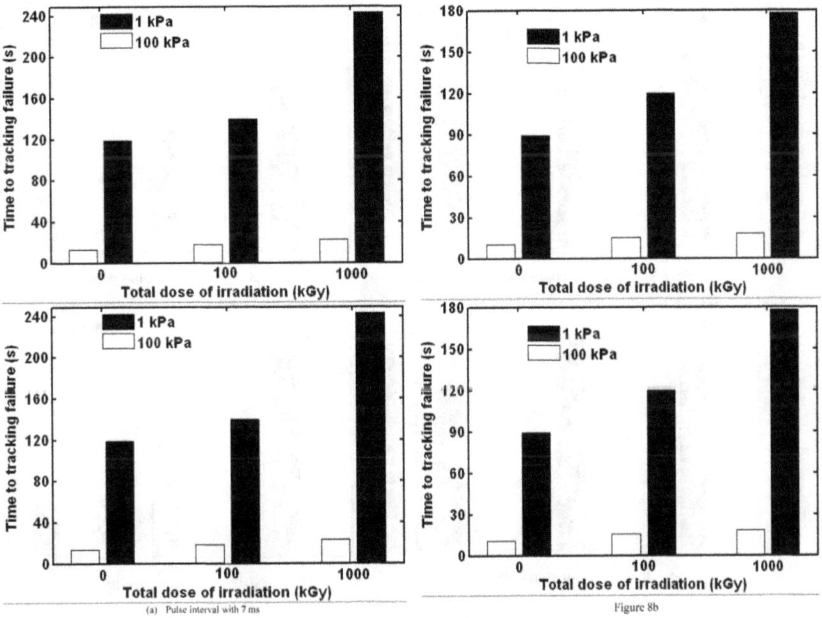

(a) Pulse interval with 7 ms

Figure 8b

Figure 9. Relation between the time to tracking failure and the total dose of irradiation for PET

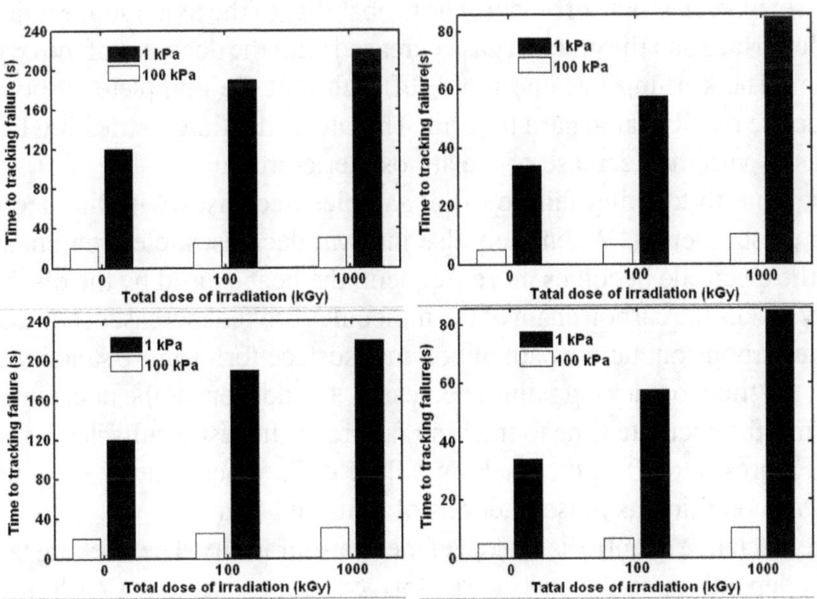

Figure 9a

Figure 9b

Figure 10. Relation between the time to tracking failure and the total dose of irradiation for PBT

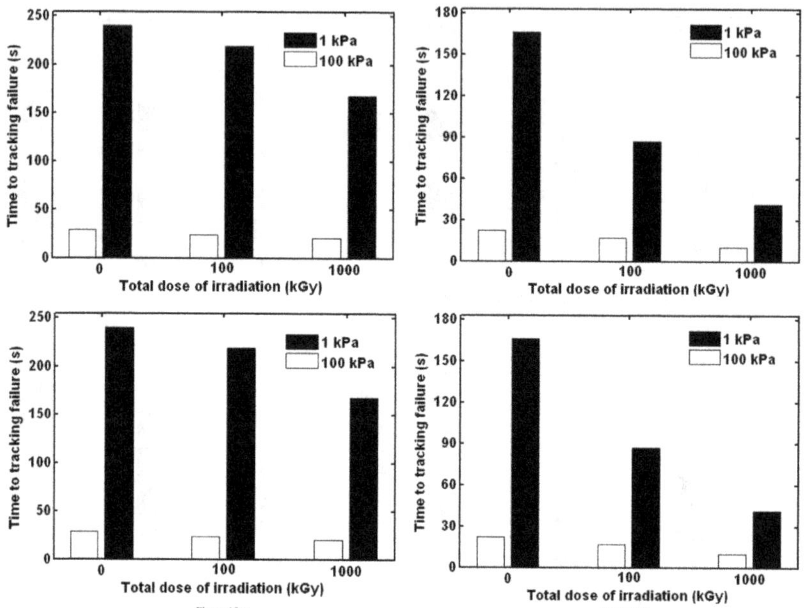

Figure 10 a Figure 10b

The time to tracking failure of all samples increased with the decrease of the atmospheric pressure. Under the reduced atmospheric pressure, the supply of oxygen is not sufficient. The probability of the oxidation reaction on sample surface and the cracked gas decreased with the decrease of the oxygen content. The burning became more difficult and the complete carbonized conductive path became hard to form. Therefore, the time to tracking failure increased with the decrease of the atmospheric pressure.

The time to tracking failure of all samples decreases with the decrease of the pulse interval. When the pulse interval decreases, electron emission from the electrode becomes more frequent, the heat caused by the discharge energy makes the carbon chain of the molecules broken increases (Du, 2001), and the carbon conductive path of polymer surface forms more steadily. As a result, the time to tracking failure decreases. In addition, at the pulse interval of 10 ms, the accurate time to tracking failure of the test results is not shown from Figures 8 to 10 as the tracking failure did not occur until 300 s under the same condition as pulse intervals of 7 ms and 5 ms.

The discharge quantity is a token of the heat-durability. The tracking failure mainly depends upon the heat energy formed by the discharge (Yoshimura et al., 1997). If discharge quantity is smaller, it suggests that the carbon chain is more difficult to be broken. The relationship between the discharge quantity and the total dose of irradiation before tracking failure for 600 discharges under 100 kPa and 1 kPa was shown in Figure 11, Figure 12 and Figure 13. For the investigated range of the irradiation, the discharge quantity decreased with PBN and PET, but increased with PBT with increasing the total dose of irradiation. From Figure 11 to 13, with the decrease of the atmospheric pressure, the discharge quantity of PBN and PET increased, but decreased with PBT.

The discharge quantity of PBN and PET decrease with the increase of the total dose, which suggests that the thermal properties of tracking resistance are improved and the tracking failure becomes more difficult. This fact is due to the occurrence of the cross-linking reaction (Du et al., 2008). With the increase of the total dose of irradiation, the discharge quantity increase with PBT, which suggests that the thermal properties are worsened and the tracking failure becomes easier. The reason for this is due to the degradation reaction by irradiation (Du et al., 2003).

The discharge quantity of PBN and PET increased with decreasing the atmospheric pressure. Under the reduced atmospheric pressure, the density of the gases was decreased, the speed of electron became fast, and the surface more readily caused the discharging. As a result, the discharge quantity increased with the decrease of the atmospheric pressure (Banford et al., 1999). The discharge quantity decreased with decreasing the atmospheric pressure for PBT. It is because under the reduced atmospheric pressure, the probability of the oxidation reaction of PBT surface and the cracked gas decreased with the decrease in the oxygen content. The carbon deposition of PBT by the ignition to the surface increased, and the discharges could not completely take place due to the decomposed carbon, so the total discharge quantity of PBT under the reduced atmospheric pressure was smaller.

From Figure 11 to 13, discharge quantity of PBN and PET increased with the decreasing the pulse interval, but the tendency is opposite for PBT. The discharge quantity is bigger in shorter pulse interval because the higher accumulative heat quantity of consecutive discharge is, the more quickly tracking failure occurred (Plessow et al., 1994). For PBN and PET, tracking failure is more difficult due to the inherent carbonization property and the discharge is continuous. The formation of carbon conductive path is gradual, and the discharge quantity is bigger at the shorter pulse interval. For PBT, the formation of carbonization and tracking failure are easier. It is observed that

Figure 11. Relation between the discharge quantity and the total dose of irradiation for PBN

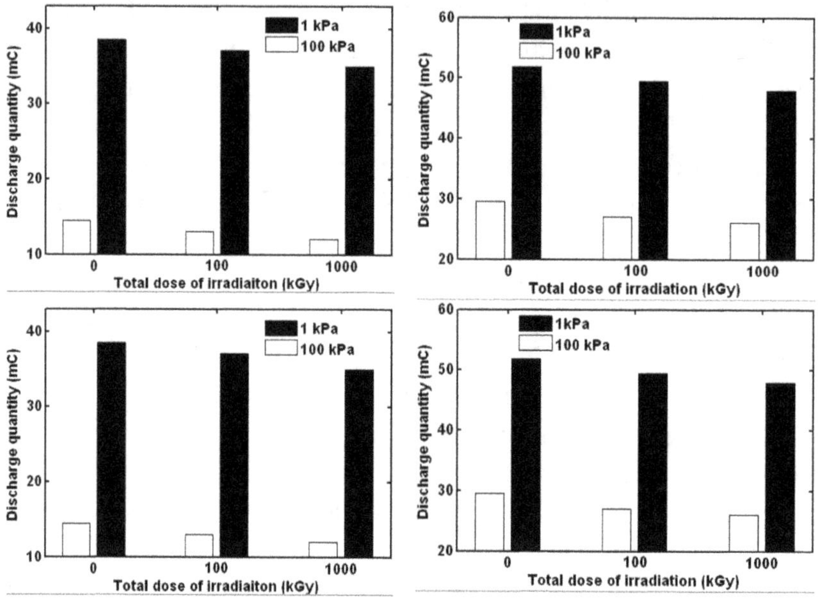

Figure 11a

Figure 11b

Figure 12. Relation between the discharge quantity and the total dose of irradiation for PET

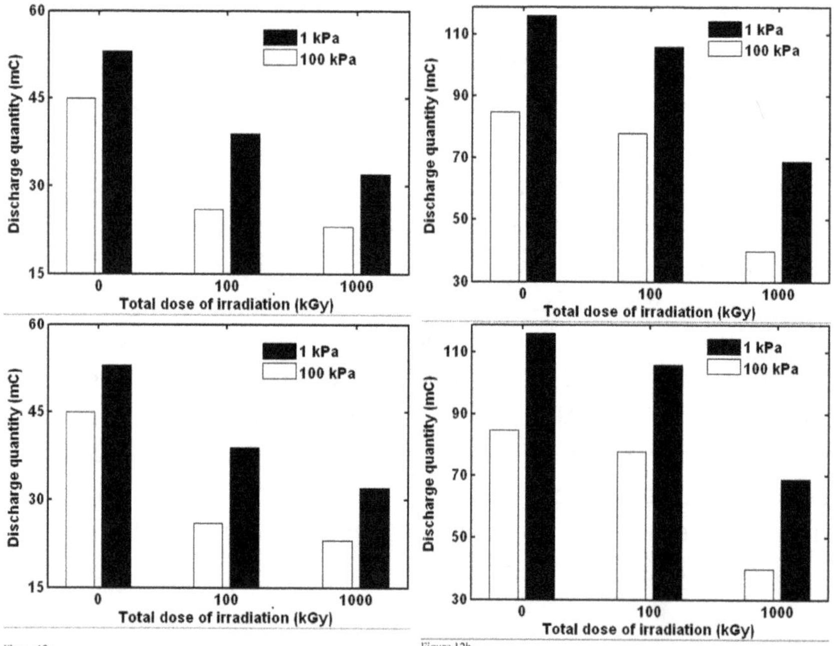

Figure 12a

Figure 12b

Figure 13. Relation between the discharge quantity and the total dose of irradiation for PBT

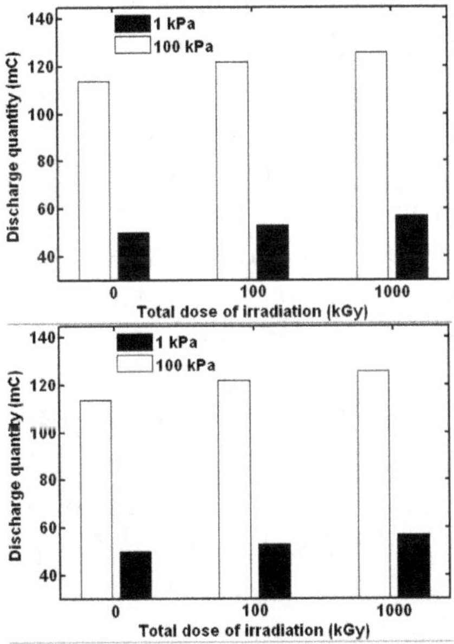

Figure 13a

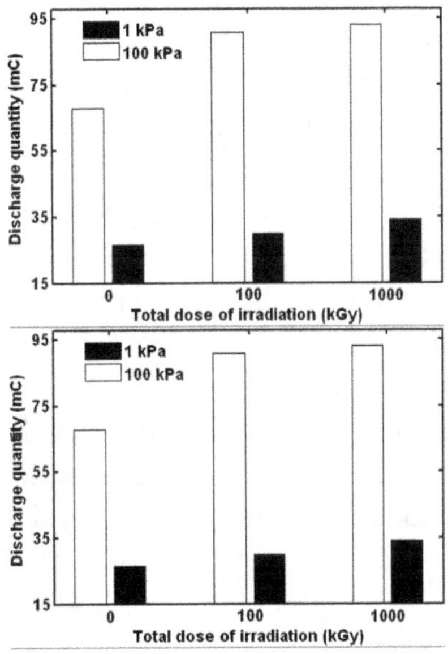

Figure 13b

short-circuit forms in a very short time, and the discharge is discontinuous. At the shorter pulse interval, the carbonization forms more quickly and the brief short-circuit occurs more frequently, which causes the discharges cannot completely take place, so the total discharge quantity is smaller.

The photographs of sample surface after the discharge for 20 s with the pulse interval of 10 ms under 100 kPa and 1 kPa are shown in Tables 1, 2 and 3. The color of sample surfaces was gradually dark according to the total dose of gamma-ray irradiation. The noticeable changes of sample surface in the tracking failure process were observed. The repetitive discharge occurred before tracking failure and the quantity of decomposition carbon increased in the area close to the electrodes. As shown in the tables, the surface was carbonized by the heat energy from the discharge. With the high repetition rate of discharges, there was a high increase of discharge quantity that affected the carbonization progress and the situation of carbon deposition. When the intensive discharge lasted for a period of time, the surface was carbonized first to the area close to the electrodes. Due to the conductive nature of carbonization products, they acted as the extension of the electrodes. After a sufficiently long time, the carbonized conduction path was finally formed over the short-circuited path.

The features of tracking failure phenomenon are different between PBN, PET and PBT. The carbonized area decreased with PBN and PET, but increased with PBT with increasing the total dose of irradiation. The difference indicates that the thermal properties of tracking resistance are improved with PBN and PET, but worsened with PBT. By comparison with the case of 1 kPa, the features of tracking failure phenomenon are different in the case of 100 kPa. The oxidation reaction, which took place on the electrodes, was an exothermic reaction initiated by scintillation discharge and increased the intensity of the discharge. In the case of 100 kPa, there was enough aerial oxygen for the samples to be oxidized completely, which shows more carbonization area in the second columns of Table 1, Table 2 and Table 3. In the case of 1 kPa, there was not enough aerial oxygen for the samples to be oxidized completely and the oxidation product was largely decreased, which shows less carbonization points in the third columns of Table 1, 2, and 3. The differences of shape and

Table 1. Sample surface after the carbonization with PBN

Total dose	100 kPa	1 kPa
0 kGy	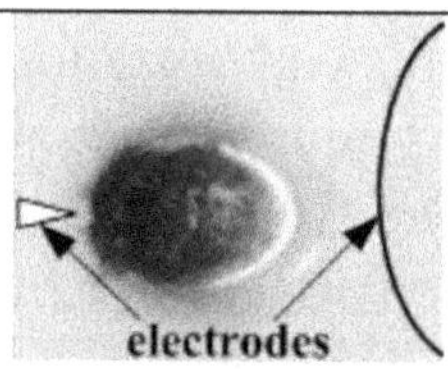	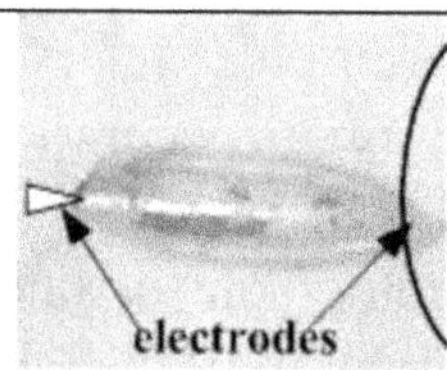
100 kGy	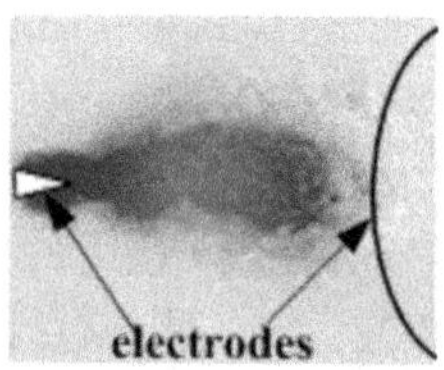	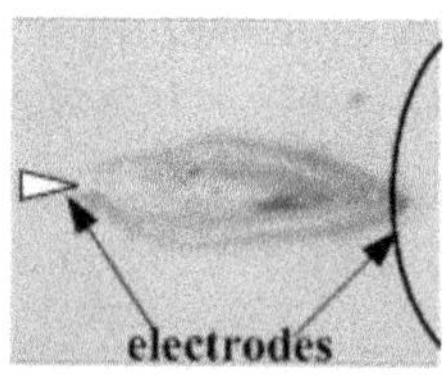
1000 kGy	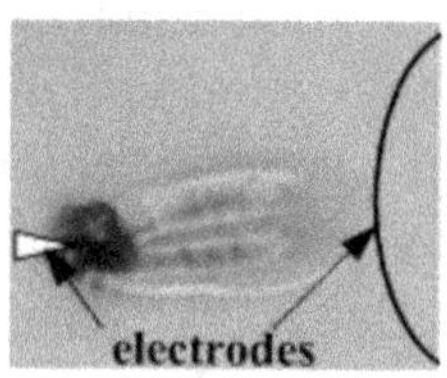	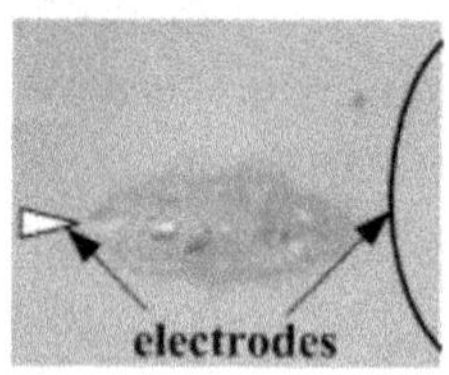

Table 2. Sample surface after the carbonization with PET

Total dose	100 kPa	1 kPa
0 kGy	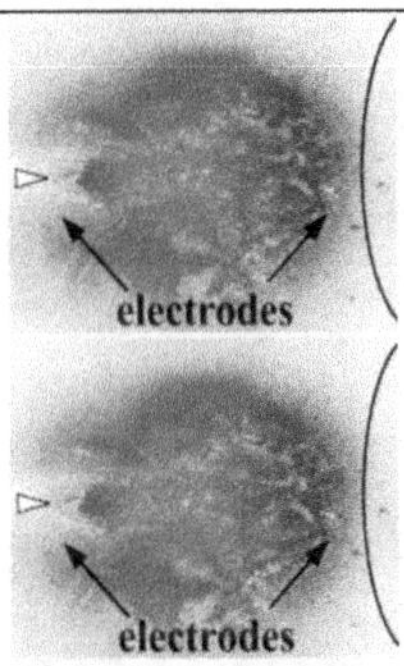	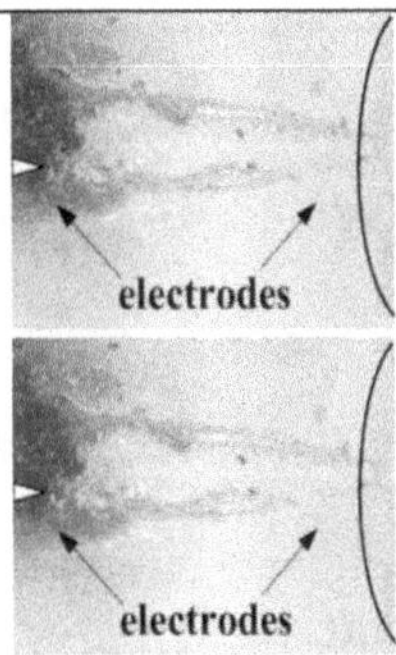
100 kGy	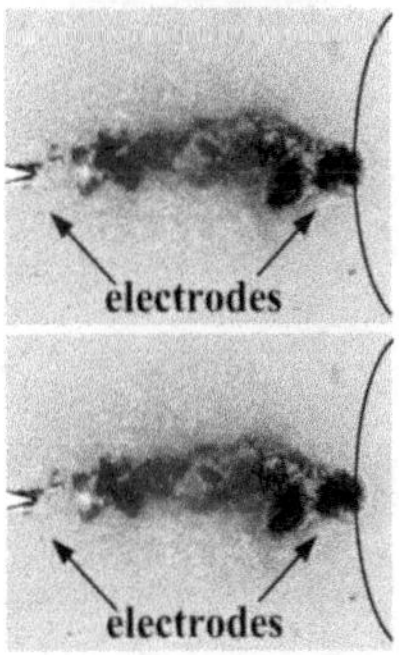	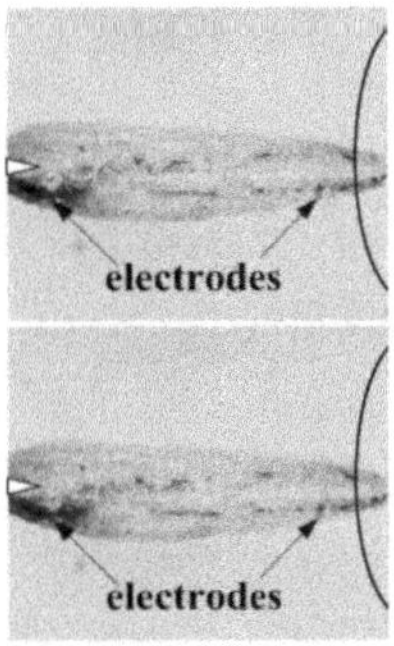
1000 kGy	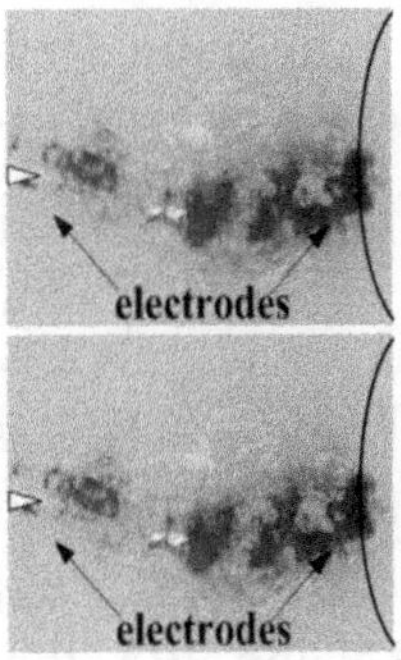	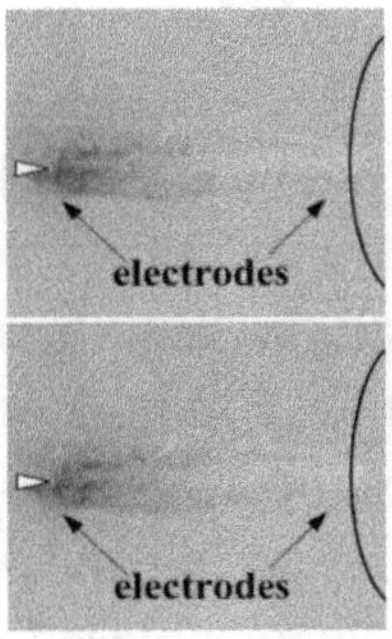

area of carbonized resultants are independent of the carbonization process with reducing the atmospheric pressure.

The tracking failure properties are improved for PBN and PET but worsened for PBT by gamma-ray irradiation. This difference is attributed to

Table 3. Sample surface after the carbonization with PBT

Total dose	100 kPa	1 kPa
0 kGy	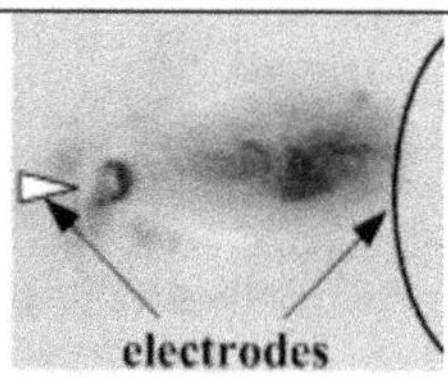	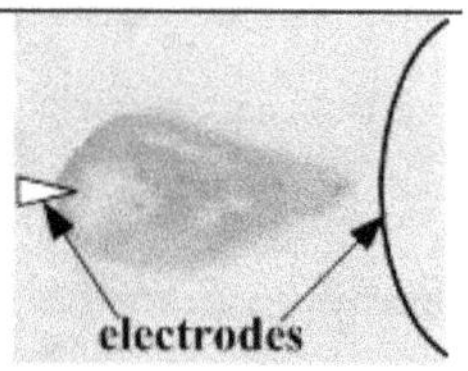
100 kGy	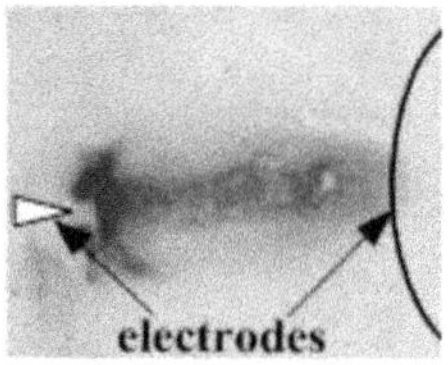	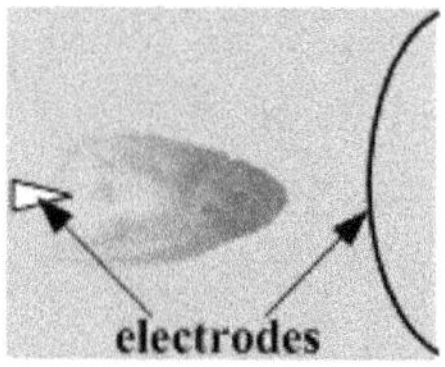
1000 kGy	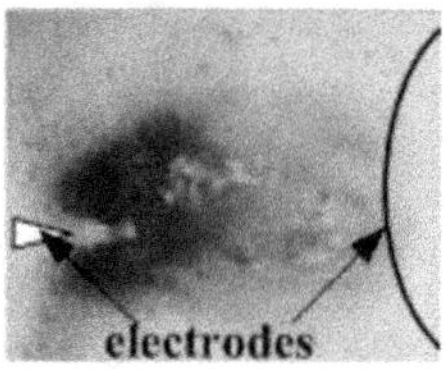	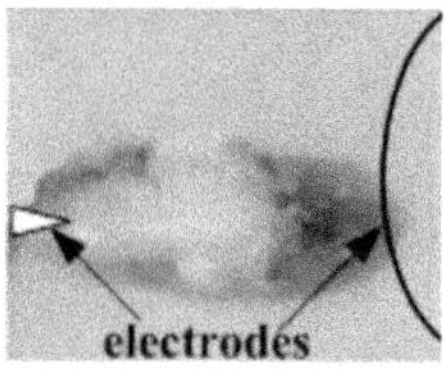

the radiation-induced cross-linking and degradation. The effect of the total dose on electrical properties is markedly different, depending on the chemical structure of the base polymer (Kuriyama et al., 1979). The radiation adds a particular dimension to the aging problem, because it interacts strongly with materials in general and brings about structural changes that alter their properties (Blodgett et al., 1969). This is because that it can alter the macroscopic properties of polymeric materials through mechanisms like chain scission, cross-linking and oxidation (Chen et al., 1999). By the comparison of the molecule formulas of PBN and PBT, there are two phenyls in the main chain of PBN but one for PBT. The amount of phenyl in the main chain plays a main role in the result of the radiation reaction, which is consistent with the result of Hanai (Hanai, 1982). In another word, the resistance to the tracking failure of polybutylene polymers was improved for PBN which contains more combined phenyls in the main chain (Banford et al., 1999). Comparing the molecular structures of PET and PBT, there are four methylene groups in the chain of PBT but two for PET. The four methylenes increase the length of the thinner, less bulky, portion of the molecular chain, resulting in easier bending (Du et al., 2008).

TRACKING FAILURE OF GAMMA-RAY IRRADIATED POLYBUTYLENE POLYMERS UNDER MAGNETIC FIELD

The Analysis Methods of Magnetic Field and Discharge Quantity

A permanent magnetic field with flux density of 250 mT was placed perpendicular to the electric field, the relative angle between $\vec{E}\times\vec{B}$ and surface normal vector $\vec{n}$ was set at 0°, 90° and 180°, which represented that the direction of $\vec{E}\times\vec{B}$ was away from, parallel to and into the surface, respectively, as shown in Figure 14. The sample was stressed electrically by applying the pulse voltage with amplitude of - 30 kV between the electrodes. The pulse width was 1 ms, and the time interval between each two pulses was 5 ms. The time from discharge inception to the formation of carbonized conductive path that bridged the electrodes was measured and defined as the time to tracking failure. Discharge quantity was calculated by the following equation,

$$Q = \int_{0}^{\tau} \left|i(t)\right| dt \qquad (1)$$

where $\left|i(t)\right|$ was the absolute value of discharge current and τ was the integral interval that could be chosen based on the demand of calculation. We calculated the discharge quantity by using 400 times of discharge pulses which corresponded to τ of 2 s. The typical surface appearance after tracking failure was captured by using a digital camera and was presented in the following section.

Effect of Gamma-Ray Irradiation

It was observed that surface discharge occurred immediately as the pulse voltage was applied between the electrodes. Subsequently, surface area close to

Figure 14. Schematic diagram of the relative angle

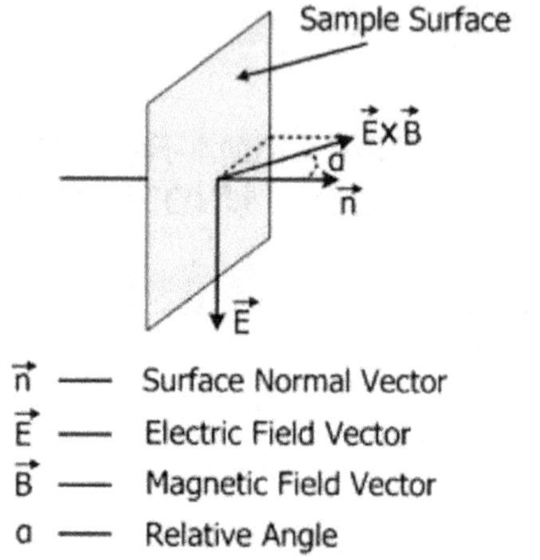

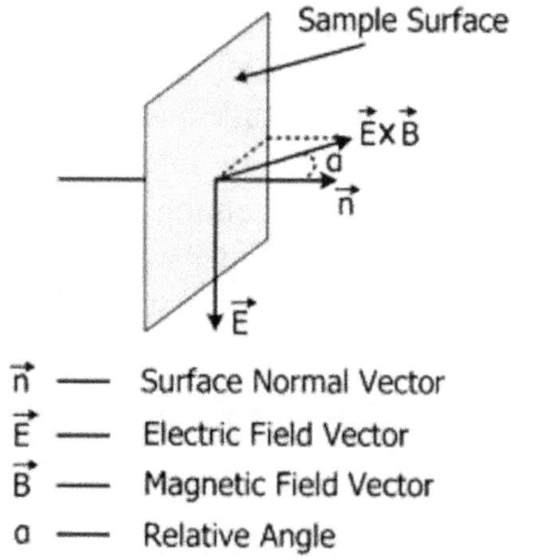

needle electrode was firstly carbonized, which appeared to propagate towards the semicircle-plate electrode. After a period of time, the carbonized path finally bridged the electrodes and the dielectric surface was breakdown. Figure 15 shows examples of PBN and PBT surfaces after the tracking failure. As can be clearly observed, the electrodes have been short-circuited by means of the carbonized track. The tracking failure is considered as one originated from the surface discharge (Sierota et al., 1988). Due to heat energy from the discharge current, molecular chains of the polymer tend to be broken (Du, 2001), which gives rise to the formation of decomposed carbonized compounds on the surface (Du et al., 2008). Since free carbons of conductive

Figure 15. Examples of sample surface after tracking failure

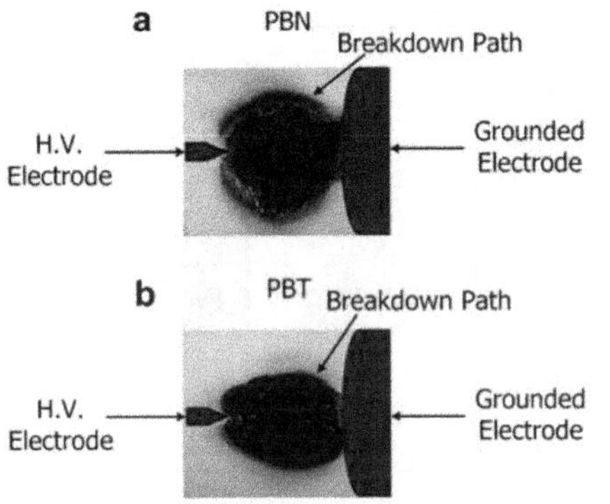

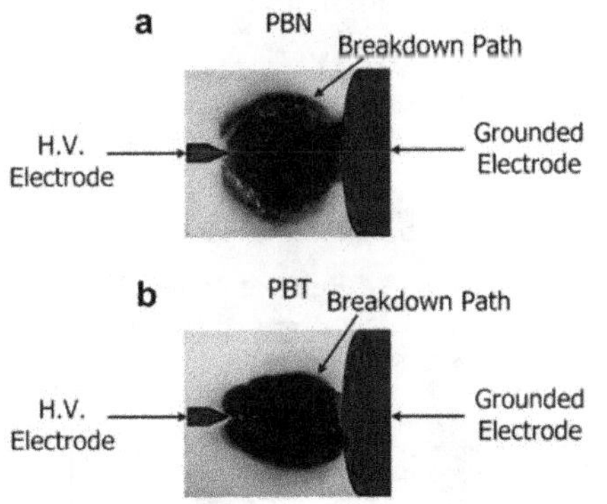

nature are involved in such compounds, they act as an extended electrode that enhances the local field and encourages more intensive discharge (Du et al., 2008). As a consequence, the deposition of carbonized compounds hence the growth of the conductive path is accelerated. As the path bridges the electrodes, tracking failure is achieved with a remarkable reduction of resistance and the perpetual loss of insulation property that can never be recovered (Du et al., 2004).

The relation between the time to tracking failure and the radiation dose for PBN and PBT is shown in Figure 16. With the increase of the radiation dose, the time to tracking failure increases for PBN but decreases for PBT. It indicates that the occurrence of tracking is becoming harder for PBN but easier for PBT. The dependence of discharge quantity for PBN and PBT upon the radiation dose is shown in Figure 17. The discharge quantity determined by $|i(t)|$ is varied as a function of surface resistance which is in response to the behavior of surface carbonization. If the discharge quantity is small, it

Figure 16. Relation between the time to tracking failure and the radiation dose

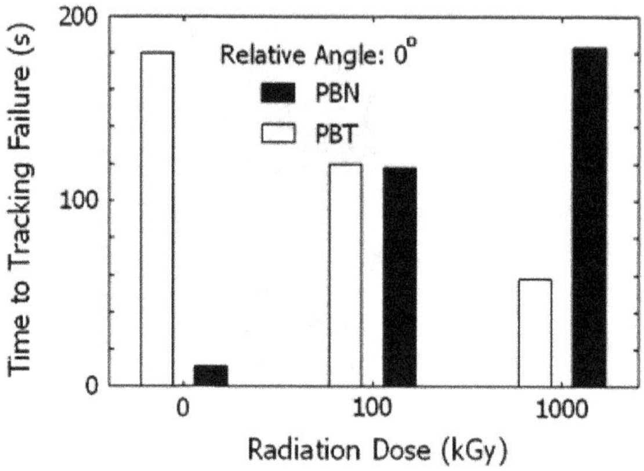

Figure 17. Relation between the discharge quantity and the radiation dose

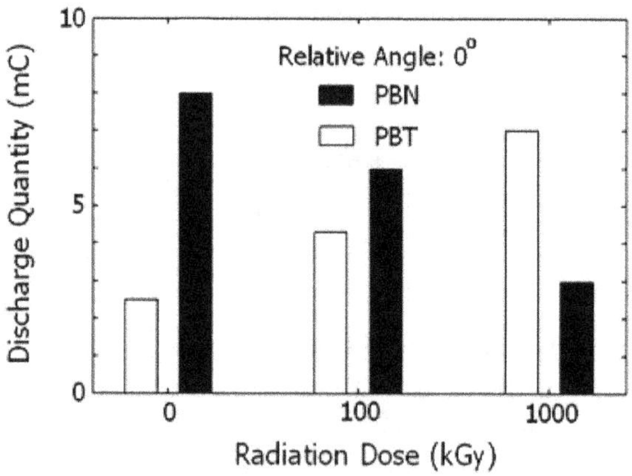

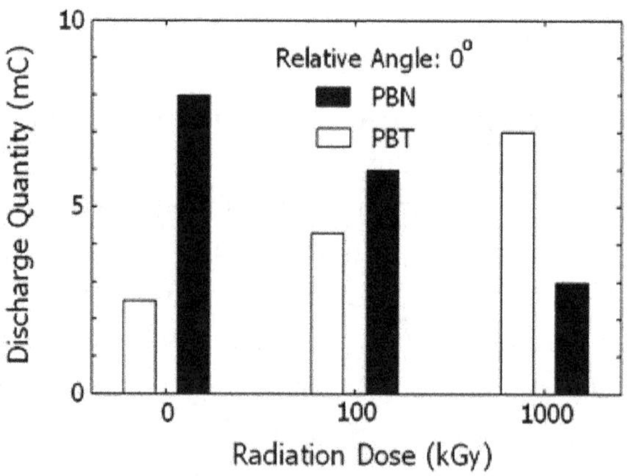

suggests that the dielectric surface is hard to be carbonized. With the growth of the radiation dose, the discharge quantity decreases for PBN but increases for PBT, which reveals that PBN surface becomes more difficult to be carbonized while PBT surface appears to be carbonized more easily. Therefore, as illustrated by Figures 16 and 17, the dielectric strength is improved for PBN but is worsened for PBT by the irradiation.

The variation of the tracking behavior can be attributed to the change of molecular structure caused by the irradiation. The atoms that make up the polymer are bounded together by weak covalent bonds that are easily disrupted by the irradiation (Laghari et al., 1990). As the bonds are broken, radicals are formed to induce chemical reactions through mechanisms like cross-linking and degradation (e.g. chain scission, oxidation) (Du, 2001). In practice, cross linking and degradation reactions occur simultaneously and the reaction result is determined by the one dominant (Kuriyama et al., 1979). As dielectric surface strength of PBN is improved by the irradiation, it is suggested that the cross-linking reaction exceeds the degradation reaction. The emergence of newly formed three-dimensional molecular chain structure tends to promote the thermal durability of the material (Scheirs et al., 1997), thus PBN surface becomes harder to be carbonized after the irradiation. As for PBT, it is proposed that the degradation reaction is superior to the cross-linking reaction. The degradation reaction leads to the formation of low-molecular-weight chain fragments and thereby the thermal durability is worsened (Du, 2001). PBT surface in turn appears to be carbonized more easily after the irradiation. The remarkable difference in the reaction results between the materials is possibly related to the fact that oxygen permeability as well as rate of bond disruption under irradiation in PBN is evidently lower than that in PBT (Scheirs et al., 1997). Since the irradiation is proceeded in air, it is considered that oxidation reaction may play a significant role in determining the reaction result for PBT but doesn't noticeably influence the reaction result for PBN. Based on the viewpoint mentioned above, it is likely to understand why the resistance to tracking is improved for PBN but is reduced for PBT by the irradiation.

Effect of Magnetic Field

The relation between the time to tracking failure and the relative angle is shown in Figure 18. With the increase of the relative angle, the time to tracking for both PBN and PBT tends to decrease. It indicates that the occurrence of tracking varies as a function of $\vec{E}\times\vec{B}$, in which the failure is easily achieved with $\vec{E}\times\vec{B}$ drift into the surface but is difficultly attained with $\vec{E}\times\vec{B}$ point away from the surface, the difficulty scale of tracking occurrence with $\vec{E}\times\vec{B}$ parallel to the surface is addressed between the two cases. Figure 19 shows the change in discharge quantity for both PBN and PBT caused by the increase of relative angle from 0° to 180°. The discharge quantity for both samples appears to increase with the relative angle, which reveals that the carbonization process is encouraged with $\vec{E}\times\vec{B}$ drift into the surface but is retarded with $\vec{E}\times\vec{B}$ drift away from the surface. Therefore, it is suggested that the dielectric strength is worsened as $\vec{E}\times\vec{B}$ is into but is enhanced as $\vec{E}\times\vec{B}$ is away from the surface.

Figure 18. Relation between the time to tracking failure and the relative angle

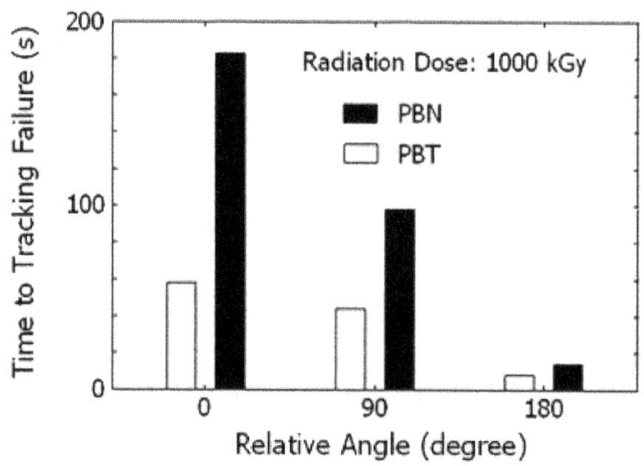

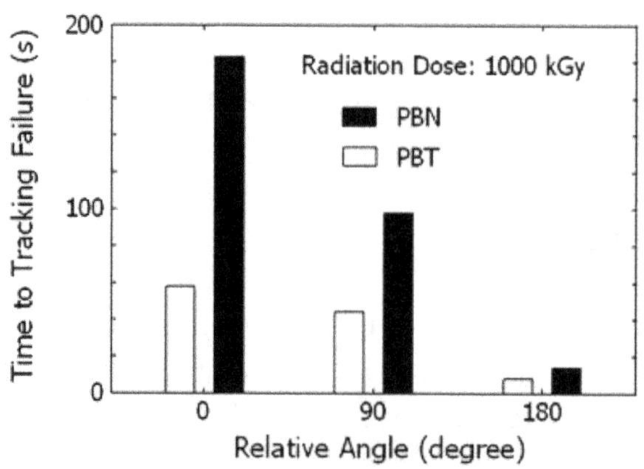

Figure 19. Relation between the discharge quantity and the relative angle

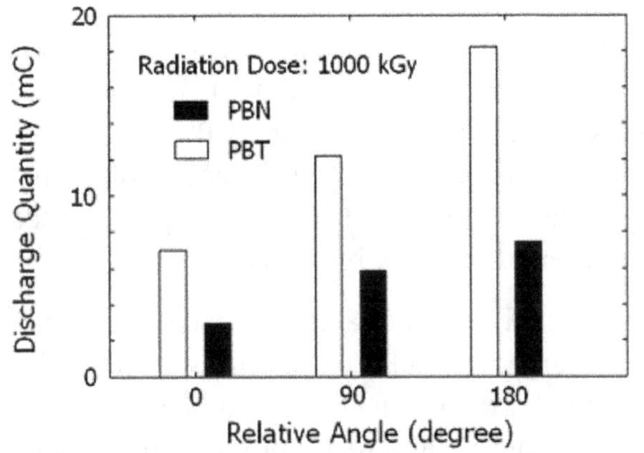

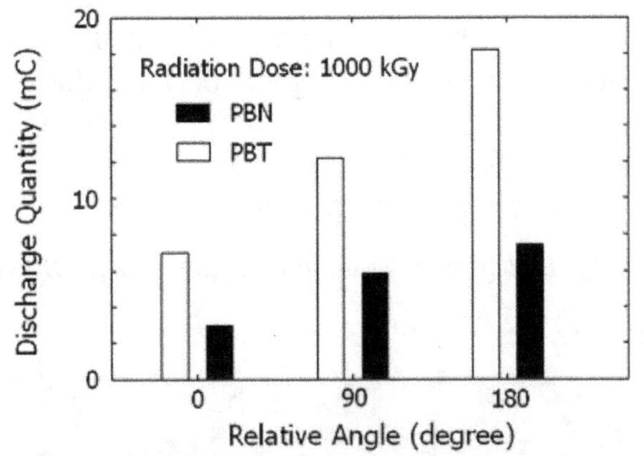

Surface discharge observed in current study is originated from electron impact ionization towards gas molecule and subsequent electron avalanche, the discharge channel is mainly composed of electrons and positive ions. With the presence of magnetic field, moving trajectory of charges in the channel is varied hence the discharge behavior is different from that without the magnetic field. Figure 20 shows schematic diagrams of charge trajectory under combined electric and magnetic field. As $\vec{E}\times\vec{B}$ drift into the sample surface, as shown in Figure 20(a), electrons as well as positive ions are driven to move towards the surface by the combined force, hence as observed in the test, the discharge channel tends to be pressed tightly onto the surface. As a consequence, heat energy caused by discharge current appears to be intense at the surface, which assists the disruption of molecular chain and the formation of carbonization path. Furthermore, it should be noticed that electron bombardment onto sample surface is likely to help breaking the chain since part of electrons in such streamer discharge channel can reach energy level as high as 5 eV (Haydon et al., 1971) that is sufficient for disrupting bands, such as C-C and C-H (Kao, 2000), whereas the disruption energy needed may vary slightly according to the specific location of the bands (Du, 2001). Therefore, the growth of carbonized conductive path is accelerated and the dielectric surface strength is weak in this case.

As $\vec{E}\times\vec{B}$ is orientated from the sample surface, as shown in Figure 20(b), charges in the discharge channel are pushed away from the surface by the combined force, thereby the channel tends to be compelled towards the air rather than onto the material, as a result of which heat energy at the surface is low. This leads to a remarkable delay for the carbonized path built-up, and the dielectric surface shows better strength than the mentioned case. As $\vec{E}\times\vec{B}$ is parallel to the surface, as shown in Figure 20(c), the charges are driven by

Figure 20. Schematic diagrams of charge trajectory under combined electric and magnetic field; (a) $\vec{E}\times\vec{B}$ drift into the sample surface; (b) $\vec{E}\times\vec{B}$ is orientated from the sample surface; (c) $\vec{E}\times\vec{B}$ is parallel to the surface

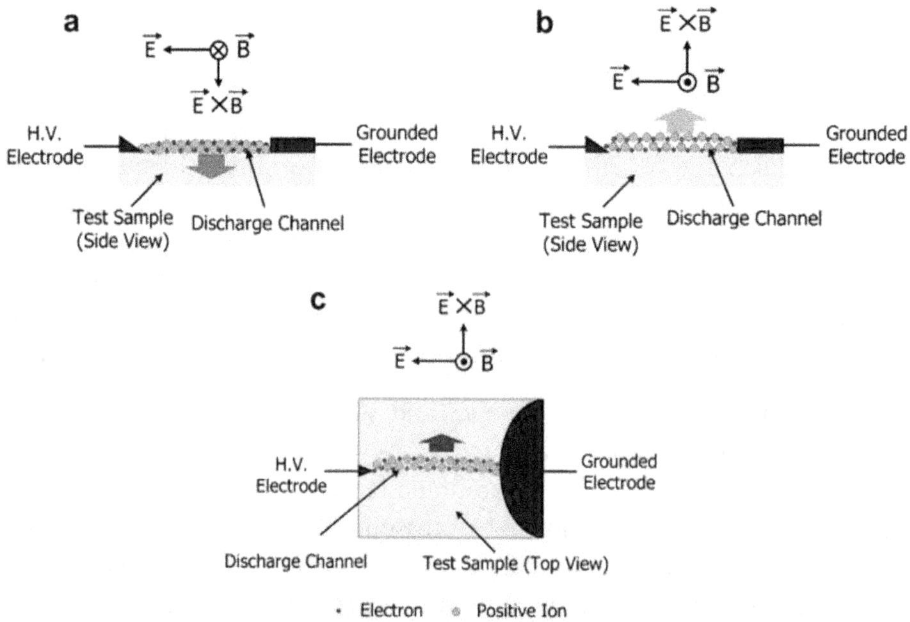

the force to deflect from the straight line between the electrodes. In other words, the force neither confines the discharge channel onto surface nor lifts it up from the material. As a result, the dielectric surface strength is found between the two cases above.

REFERENCES

Banford, H. M., & Fouracre, R. A. (1999). Nuclear Technology and Ageing. *IEEE Electr. Insul. Mag.*, *15*(5), 19–27. doi:10.1109/57.793826

Blodgett, R. B., & Fisher, R. G. (1969). Insulation and Jackets for Control and Power Cables in Heat Reactor Nuclear Generating Stations. *IEEE Transactions on Power Apparatus and Systems*, *88*(5), 529–537. doi:10.1109/TPAS.1969.292321

Chen, G., Banford, H. M., & Davies, A. E. (1998). Space charge formation in γ-irradiation low density polyethylene. *IEEE Transactions on Dielectrics and Electrical Insulation*, *5*(1), 51–57. doi:10.1109/94.660764

Chen, G., Banford, H. M., & Davies, A. E. (1998). Space charge formation in γ-irradiation low density polyethylene. *IEEE Transactions on Dielectrics and Electrical Insulation*, *5*(1), 51–57. doi:10.1109/94.660764

Chen, G., Davies, A. E., & Banford, H. M. (1999). Influence of radiation environments on space charge formation in γ-irradiated LDPE. *IEEE Transactions on Dielectrics and Electrical Insulation*, *6*(6), 882–886. doi:10.1109/94.822031

Chen, G., Davies, A. E., & Banford, H. M. (1999). Influence of Radiation Environments on Space Charge Formation in γ-irradiated LDPE. *IEEE Transactions on Dielectrics and Electrical Insulation*, *6*(6), 882–886. doi:10.1109/94.822031

Das-Gupta, D. K. (1990). Decay of electrical charges on organic synthetic polymer surfaces. *IEEE Transactions on Electrical Insulation*, *25*(3), 503–508. doi:10.1109/14.55723

Das-Gupta, D. K. (1992). Electrical properties of surfaces of polymeric insulators. *IEEE Transactions on Electrical Insulation*, *27*(5), 909–923. doi:10.1109/14.256469

Das-Gupta, D. K., & Schmidt, W. F. (1995). Electrical properties of surfaces of unirradiated and irradiated polymers in humid environments. IEEE Trans. Components. *Packaging and Manufacturing Technology*, *18*(2), 266–269. doi:10.1109/95.390301

Du, B. X. (2001). Discharge Energy and dc Tracking Resistance of Organic Insulating Materials. *IEEE Transactions on Dielectrics and Electrical Insulation*, *8*(6), 897–901. doi:10.1109/94.971443

Du, B. X., & Dong, D. S. (2008). Recurrence plot analysis of discharge currents in tracking test of gamma-ray irradiated polymers. *IEEE Transactions on Dielectrics and Electrical Insulation*, *15*(4), 974–981. doi:10.1109/TDEI.2008.4591218

Du, B. X., & Dong, D. S. (2008). Recurrence Plot Analysis of Discharge Currents in Tracking Tests of Gamma-ray Irradiated Polymers. *IEEE Transactions on Dielectrics and Electrical Insulation*, *41*(4), 974–981. doi:10.1109/TDEI.2008.4591218

Du, B. X., & Gao, Y. (2005). Dielectric breakdown of printed circuit board under magnetic field. *Inter Symp Electr Insul Materi*, *2*, 308–311.

Du, B. X., & Gao, Y. (2009). Gamma-ray irradiation inhibiting surface charge accumulation on polyethylene. *IEEE Transactions on Dielectrics and Electrical Insulation*, *16*(3), 876–881. doi:10.1109/TDEI.2009.5128529

Du, B. X., Gu, L., Dong, D. S., Zheng, X. L. (2008). Recurrent Plot Analysis of Discharge Sequences in Tracking Test of Polybutylene Polymers. *J. Phys. D: Appl. Phys.*, *41*(19).

Du, B. X., & Kobayashi, S. (2001). Wavelet analysis of scintillation discharge current on dc tracking resistance of gamma-ray irradiation polyethylene and polycarbonate. *Radioiso*, *50*(1), 1–11. doi:10.3769/radioisotopes.50.1

Du, B. X., & Kobayashi, S. (2003). Environmental Factors Affecting dc Resistance to Tracking of Polyethylene. *IEEE Transactions on Dielectrics and Electrical Insulation*, *10*(2), 271–277. doi:10.1109/TDEI.2003.1194110

Du, B. X., Liu, H. J., & Liu, Y. (2007). Effects of gamma-ray irradiation on dielectric surface breakdown of polybutylene polymers. *IEEE Transactions on Dielectrics and Electrical Insulation*, *14*(3), 696–701. doi:10.1109/TDEI.2007.369533

Du, B. X., Liu, Y., & Liu, H. J. (2008). Effects of low pressure on tracking failure of printed circuit boards. *IEEE Transactions on Dielectrics and Electrical Insulation*, *15*(5), 1379–1384. doi:10.1109/TDEI.2008.4656247

Du, B. X., Suzuki, A., Kishi, K., & Kobayashi, S. (1995). Effects of atmospheric pressure on tracking failure of organic insulating materials. *Trans IEE Japan*, *115-A*(12), 1284–1293.

Du, B. X., & Zhang, X. J. (2004). Discharge energy and dielectric breakdown of gamma-ray irradiated polybutylene naphthalate. *IEEE Int Conf Solid Dielectr*, *2*, 947–950.

Faircloth, D. C., & Allen, N. L. (2003). High resolution measurements of surface charge densities on insulator surfaces. *IEEE Transactions on Dielectrics and Electrical Insulation*, *10*(2), 285–290. doi:10.1109/TDEI.2003.1194112

Fang, Z., Fouracre, R. A., & Farish, O. (1996). Investigations of surface charging of DC insulator spacers. *IEEE Annual Report of CEIDP*, *1*, 20–23.

Garrett, H. B., & Whittlesey, A. C. (2000). Spacecraft charging, an update. *IEEE Transactions on Plasma Science*, *34*(5), 2017–2028. doi:10.1109/27.902229

Goetzel, C. G., Rittenhouse, J. B., & Singletary, J. B. (1965). *Space Materials Handbook*. Reading, MA: Addison-Wesley.

Hanai, S. (1982). Irradiation and Insulating Materials. *J. Inst. Electrostatics Japan*, *6*(2), 89–95.

Haydon, S. C., Mcintosh, A. I., & Simpson, A. A. (1971). The effect of magnetic field on electron energy distribution functions and derived collision frequencies. *Journal of Physics. D, Applied Physics*, *4*(9), 1257–1265. doi:10.1088/0022-3727/4/9/304

Jaitly, N. C., & Sudarshan, T. S. (1988). In-situ insulator surface charge measurements in dielectric bridged vacuum gaps using an electrostatic probe. *IEEE Transactions on Electrical Insulation*, *23*(2), 261–273. doi:10.1109/14.2362

Kao, K. C. (2000). Electrical conduction and breakdown in insulating polymers. *International Conference on Properties and Applications of Dielectric Materials*, 1-17. doi:10.1109/ICPADM.2000.875620

Kostoski, D., Dojčilović, J., Novaković, L., & Suljovrujić, E. (2006). Effects of charge trapping in gamma irradiated and accelerated aged low-density polyethylene. *Polymer Degradation & Stability*, *91*(9), 2229–2232. doi:10.1016/j.polymdegradstab.2005.12.016

Kumada, A., Okabe, S., & Hidaka, K. (2004). Resolution and signal processing technique of surface charge density measurement with electrostatic probe. *IEEE Transactions on Dielectrics and Electrical Insulation*, *11*(1), 122–129. doi:10.1109/TDEI.2004.1266325

Kuriyama, I., Hayakawa, N., Nakase, Y., Kawawata, S., Ogura, J., Kasai, K., & Onishi, T. (1978). Radiation Resistance of Cable Insulating Materials for Nuclear Power Generating Stations. *IEEE Transactions on Electrical Insulation*, *13*(3), 164–171. doi:10.1109/TEI.1978.298125

Kuriyama, I., Hayakawa, N., Nakase, Y., Ogura, J., Yagyu, H., & Kasai, K. (1979). Effect of Dose Rate on Degradation Behavior of Insulating Polymer Materials. *IEEE Transactions on Electrical Insulation*, *14*(5), 272–277. doi:10.1109/TEI.1979.298231

Kyoto, M., Chigusa, Y., Ohe, M., Go, H., Watanabe, M., Matsubara, T., & Okamoto, S. et al. (1992). Gamma-ray radiation hardened properties of pure silica core single-mode fiber and it data link system in radioactive environments. *Journal of Lightwave Technology*, *10*(3), 289–294. doi:10.1109/50.124489

Laghari, J. R., & Hammoud, A. N. (1990). A brief survey of radiation effects on polymer dielectrics. *IEEE Transactions on Nuclear Science*, *37*(2), 1076–1083. doi:10.1109/TNS.1990.574201

Larbalestier, D., Gurevich, A., Feldmann, D. M., & Polyanskii, A. (2001). High-T_c superconducting materials for electric power applications. *Nature*, *414*(6861), 368–377. doi:10.1038/35104654 PMID:11713544

Lorenzi, D. A., Grando, L., Pesce, A., Bettini, P., & Specogna, R. (2009). Modeling of epoxy resin spacers for the 1MV dc gas insulated line of ITER neutral beam injector system. *IEEE Transactions on Dielectrics and Electrical Insulation*, *16*(1), 77–87. doi:10.1109/TDEI.2009.4784554

Nakamura, K., Saito, K., & Ichinokura, O. (2003). Dynamic analysis of interior permanent magnet motor based on a magnetic circuit model. *IEEE Transactions on Magnetics*, *39*(5), 3250–3252. doi:10.1109/TMAG.2003.816738

Ootera, H., & Nakanishi, K. (1988). Analytical method for evaluating surface charge distribution on a dielectric from capacitive probe measurement-application to a cone-type spacer in ±500 kV DC-GIS. *IEEE Transactions on Power Delivery*, *3*(1), 165–172. doi:10.1109/61.4242

Plessow, R., & Pfeiffer, W. (1994). Influence of the Frequency on the Partial Discharge Inception Voltage. *IEEE Conf. Electr. Insul. Dielectr. Phenom.*, 97-102. doi:10.1109/CEIDP.1994.591699

Scheirs, J., & Cardette, J. L. (1997). Photo-oxidation of poly (butylene naphthalate). *Polymer Degradation & Stability*, *56*(3), 351–356. doi:10.1016/S0141-3910(96)00200-5

Sierota, A., & Calderwood, J. H. (1988). Degradation and breakdown of solid dielectric materials resulting from surface discharge in air and in insulating liquids. *IEEE Transactions on Electrical Insulation*, *23*(6), 993–998. doi:10.1109/14.16525

Stark, K. H., & Carton, C. G. (1955). Electric strength of irradiated polyethylene. *Nature*, *176*(4495), 1225–1226. doi:10.1038/1761225a0

Takada, T., Miyake, H., & Tanaka, Y. (2006). Pulse acoustic technology for measurement of charge distribution in dielectric materials for spacecraft. *IEEE Transactions on Plasma Science*, *34*(5), 2176–2184. doi:10.1109/TPS.2006.883289

Takuma, T., Yashima, M., & Kawamoto, T. (1998). Principle of surface charge measurement for thick insulating specimens. *IEEE Transactions on Dielectrics and Electrical Insulation*, *5*(4), 497–504. doi:10.1109/94.708264

Toomer, R., & Lewis, T. J. (1980). Charge trapping in corona-charged polyethylene films. *Journal of Physics. D, Applied Physics*, *13*(7), 1343–1356. doi:10.1088/0022-3727/13/7/031

Yoshimura, N., Kumagai, S., & Du, B. X. (1997). Research in Japan on the tracking phenomenon of electrical insulating materials. *IEEE Elect Insul Mag*, *13*(5), 8–19. doi:10.1109/57.620513

Yoshimura, N., Kumagai, S., & Du, B. X. (1997). Research in Japan on the Tracking Phenomenon of Electrical Insulating Materials. *IEEE Electr. Insul. Mag.*, *13*(5), 8–19. doi:10.1109/57.620513

Chapter 10
Polymer Insulation for Superconductive Application

ABSTRACT

With the research and development of high temperature superconducting technology, superconducting insulating materials under liquid helium and nitrogen temperature have been gradually taken seriously. Considering the unique operating environment, epoxy resin and PI face the challenge of low temperature. Electrical tree is one of the aging failure phenomena occurring in solid dielectrics. These imperfections could cause the field concentration with the application of high voltage, which results in partial discharges (PD). PD testing is an important quality check for the insulation of HTS cable. This chapter presents a study aimed at clarifying the influence of low temperature, pulse frequency and pulse duration on the electrical tree characteristics in epoxy resin, as well as PD characterization of PI film in LN2. The results show that the number of discharges and the discharge quantity in PI films increase with the increasing of the applied voltage and the defect size. The PD inception voltage decreases when the void defect diameter in PI enlarged and it is higher in LN2 than that at room temperature.

INTRODUCTION

With the research and development of high temperature superconducting technology, superconducting insulating materials under liquid helium and nitrogen temperature have been gradually taken seriously (Ivanov et al.,

DOI: 10.4018/978-1-5225-2309-3.ch010

2013). The international thermonuclear experimental reactor (ITER) project and High temperature superconducting (HTS) cable develop very fast. Since the device operates in the liquid He environment insulating materials in the superconducting magnet system and HTS cable endure the extremely low temperature.

Considering the no-polluting and high-energy characteristics of nuclear fusion energy, the international thermonuclear experimental reactor (ITER) project has attracted much attention in China, India, EU, USA, Japan, Russian Federation and South Korea (Hong et al., 2014). During the operation of ITER, superconducting magnet insulators take the responsibility of the insulation for the liquid helium and nitrogen channels (Boyer et al., 2014). Moreover, fusion process is driven by radio frequency (RF) and pulse power. ITER superconducting coils face the challenge of pulse voltage during superconducting coil current quench, plasma startup and disruption. Epoxy resin is used as the adhesives and electrical insulation in the superconducting magnet system because of its excellent performance in many aspects (Bittner-Rohrhofer et al., 2003). However, in the fabricated process, it is inevitable to bring some defects into the insulating materials, such as voids and impurities, which result in the uneven electrical field strength. The defect regions may cause the stress concentration and become the weak points for partial discharges (Danikas et al., 1997). Therefore, the defects have a close connection with the electrical tree initiation and breakdown process of electrical insulation systems (Li et al.,2012).From the view of reliability of ITER system, it is very important to investigate the treeing mechanism in epoxy resin under the low temperature.

Many researches have been carried out to investigate the tree growth mechanism in epoxy resin. Bahadoorsingh et al researched the relation between the harmonics and the electrical tree growth in epoxy resin. It was indicated that electrical trees in a harmonic distortion environment would reduce the lifetime of electrical insulation (Bahadoorsingh et al., 2010). Chen et al investigated the tree initiation time of pure epoxy resins and found that partial discharge processes are involved in tree initiation, especially at low electrical field strength (Chen et al., 2010). Stone et al studied the effect of pulse voltage on epoxy resin and revealed that the lifetime of epoxy resin decreased with the increase of pulse voltage, and that the insulator had a shorter lifetime with positive pulse than with negative pulse voltage (Stone, 1992). These experiments were carried out above the room temperature. Tree characteristics in silicone rubber under the low temperature were analyzed and three typical profiles of electrical trees could be observed (Du et al., 2013; Du et al., 2014). However, few papers studied the effect of low temperature on electrical tree and breakdown characteristics in epoxy resin with different pulse frequencies and durations (Chowdhuri, 1981).

Firstly, considering the operating environment of ITER device, this chapter will investigate the treeing process of epoxy resin with different pulse frequencies and durations under low temperature.

Superconducting magnetic energy storage systems seem promising when used for grid oscillation damping. Power transmission capacity may also be considerably improved by this technique which may be less expensive than increasing the voltage level. The rapid development of HTS cable has now reached the stage where such applications in power system equipment are possible (Honjo et al., 2003). In metropolitan areas the electricity demand is extremely high; therefore the HTS cable is an outstanding way to increase the transmission capacity of the electricity grid. The superconducting cable cooling system which uses LN_2 without insulating oil or SF6 will not cause environmental pollution, and has the characteristics of being flame and explosion proof. Polyimide film has good insulation performance at cryogenic temperature. It could prevent crack propagation and premature quench of coil effectively when the crack appears in the matrix material of the superconducting magnet coil. In recent decades, much effort has been devoted towards the PD characterization of Polypropylene laminated paper (PPLP) in LN_2 (Lee et al., 2013; Liu et al., 2013; Du et al., 2008; Champion et al., 2001). However, the PD characterization of polyimide film is not clear in LN_2. Secondly, this chapter will investigate PD characterization of polyimide film in LN_2 and at room temperature.

EFFECT OF PULSE DURATION ON ELECTRICAL TREE AND BREAKDOWN PROCESS OF EPOXY RESIN IN LN_2

Tree Growth Characteristics

With the change of pulse duration and ambient temperature, tree structures may be completely different. By using repetitive pulse power, the experiment took into consideration how the electrical tree structure is influenced by the pulse duration and ambient temperature. The typical structures of electrical trees are shown in figure 1.These electrical tree images are selected from different pulse duration under the low temperature environment. As shown in figure 1, when the temperature and treeing time is same respectively, with increasing the pulse duration from 50 to 220 μs, the structures of electrical tree show a significant difference. Taking -196 °C for an example, when the sample is stressed with pulse voltage for 200 min, the tree lengths at pulse duration of 50 and 110 μs are only 400 and 800 μm, respectively. The number of side tree branches is relatively small. These branches grow along the electric field direction with less overlapping. While the electrical tree can grow up to 1900 μm and tree branches become denser with a larger number of small branches generating from the main branches at 220 μs. These side branches are overlapping with each other, causing the tree structure become complex. It means that the pulse duration has an influence on the tree morphology.

Some experiments have concluded that the number of high-energy charges increases with increasing the treeing time (Dissado, 2002; Du, Ma & Gao, 2009). At a fixed frequency, the increase of single pulse duration implies the longer treeing time. Therefore, considering the difference of charging time for different pulse duration, it could be inferred that with the larger pulse duration, much more high-energy charges may be injected into the polymer and collide with the molecular chain, which promotes the degradation of dielectric materials. Therefore, the electrical tree tends to grow at a faster speed with the larger pulse duration.

Besides the pulse duration, tree structures can be affected by the low temperature. With the decrease of temperature, it needs longer time to make tree branches to achieve the same length. Moreover, there are only a few runaway branches at -30 °C, which means fewer branches reach the ground electrode. The number of small tree branches is relatively small. These branches grow along the electric field direction with less overlapping. When tested at -90 and -196 °C, many small tree branches are generated from the main branches.

These branches are overlapping with each other. The lower temperature resists the partial discharge activity. The heat and gas pressure inside the insulating material is not enough to break the molecular chain of polymer. The oxidation and erosion are also weakened by the lower temperature. These factors result in the change of tree structure.

The length of electrical tree (L) is defined as the distance from the furthest tree extent to the needle electrode along the electric field direction. The average value of experiment data was used in the figure. It can be seen from figure 2 that at the same treeing time, the tree length shows an obvious difference with the increase of pulse duration. It means that the growth speed of electrical

Figure 1. Typical profiles of electrical trees at different pulse duration under the low temperature (the tree lengths in each column are about the same)

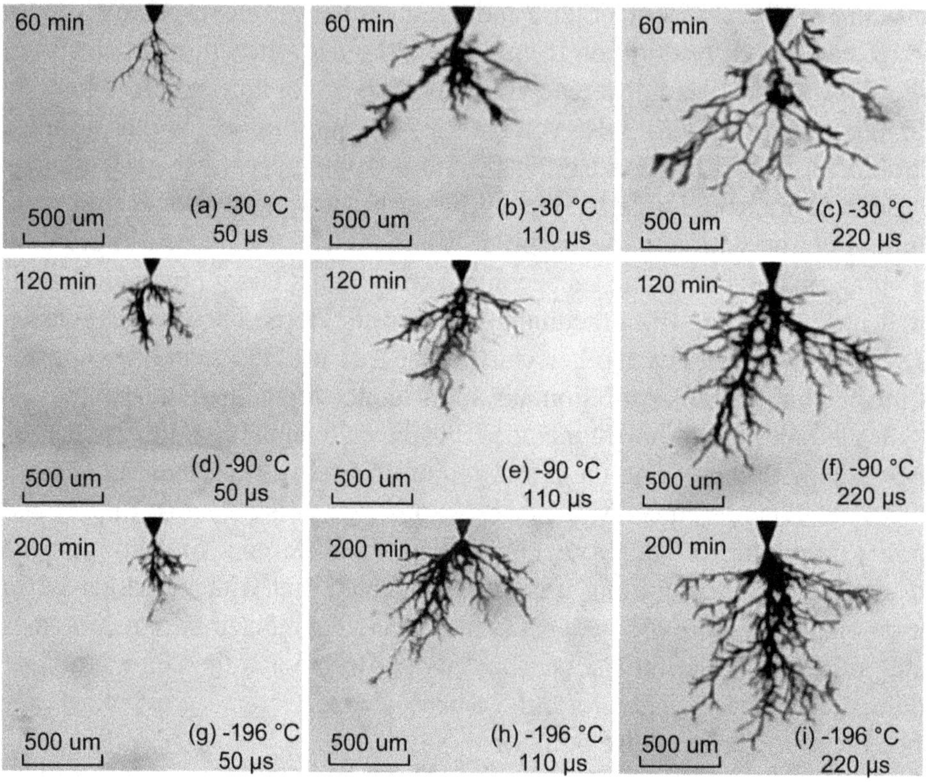

tree is different with changing the pulse duration. Figure 2 shows the relation between the tree length and treeing time with different pulse duration. The confidence bounds are set to 90%. It is found that if the samples are tested with the equal treeing time and the same temperature, the tree length with pulse duration of 220 μs is much longer than that with 50 and 110 μs. Compared with the smaller duration, electrical tree grows at a fast speed with 220 μs. However, the tree length increases slowly or even stops increasing as the tree branches reach to a certain length at pulse duration of 50 and 110 μs, whereas the speed maintains a high level at 220 μs with time elapsing.

The charging time of the sample becomes longer when the duration of the pulse voltage increases. If the charging time of samples is reduced, the charges could not obtain the enough energy to inject into the polymers. Therefore, the collision between the charges and polymers is weakened, which makes it difficult to form the tree channel. As the tree channel reaches to a certain

Figure 2. Relation between the tree length and the treeing time with different pulse duration under the low temperature

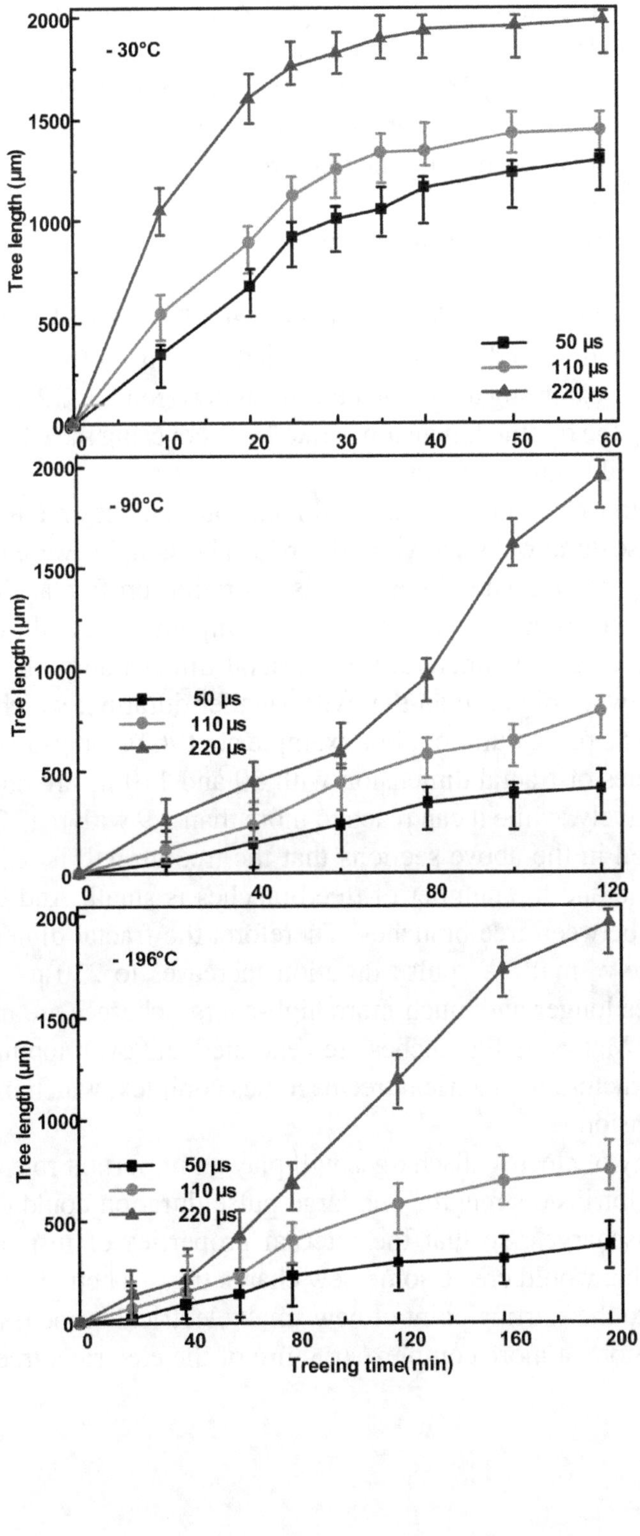

length, the electrical field strength at the tree tip is lower than at the needle tip. The partial discharges are reduced. The growth of electrical tree almost stagnates. When the pulse duration is large enough, the partial discharge would be enhanced and the number of high-energy charges would also increase. The collision becomes strong. More serious partial discharge occurs and it makes the electrical tree grow at a faster speed. In this case, the pulse duration plays an important role in promoting the tree growth.

Tree Structure Characteristics

In order to define the electrical tree characteristics in the detail, the fractal dimension is introduced. Box-counting method is used to calculate fractal dimension. Research suggests that the value is between 1 and 2. In the process of calculating the fractal dimension, a fixed size covering the whole electrical tree, is selected from the images.

The relation between the fractal dimension and the treeing time with different pulse duration is shown in figure 3. The samples were tested under the low temperature. The effect of pulse duration on fractal dimension is similar with each other under different low temperature. As shown in figure 3, when at the same temperature, the fractal dimension with larger pulse duration is always higher than that with smaller duration, which is obvious with the 220 μs pulse duration. For example at -196 °C and stressed for 200 min, the values of fractal dimension with 50 and 110 μs are only 1.45 and 1.60, respectively, while it can reach to more than 1.9 with the 220 μs. It has been analyzed in the above sections that the tree growth is resisted in the small duration and the number of tree branches is small. And there is less overlapping between tree branches. Therefore, the fractal dimension tends to be relatively small. As pulse duration increases to 220 μs, the treeing time becomes longer and much more high-energy charges are injected into the polymer. Many small branches are generated and overlapping with each other. The structure of electrical tree becomes complex, which increases the fractal dimension.

The energy of electric discharges will play an important role in attaining the value of intrinsic strength. The large pulse duration could enhance the energy of discharges, so that the internal properties of insulation would change and that would create some new charge traps. Then, the new created traps turn into the starting point of new small electrical branches (Wu et al., 2005). Therefore, a more complex structure of the electrical tree caused by

Figure 3. Relation between the fractal dimension and the treeing time with different pulse duration under the low temperature

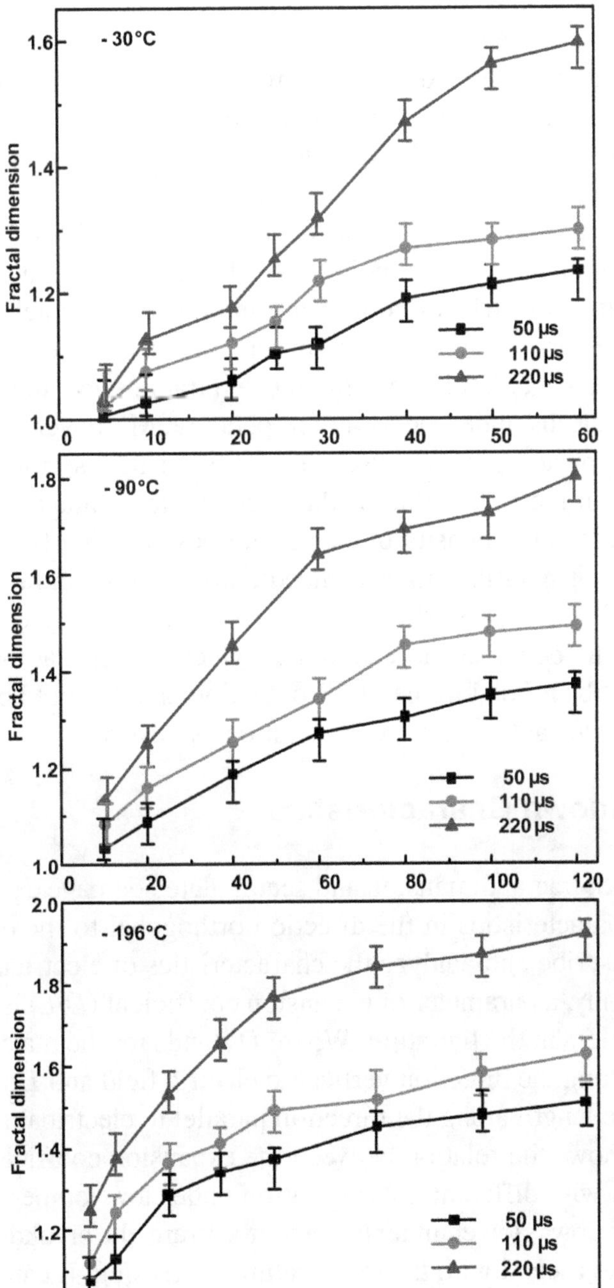

larger amount of small branches is responsible for the large fractal dimension at larger pulse duration.

Accumulated damage describes the damage extent of electrical tree. It can measure the degree of destruction to polymer materials on account of electrical tree aging. The fixed size of area (500 pixels x 500 pixels) covers electrical tree images. The relation between the accumulated damage and the treeing time with different pulse duration under low temperature is shown in figure 4. It is clearly that the value of accumulated damage becomes bigger with the larger pulse duration at the same temperature. More tree branches means the degradation zone in polymers is lager and it can be seen that the pixels covered by electrical tree increase more greatly, which is consistent with the fractal dimension. Accordingly, larger pulse duration leads to bigger accumulated damage. With the same pulse duration, accumulated damage will be more serious under the lower temperature. Moreover, more runaway branches forming at the larger pulse duration cause more serious destruction, which results in the high value of the accumulated damage. As the pulse duration increases, the density of electrical trees is higher because of more small branches generating from the needle tip or the main branches. Thus, compared with small pulse duration, the destructive area grows larger and the tree structure becomes more complex, which means that the large pulse duration is not beneficial to the ITER operation and it should be avoided to operate the device at the large pulse duration as possible.

Tree Breakdown Characteristics

The tree length, fractal dimension and accumulated damage cannot describe the growth characteristics in the direction orthogonal to the electric field. In order to describe and analyze the characteristics of electrical trees more comprehensively, a parameter of expansion coefficient (*D/L*) is proposed to solve the problem in the literature. Where *D* stands for the maximum length of branches along the direction vertical to electric field and *L* stands for the maximum tree length along the direction parallel to electrical field.

Figure 5 shows the relation between the expansion coefficient and time to breakdown with different pulse duration under low temperature. As the electrical tree grows slower under lower temperature, the breakdown time has the tendency to ascend with the temperature decreasing. It can be observed that the lower the temperature is, the larger the expansion coefficient will be when applied with the same pulse duration. Thus, it implies that the tree

Figure 4. Relation between the accumulated damage and the treeing time with different pulse duration under the low temperature

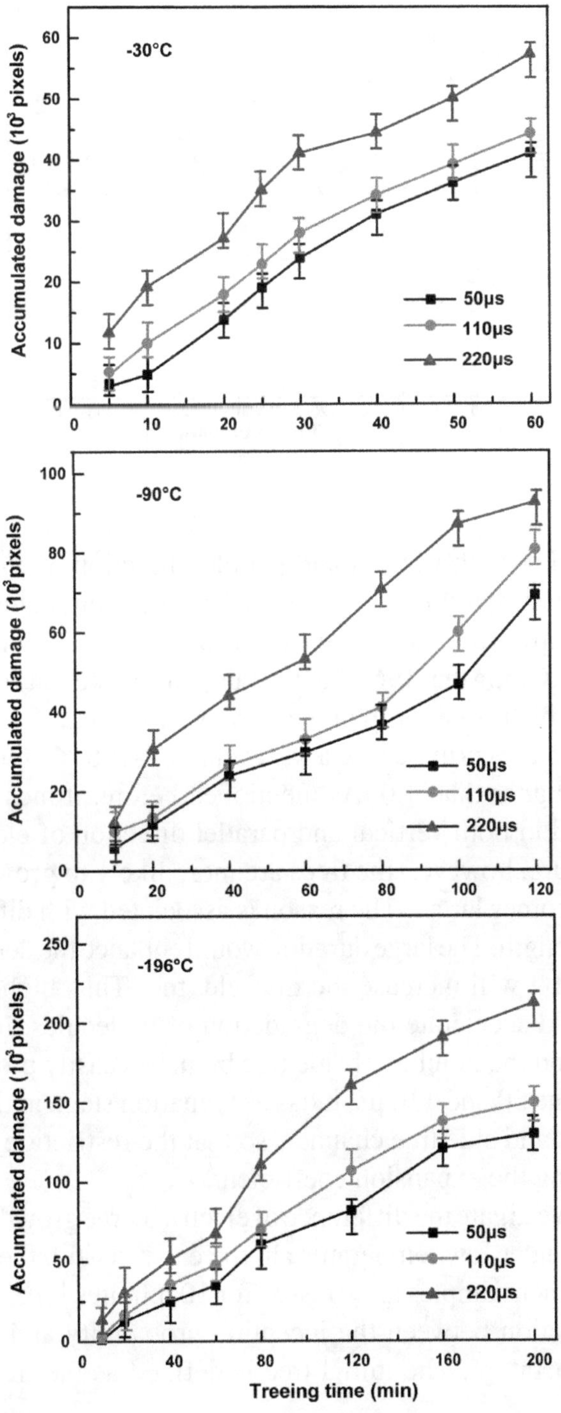

Figure 5. Relation between the expansion coefficient and time to breakdown with different pulse duration under low temperature

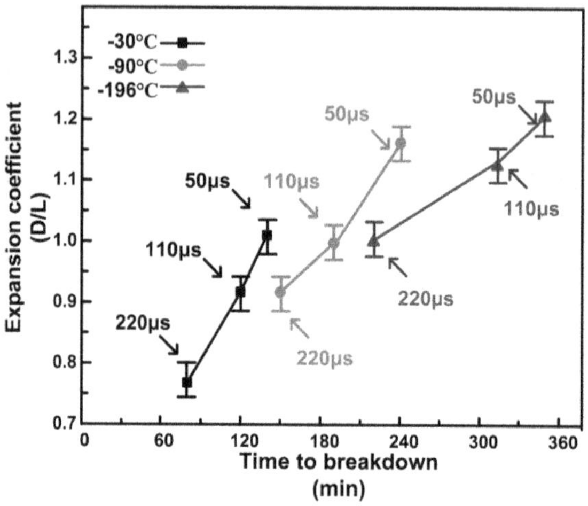

branches extend in both vertical and parallel direction of electrical field at lower temperature. On the other hand, when the temperature is the same, there exists an upward trend of the expansion coefficient with the duration becoming larger. Citing the -196 °C for instance, the value can be approximately 1.2 at 220 μs, which is larger than the ones of 1.0 and 1.15 at 50 and 110 μs respectively. Moreover, under the temperature of -196 °C, the values of D/L stay at a level higher than 1.0. As mentioned before, it means that the tree branches extend in both vertical and parallel direction of electrical field at lower temperature, however, the trees are more likely to grow parallel when the duration becomes larger. The reason is associated with differences of the electric field strength. The large duration would enhance the density of charges more easily and it will increase the tip field, too. Thus high energy partial discharges would accelerate the degradation of molecules. It illustrates the electrical tree and the number of side tree branches easily propagates at 220 μs, while as to the 50 and 110 μs, it has a stagnation stage in the propagation of tree branches and side tree channels so that the restriction of the tree is a factor in reducing the expansion coefficient.

In order to investigate the difference of electrical tree growth with different pulse duration under low temperature in the early period, the probability of tree inception when samples are stressed for 10 minutes is calculated. Figure 8 shows the relation between the inception probability and pulse duration under low temperature. The initial tree is defined as the tree grows to the length of 30 μm. The initial probability refers to the proportion of the tested samples which have grown initial trees in all 20 samples in 10 minutes. The samples without trees are eliminated out of the calculation of experiment data. It can be seen from figure 6 that with the decrease of the temperature from -30 °C to -196 °C, the probability of tree inception presents a decrease trend obviously. Taking the 220 μs for an example, the inception probability is only 35% at -196 and 50% at -90 °C respectively, while the value can be 80% when the temperature is up to -30 °C. It implies that the lower temperature contributes to reduce the insulation aging, which can extend the life length of the epoxy resin.

When under the same low temperature, there exists obvious difference of the tree inception probability with the pulse duration varying from 50 to 220 μs, as shown in figure 6. The rising pulse duration would result in the reduction in the lifespan of the epoxy resin in ITER device. Taking the -30 °C for example, the tree inception probability can be up to 80% with the duration of 220 μs, which is twice higher than that with 50 μs for contrast. The probability could maintain a relative low value at the small duration. Thus, the tree is more likely to grow at the large pulse duration, which means it should be avoided to operate the ITER device at the large pulse duration environment as soon as possible.

The effects of pulse duration on the time to breakdown (TTB) with different low experimental temperature are also shown in Figure 8. Differences in pulse duration could cause a variation in TTB. It can be seen clearly that

Figure 6. Relation between the time to breakdown and the pulse duration under different temperatures

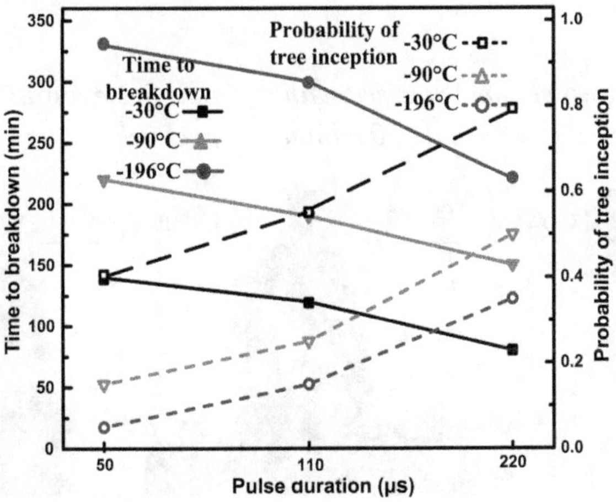

the TTB has a significantly increasing tendency with the pulse duration getting larger, indicating the large duration could accelerate the speed of breakdown in epoxy resin. Taking -30 °C for example, with the application of 50 μs pulse duration, the TTB is approximately 140 min, while the value of 110 and 220 μs is respectively 120 and 80 min as contrast. The increase of the time to breakdown is associated with the effect of partial discharges in electrical tree channels.

At the same pulse duration, with temperature increasing from -196 to -30 °C, the probability of breakdown and the probability of the tree inception all present an increase trend, and what can be seen clearly is that the initial tree is much harder to grow at the liquid nitrogen temperature than the others. Similarly, it becomes more difficult for the electrical breakdown to take place at -196 °C.

The tree breakdown occurs when one or some branch channels reach the ground electrode, which attached with a 100 μm thick copper foil (making a close contact between the sample and ground electrode) in this experiment. Apparently, the increasing pulse duration would strengthen the energy of partial discharge. Because of the high gas pressure, as well as the high temperature that caused by partial discharge in branch channels, the electrical tree would be easy to propagate along the micro pores. In this way, field moderation and tree propagation come into being with more high energy space charges injected from the needle electrode. Consequently, electrical deterioration probability increases. So the TTB indicates the ability of inhibiting the tree propagation, thus, as mentioned above, the tree is more likely to be inhibited at the small pulse duration, which indicates the large pulse duration will make the epoxy resin insulation weaker and result in a greater deterioration.

Figure 7. Breakdown channel stressed with different pulse duration at -90 °C (the breakdown time is 220, 187 and 150 minutes respectively)

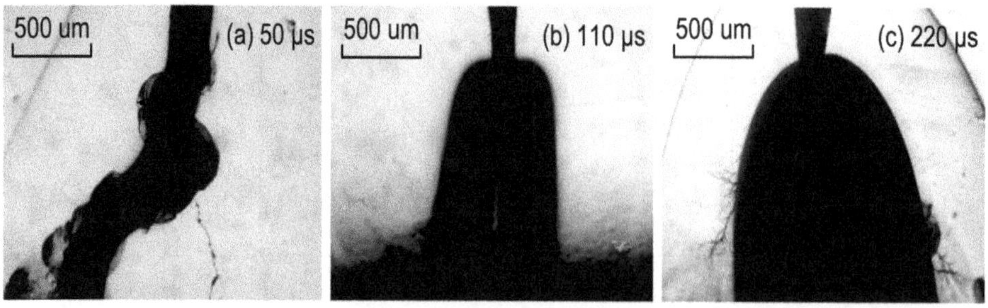

Figure 7 shows the differences among breakdown channels at -90 °C with the pulse duration of 50, 110 and 220 μs. Clear of the stagnated trees, the figures illustrate that the width of breakdown channels increases with the rise of pulse duration. It can also be observed the electrical tree grows denser at larger pulse duration, which causes a larger damaging region. The same conclusions also exist at -30 °C and -196 °C. Such phenomenon is associated with partial discharge intensity, which is closely connected with the pulse duration. Although the breakdown happens in a flash so that oxygen and carbon have no time to combine together, and this leads to differences between breakdown path and electrical tree branch, large pulse duration still has a promotional function in both of them. Moreover, high intensity partial discharge makes it easy to break molecular chains in epoxy resin and bring about a large damage.

From the above analysis, it can be inferred that larger pulse durations could result in larger damaging regions, contrasting the smaller ones. The phenomenon means that the insulation in the superconducting magnet system will get a greater damage with application of large pulse duration.

Tree structures vary obviously with both frequency and low temperature (Du et al., 2016). With the decrease of temperature, the color of branches becomes darker. Whether the decrease of temperature or the increase of frequency, the branches appear to be dense and a large number of small branches are overlapping together. The tree growth characteristics and density of branches under low temperature are analyzed. Fractal dimension is introduced to measure the complexity of trees. It is revealed that the low temperature can reduced the tree growth rate (Du et al., 2015). However, the fractal dimension and accumulated damage has an opposite trend. When the samples are tested under lower temperature, fractal dimension is larger which means the higher density of branches and larger damaging area.

EFFECT OF PULSE FREQUENCY ON TREE CHARACTERISTICS IN EPOXY RESIN UNDER LOW TEMPERATURE

Tree Structure

Electrical tree aging is a complex electrical, physical, and chemical process. Tree initiation and propagation depend on partial discharge activity, oxidization and chemical decomposition in the micro tree channels (Zheng et al., 2008). With the change of ambient temperature and pulse frequency, tree structures may be completely different. The typical tree structures with the applied different frequencies under the low temperature are shown in figure 8. The amplitude of pulse voltage is 14 kV with the positive polarity. The selected results from 300, 400 and 500 Hz are reported here since the changing trend of tree structures from 100 and 200 Hz is similar to that from 300 to 500 Hz. Here, the tree images from pre-breakdown are selected to obtain the whole tree structures.

As is shown in figure 8, at the same pulse frequency, the density of electrical tree is significantly larger with the decrease of temperature. Only a few slim

Figure 8. Typical profiles of electrical tree with different pulse frequencies and different temperatures at 14 kV

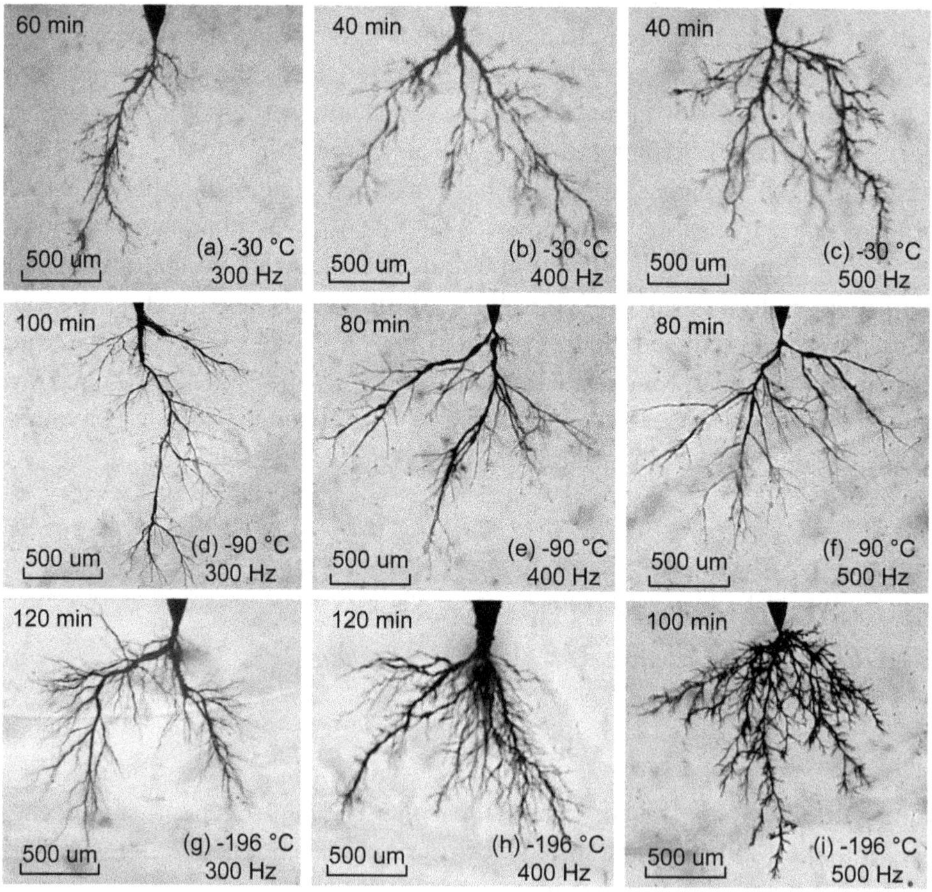

main branches extend towards the electrical field direction at -30 °C. As the experimental temperature decreases, tree branches become denser and a larger number of small branches generate from the main branches, the phenomenon of which is especially obvious at the liquid nitrogen temperature. Moreover, the branches become darker and clearer at the lower temperature. Considering the wide range of experimental temperature, it can be concluded that when the temperature is down to the liquid helium temperature, the electrical tree tends to be much denser and darker than the above-mentioned temperature. And the number of tree branches appears to be larger at the liquid helium temperature. While under the same low temperature, the tree morphology with the change of pulse frequency shows little or no significant difference, but the difference of tree density can be observed. The number of main branches and side branches gets larger and more tree channels are overlapping together with the higher frequency, while there are only a few runaway branches with the low frequency. In addition, when applied with the frequency of 300 Hz, the runaway branches develop mainly along the direction of electrical field. While there are more side branches extending in a wider arc angle direction around the needle tip, even perpendicular to the electric field direction with the frequency of 400 and 500 Hz.

The growth of electrical tree is involved in partial discharge activity in micro channels. High frequency promotes the partial discharges activity. Thus, the number of space charges produced by PD becomes larger than that with the low frequency. Space charges injected from tip electrode collide with the polymer molecules causing the degradation of molecular chains. As a result, there are a large number of small branches generating from the main branches.

Growth Characteristics

In general, the growth speed of electrical tree could be expressed by the tree growth rate. Growth rate can be calculated by the following equation (Chen et al., 2009):

$$\text{Growth Rate} = \frac{L_2\text{-}L_1}{T_2\text{-}T_1} \tag{1}$$

where L is referred to the maximum tree length from the needle tip to tree tip. T_1 and T_2 are the treeing time. L_1 and L_2 are responded to the application of pulse voltage for T_1 and T_2 respectively. In general, T_1 is set to 0 minutes.

Figure 9. Growth process of the electrical tree with treeing time at liquid nitrogen temperature

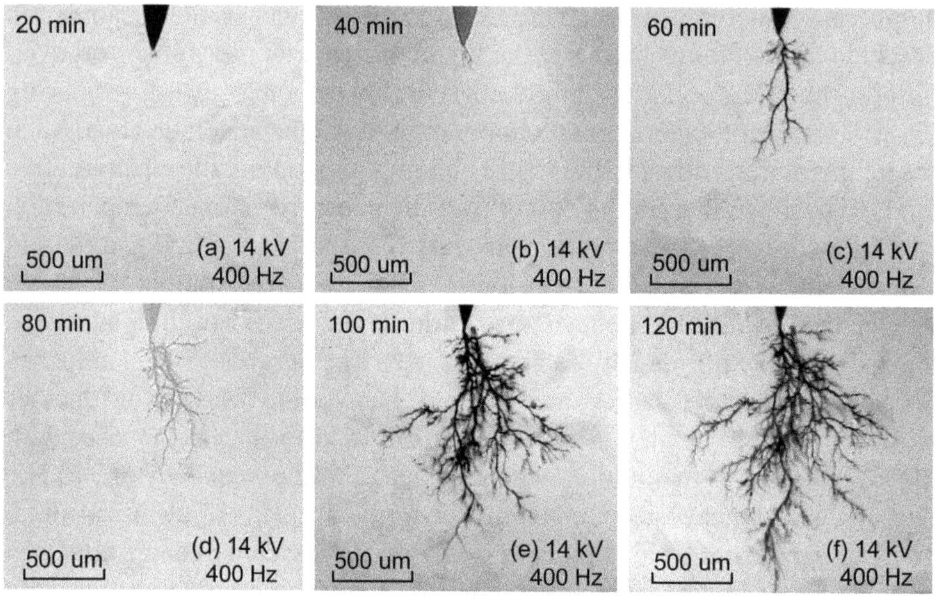

In addition, the tree length as a function of treeing time can be also used to describe the tree growth speed. Here, the second method is selected to analyze the tree growth characteristics.

Figure 9 shows the growth process of the electrical tree with time. The selected results from liquid nitrogen temperature with 400 Hz and 14 kV are reported here. In the first 60 seconds, the electrical tree grows very slowly and there are only a few tree branches. When the time is over 60 minutes, the tree branches expends at a relatively high speed. Moreover, many small branches generate from the main branches. The number of branches increases rapidly.

Figure 10 shows the relation between the tree length and the treeing time with different pulse frequencies under low temperature. The amplitude of pulse voltage is 14 kV with the positive polarity. It can be revealed that the length of electrical tree increases at a higher speed with the increase of frequency at the same low temperature. This is due to the difference of partial discharges caused by the frequency variation. When the electrical energy is adequate to stimulate the charges inject from the tip to the polymers (Chowdhuri, 1981), the number of PD activity increases with the increase of pulse frequency, which accelerates the degradation of epoxy resin molecules. As the frequency increases from 300 Hz to 500 Hz, the repeated charge

Figure 10. Relation between the tree length and the treeing time with different pulse frequencies at 14 kV

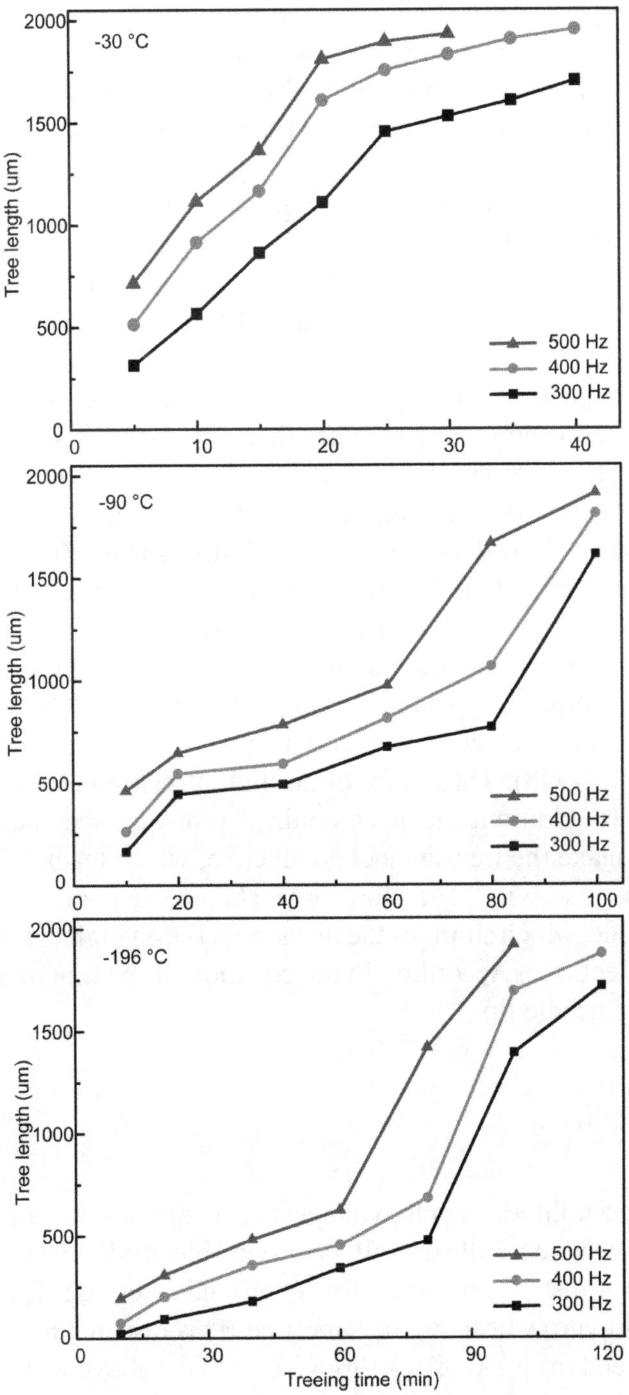

injection process between needle tip and polymers becomes more frequent. In this case, compared with the low pulse frequency, electrical tree length tends to be longer with the higher frequency.

As is shown in figure 10, the electrical tree grows at an obviously slower speed under the lower temperature. The higher density of crystalline regions under lower temperature makes it difficult to break the molecular chai. Besides, the oxidization is weakened by the low temperature. Both two factors cause the slow growth speed. In addition, the growth patterns of electrical tree are different with the variation of temperature. Tree growth process includes the initiation and propagation period. When the samples are tested at -30 °C, tree length increases at a high speed in the initial period. As the tree length is up to 1700 μm, electrical tree growth becomes slower. With the temperature of -90 and -196 °C, it needs longer time to form the initial tree or tree bub. Then, when the treeing time is long enough, there is an obvious increase in tree length. The different growth patterns can be explained as follows. Compared with extremely low temperature, the motion of charges and oxidization in the tree channels (Chowdhuri, 1981) is much stronger at -30 °C, which could promote the tree growth in the initial period. As the temperature decreases, such as -90 and -196 °C, the energy is not sufficient to form the initial tree. However, when the samples are subjected to pulse voltage for a long time, the local high temperature and high pressure accumulated during the partial discharges would accelerate the degradation of epoxy resin molecules (Shimizu et al., 1998). Thus, a followed high growth rate of electrical tree is observed. In addition, much carbonized products accumulated in this environment make the tree channel conductive, which leads to the increase of channel conductivity. Therefore, tree channel becomes an equivalent needle electrode, which shortens the distance between high voltage electrode and ground electrode. According to the equation of maximum electric field strength at the needle tip (E):

$$E = \frac{2U}{r\ln(1 + 4R / r)} \tag{2}$$

where U is the amplitude of pulse voltage; r is the tip curvature radius; R is the distance between the needle tip and the ground electrode. E would be higher with the shorter distance between tip and ground electrode. The two factors promote trees to grow faster in the later time. This research has analyzed the tree growth trend from -30 °C to -196 °C. Due to the above wide temperature range, it is reasonable to deduce that the similar trends from the experimental temperature are also suitable for the lower temperature.

Fractal Dimension and Accumulated Damage

Tree length can only express the growth characteristics of electrical tree along the direction of the applied electric field. The fractal dimension is introduced as a parameter which provides a measure of the branches complexity. By using the box-counting method, the value of fractal dimension varies between 1 and 2, which is consistent with the previous literatures. Here, the value of fractal dimension is selected from the tree images of pre-breakdown.

The relation between the fractal dimension and the time to breakdown (TTB) with different pulse frequencies and low temperatures is shown in figure 11. The pulse frequency varies from 100 to 500 Hz with magnitude of 14 kV. The experiment is carried out at sub-zero temperature. The growth of electrical tree is much slower under lower temperature. Thus, the TTB becomes longer with the decrease of temperature. In addition, it has been analyzed from figure 10. That high frequency accelerates the tree growth when under the same low temperature. Therefore, the TTB tends to be shorter with the increase of pulse frequency, while the fractal dimension shows the opposite tendency. As the frequency increases, numerous charges may collide with polymers, breaking the molecular chain. Thus, the new degradation zone forms (Wu et al., 2005). There are more side branches generating from the main branches and these small branches may be overlapping with each other during their extension process. This makes the tree branches packed and the complexity of electrical tree increases accordingly.

In addition, the effect of frequency on fractal dimension and the time to breakdown shows significant difference with the decrease of temperature. With the frequency varying from 100 to 500 Hz, the time to breakdown decreases from 80 to 42 minutes at -30 °C, while the value of TTB is from 240 to 115 minutes at -196 °C. There is a bigger time gap in breakdown with the decrease of temperature. It can be speculated that pulse frequency appears to have greater influence on TTB at the lower temperature, which is suitable for the liquid helium temperature. At the liquid helium temperature, TTB tend to be a wider range with the increase of pulse frequency.

Accumulated damage indicates the degree of destruction to polymer materials dueto electrical tree aging. A fixed size of area (500 pixels x 500 pixels) is selected from the electrical tree images. The maximum value of

Figure 11. Fractal dimension as a function of TTB with different frequencies under low temperature at 14 kV

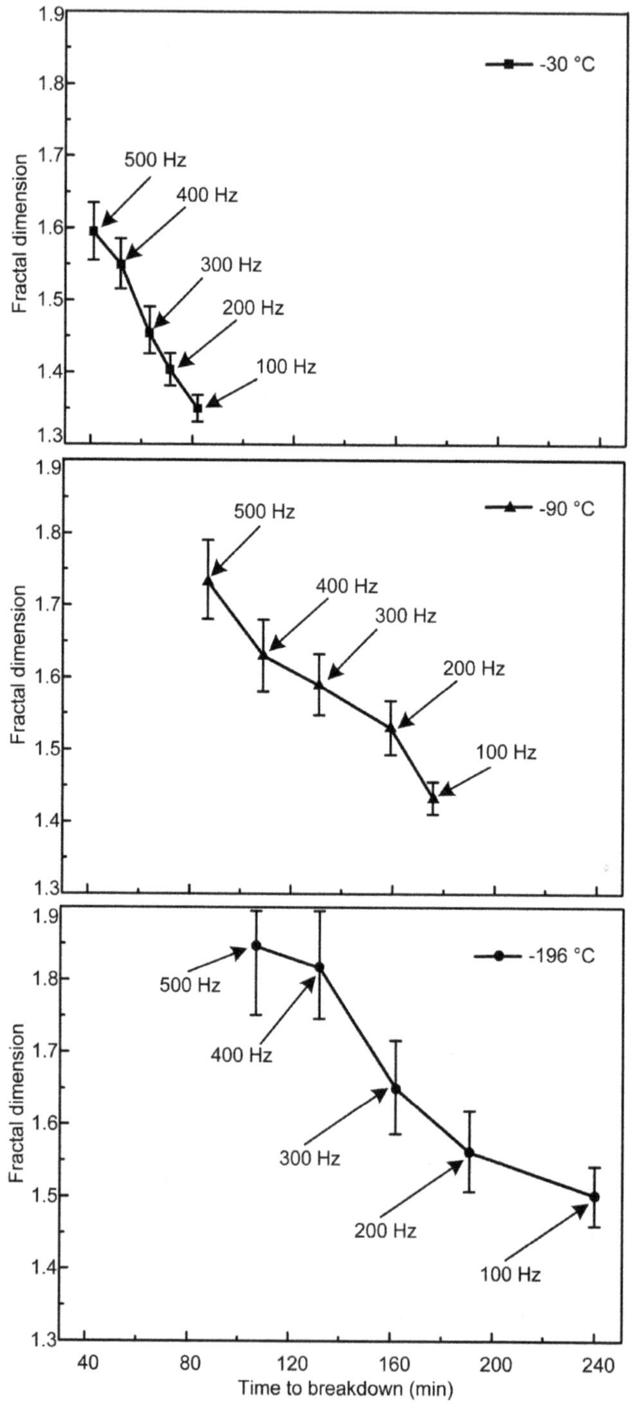

Figure 12. Accumulated damage as a function of TTB with the different frequencies under low temperature at 14 kV

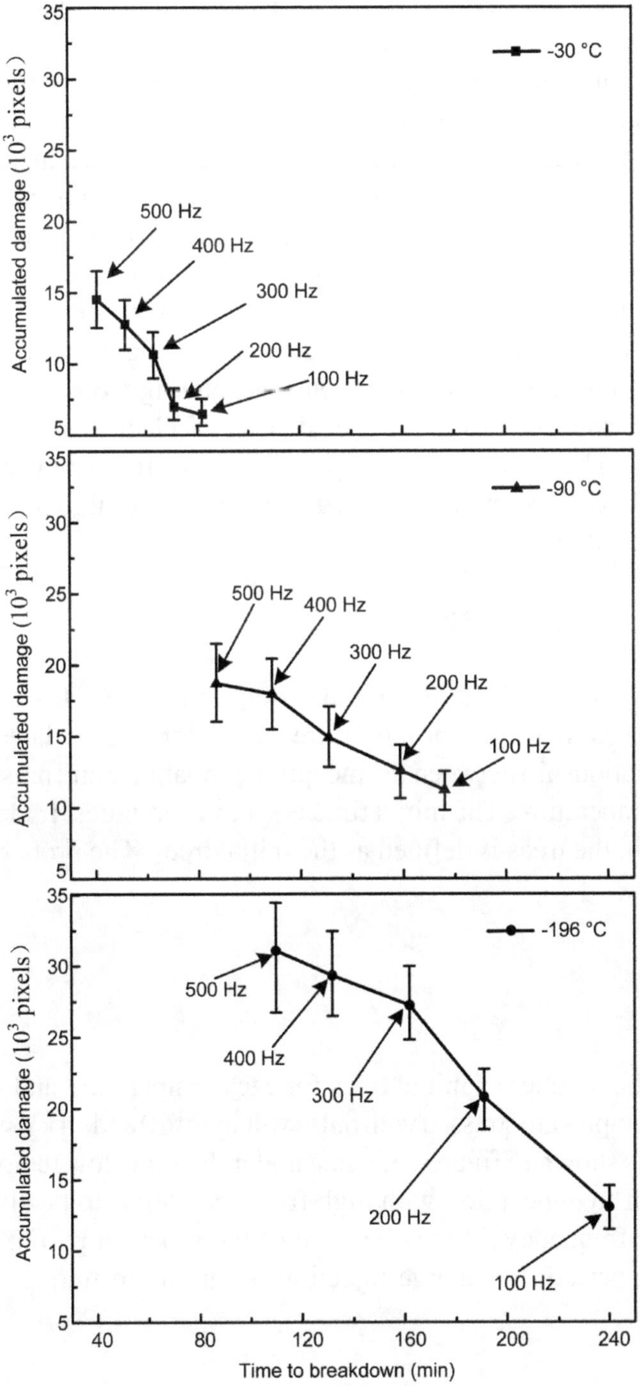

accumulated damage is selected Accumulated damage is calculated according to the area covered by tree branches in pixels.

Figure 12 shows accumulated damage as a function of time to breakdown with different frequencies under low temperature. It can be seen that the value of accumulated damage becomes bigger under the lower temperature. More tree branches under the lower temperature means the larger degradation zone in polymers and the pixels covered by electrical tree increase greatly. Therefore, lower temperature lead to bigger accumulated damage. When under the same low temperature, accumulated damage tends to be more serious with the higher frequency, which is consistent with the fractal dimension. Fewer runaway branches forming at the lower frequency cause less destruction, which results in the low value of accumulated damage. With the increase of pulse frequency, many small branches generating from the needle tip or the main branches lead to the higher density of electrical trees. Thus, the destructive area becomes larger compared with low frequency, which means that the ITER device should be avoided to operate at the high pulse frequency as soon as possible.

Tree Initial Characteristics

In order to investigate the effect of pulse frequency on electrical tree growth in the early period, the probability of tree inception is calculated. Figure 13 shows the relationship between the inception probability and pulse frequency under low temperature. The initial time is set to 10 minutes. If the tree length gets to 30 μm, the trees is defined as the initial trees. The probability of tree inception is:

$$P = \frac{N}{20} \times 100\% \tag{3}$$

where N is the number of initial trees for each temperature and frequency.

And the samples are stressed with pulse voltage of 10 and 14 kV with positive polarity. As is shown in figure 13, when under the same low temperature, the probability of tree inception with high frequency tends to be always higher than the low frequency. This is because of the repeated partial discharges. In the initial period, the charge injection occurs more frequently with the

Figure 13. Relation between the inception probability and the low temperature with different frequencies at10 minutes

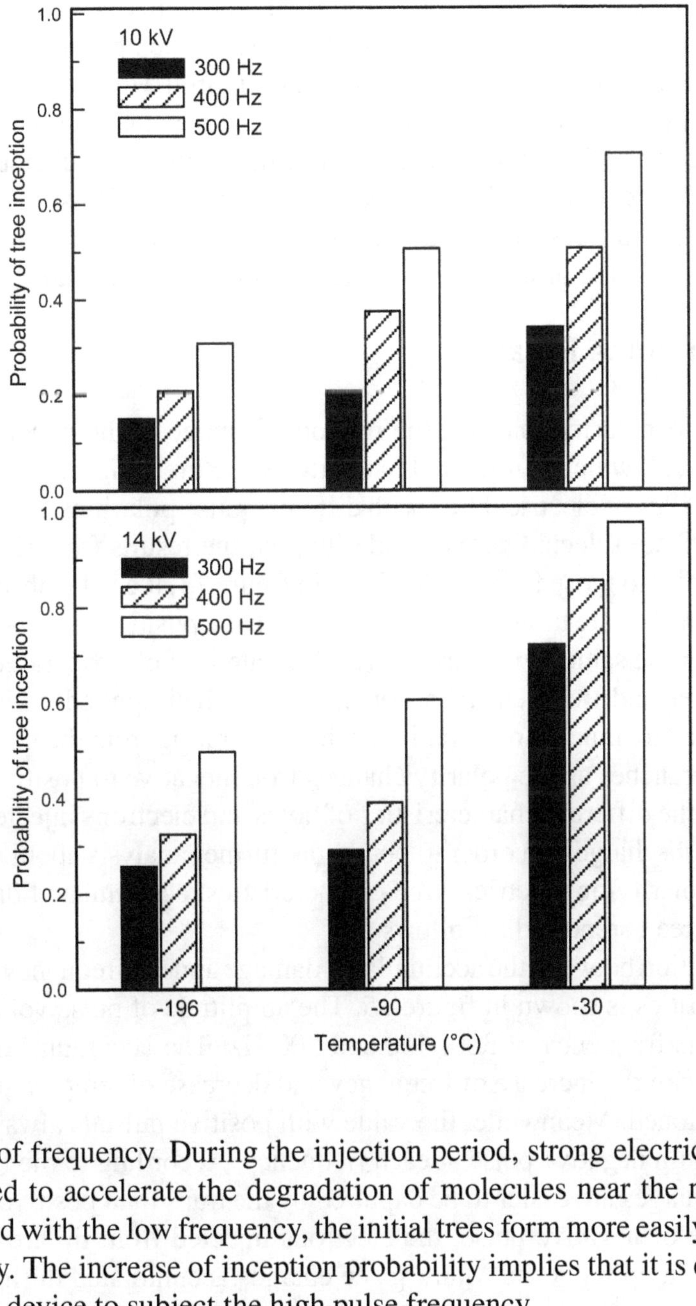

increase of frequency. During the injection period, strong electrical energy is released to accelerate the degradation of molecules near the needle tip. Compared with the low frequency, the initial trees form more easily at higher frequency. The increase of inception probability implies that it is dangerous for ITER device to subject the high pulse frequency.

When at the same frequency, there is obvious difference of the tree inception probability with the temperature varying from -30 to -196 °C. Taking the 14 kV and 500 Hz for an example, the inception probability is only 50 and 60% at -196 and -90 °C respectively, while the value can be up to 100% when the temperature is -30 °C. It has been analyzed in figure 10 that low temperature can resist the tree growth. The rigidity of epoxy resin is supposed to be higher due to the decrease of temperature, which makes it difficult for the discharge to break the molecular chain of epoxy resin. Thus, it is difficult for the electrical trees to generate from the needle tip at the extremely low temperature. So the probability of tree inception decreases significantly.

Effect of Pulse Polarity

As the pulse polarity changes, the motion of charge in the materials tends to be different, which is reflected in the tree structures. Figure 14 shows the typical profiles of electrical tree with different pulse polarities. Results from the -196 °C are selected here considering that the results from -30 and -90 °C are similar to those from -196 °C. From figure 14, it can be observed that the tree branches are more easily to extend with the positive than the negative pulse under the same low temperature. The length of electrical tree appears to be longer and tree density becomes larger when applied with positive pulse. There are much more small branches generating from the needle tip or the main branches as the polarity changes from negative to positive. This is caused by the different characteristics of holes and electrons injected during the partial discharges. In order to obtain the further analysis about the effect of pulse polarity on electrical tree characteristics, accumulated damage of electrical tree is reported as follows.

The relation between the accumulated damage and the frequency with two pulse polarities is shown in figure 15. The amplitude of pulse voltage is 14 kV with the frequency of 300, 400 and 500 Hz. The accumulated damage increases with the increase of frequency and decrease of temperature as has been mentioned. Meanwhile, the value with positive pulse is always higher than that with negative pulse at each frequency. According to the literature, negative charges are easier to be captured by the traps than positive charges. In the case of negative pulse, the electrons injected from tip are easier to be trapped in epoxy resin. Many space charges accumulated during partial

Figure 14. Typical profiles of electrical tree with different pulse polarities at liquid nitrogen temperature

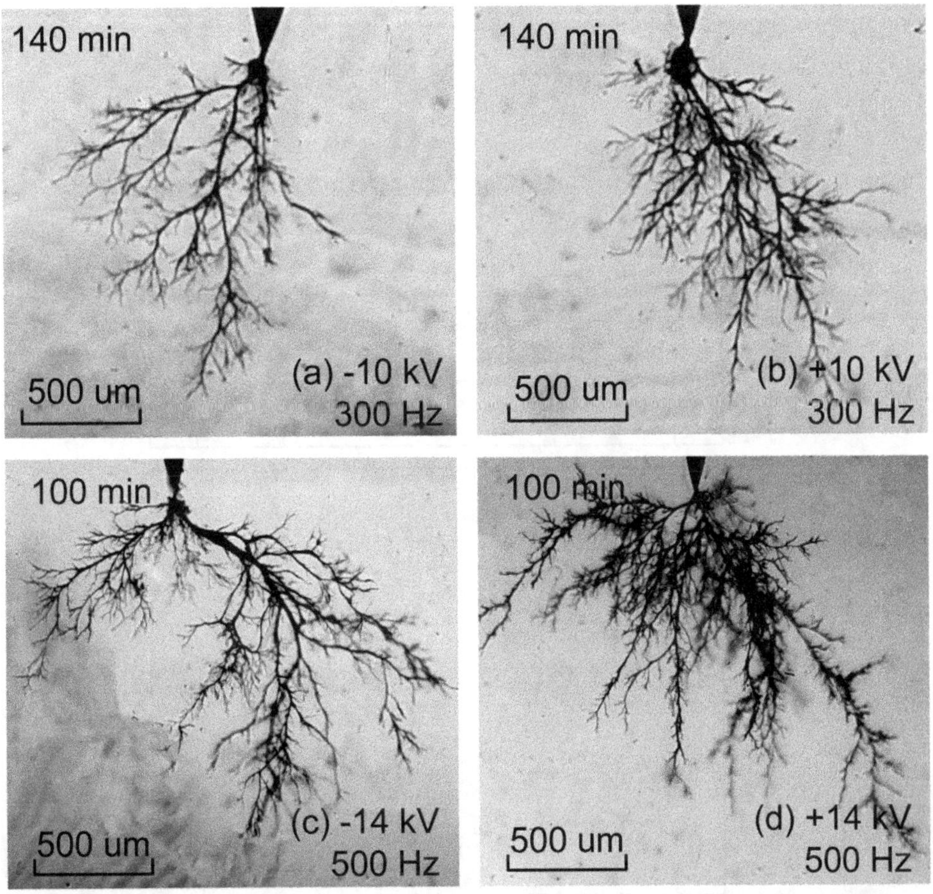

discharge form a shielding layer around the needle and tree tips, which causes the electrical field distribution more uniform and reduces the electrical field. The weakened electrical field makes it difficult for small branches to initiate. Therefore, the destructive area becomes smaller with the negative pulse. However, the motility of holes produced by positive pulse is relatively poor. The strong electrical field near the tip encourages the formation of small branches, resulting in the serious destruction in epoxy resin.

Figure 15. Relation between the accumulated damage and the frequency with positive and negative pulse polarity at 14 kV

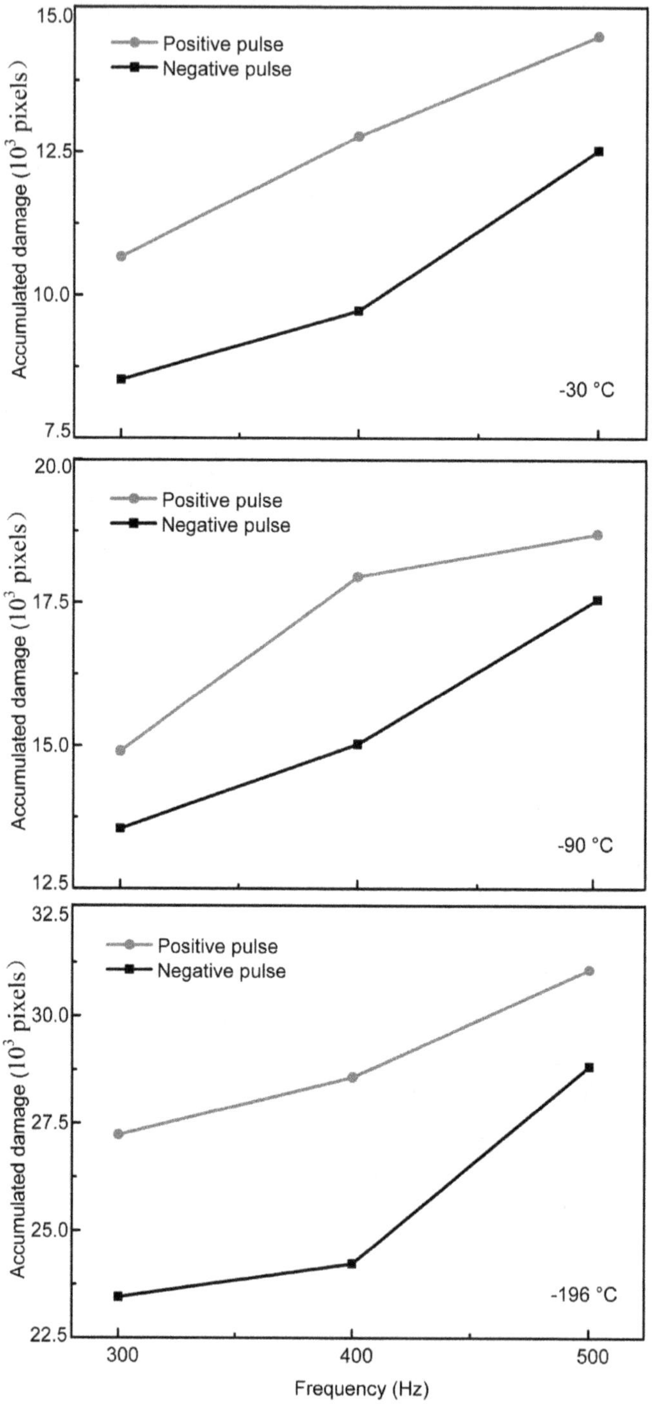

It has obvious deficiency to use a single parameter to describe the characteristics of electrical trees. Tree length, fractal dimension and accumulated damage can't describe the growth characteristics in the perpendicular direction of electrical field. A parameter of expansion coefficient (D/L) is proposed to solve the problem in the literature (Zheng et al., 2008). Where D is the maximum length of branches along the direction perpendicular to the electrical field and L is the maximum tree length along the direction parallel to electrical field.

Figure 16 shows the relation between the expansion coefficient and treeing time with two polarities under the low temperature. The pulse voltage is 14 kV and 500 Hz. At the same temperature, there is little or no significant difference in D/L with negative and positive pulse. However, it can be observed that the low temperature has great influence on the value of D/L. With the decrease of temperature, the D/L appears to rise, which implies that the tree branches extend in both perpendicular and parallel direction of electrical field at lower temperature. The D/L of 0.7 at -30 °C means that branches grow mainly towards the electrical field direction.

The effects of low temperature on breakdown characteristics are obvious. It takes longer time to breakdown under lower temperature. In addition,

Figure 16. Relation between the expansion coefficient and treeing time with negative and positive pulse polarity

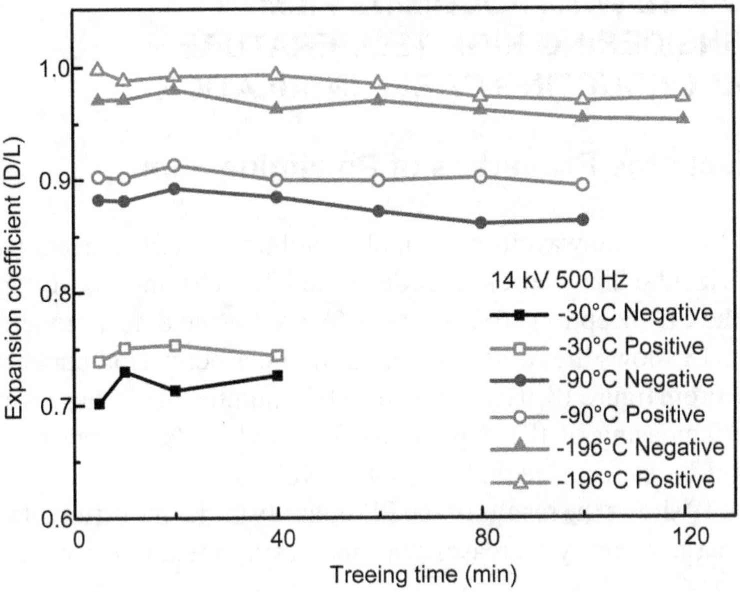

compared with the room temperature, the diameter of breakdown channels has a significant increase with the decrease of temperature (Du et al., 2015). This means that the low-temperature insulation breakdown will lead to great destruction in the superconducting magnet system. High frequency promotes the tree initiation and propagation. Thus, the value of TTB decreases obviously, which is adverse to the operation of ITER device. Moreover, with the decrease of temperature, the effect of pulse frequency on TTB is more significant. And the TTB tends to be much longer at the lower temperature. At the same temperature, there is a higher probability of tree inception with the higher pulse frequency, which is related with the frequent partial discharges and charge injection. Besides the pulse frequency, the pulse polarity and the amplitude of pulse voltage also has effect on electrical tree characteristics under the low temperature. It is revealed that compared with the negative pulse, it is easier for tree branches to grow at the positive pulse under the same low temperature. And the accumulated damage becomes larger with the positive than negative pulse. The reason is the difference of electrons and holes produced by negative and positive pulse respectively under low temperature. The tree inception probability is higher with application of higher pulse voltage. Moreover, the electrical tree growth rate increases when the pulse voltage ranges from 10 kV to14 kV.

CHARACTERIZATION OF PARTIAL DISCHARGE WITH POLYIMIDE FILM IN LN_2 CONSIDERING HIGH TEMPERATURE SUPERCONDUCTING CABLE INSULATION

The Discharge Properties of Polyimide Film

Figure 17 shows the waveform of applied voltage and PD current pulses for 1 mm horizontal diameter circular defect in LN_2. The applied voltage was 2 kV and the PD inception voltage was 1.58 kV for the defect model.

Figure 18 shows the results of the PD number per period under different voltages. From figure 18, it can be seen that the number of discharges increases with the increasing of the applied voltage, and it also increases with the addition of the defect size under the same voltage.

Figure 19 shows the results of the PD quantity under the different voltages. The discharge quantity increases with increasing the applied voltage at the

Figure 17. Applied voltage and PD current waveform

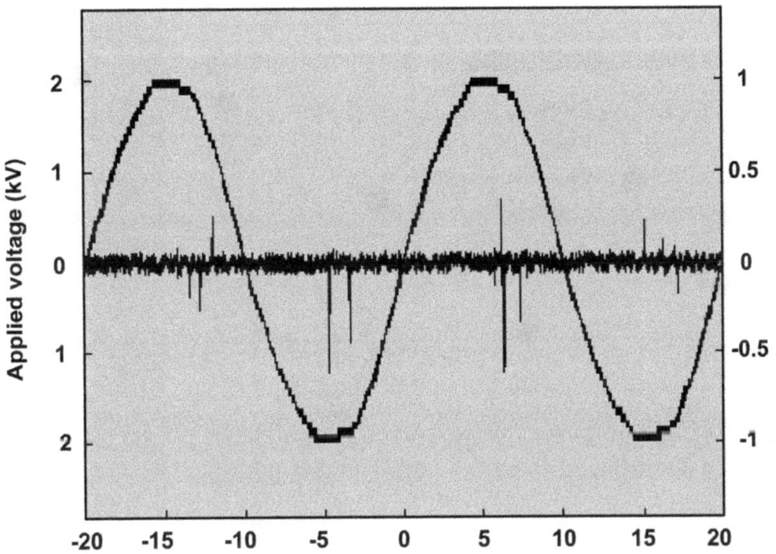

Figure 18. Relation between the number of discharges and the applied voltage in different voids per period

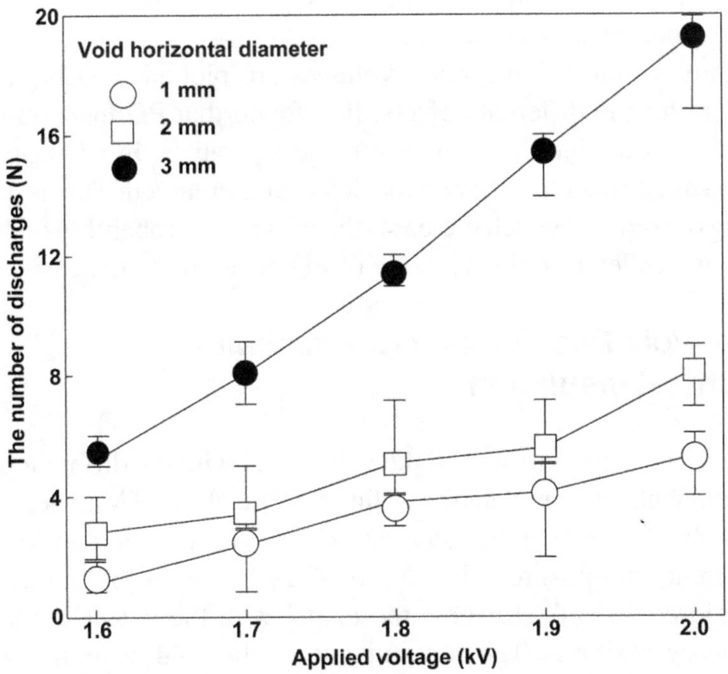

Figure 19. Q-E characteristic in the different voids

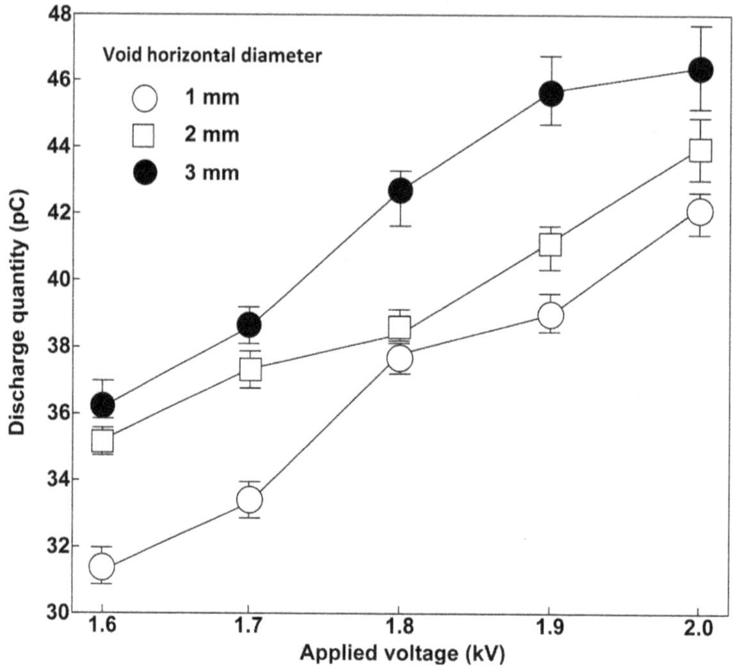

same void. In addition, when the void defect is larger, the discharge quantity is greater under the applied voltage.

In figure 20, the PD inception voltages are plotted for LN_2 and room temperature for the different defects. It is found that PD inception voltage in LN_2 is about double than that at room temperature. In addition, the PD inception voltage decreased when the defect size enlarged. This is due to the increasing of void size which increases the formation probability of discharge channels, which leads to the decrease of PD inception voltage.

Effect of Void Position and Contaminant on Electrical Insulation

The electric field strength around the void defect with the different positions and contaminants was simulated by 2-dimension analysis of Maxwell software. Figure 21 shows three simulation models of 3-layer polyimide film samples with different void positions in LN_2. Figure 21a shows the model without defect. Figure 21b and 21c show the model with the defect in the middle and boundary positions. The contaminants in the void were nitrogen, dust

Figure 20. Relation between the PD inception voltage and the void size in the different temperatures

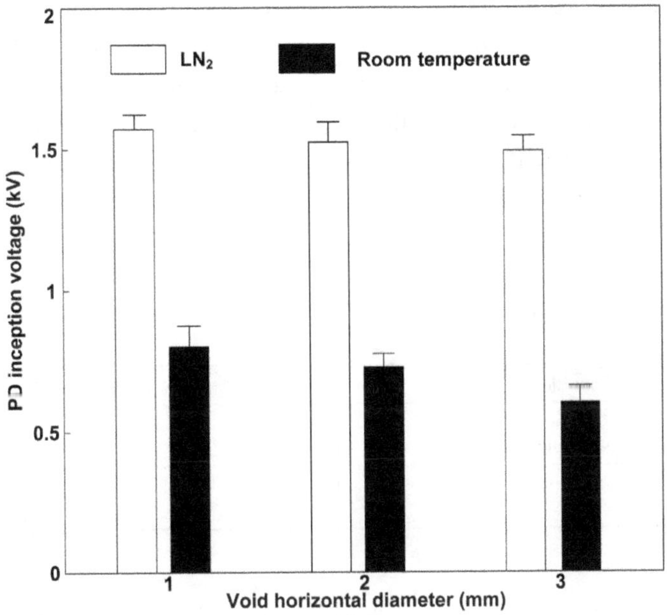

Figure 21. (a) Without defect, (b) Defect in the middle, (c) Defect in the boundary region

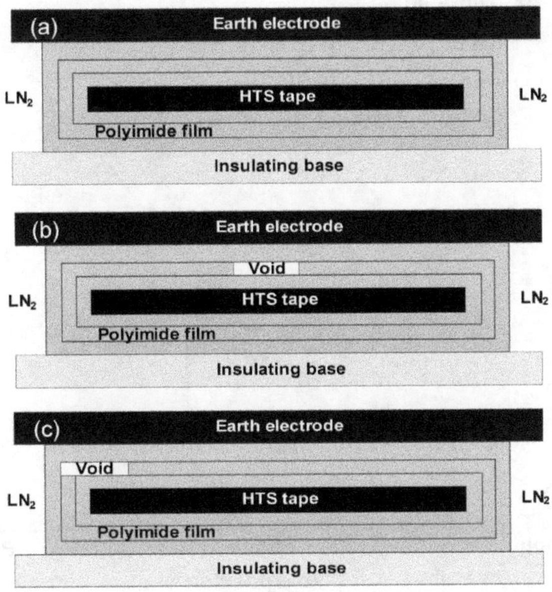

and ice respectively. The electric field strength was calculated by the applied voltage and the relative permittivity listed in table 1(Rezaeifar et al., 2010).

Figure 22 shows the maximum electric field strength of simulation models under AC 1.5 kV in LN_2. It was found that due to the existence of the defect, the maximum field strength increases significantly. The maximum field strength in the model with middle void is 24% higher than that without the defect. The impact of different contaminant components on the maximum field strength is small when the void is in the middle, but it is apparent when the

Table 1. Relative Permittivity

Condition	Relative permittivity
N_2(Gas) N_2(Liquid)	1.0 1.4
Dust Ice	3 4

Figure 22. Relation between the maximum electric field strength and the void position with different contamination

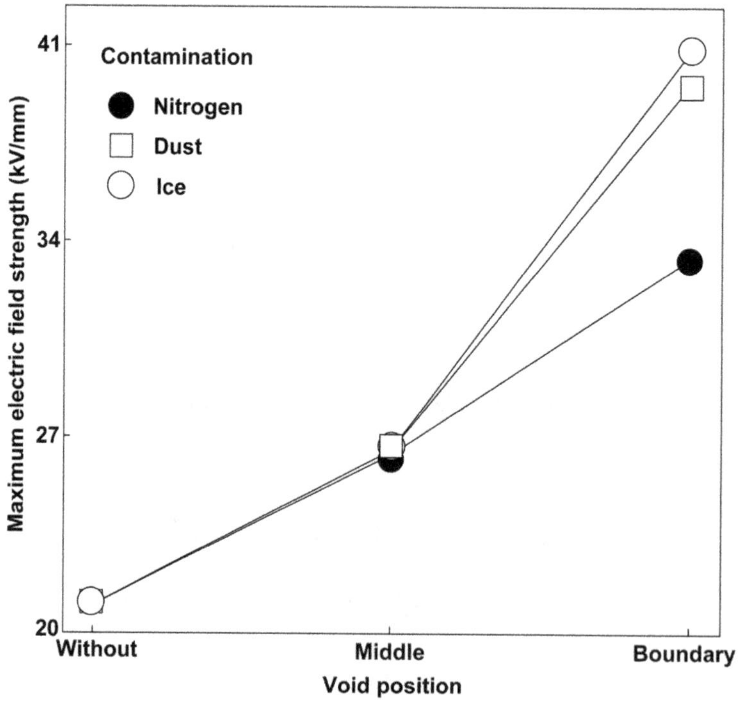

void is in the boundary region. The insulation in HTS apparatus is complex, which consist of non-uniform field configuration, interfaces between solids and LN_2, nitrogen-filled voids in imperfect solids and other contaminants (Du et al., 2016). The monitoring of the insulation condition should take all of these factors into account.

REFERENCES

Bahadoorsingh, S., & Rowland, S. M. (2010). Investigating the Impact of Harmonics on the Breakdown of Epoxy Resin Through Electrical Tree Growth. *IEEE Transactions on Dielectrics and Electrical Insulation*, *17*(5), 1576–1584. doi:10.1109/TDEI.2010.5595560

Bittner-Rohrhofer, K., Humer, K., Wang, Z. D., Weber, H. W., Fabian, P. E., & Munshi, N. A. (2003). Radiation Hardness of Newly Developed ITER Relevant Insulation Systems. *Fusion Engineering and Design*, *66-68*, 1209–1213. doi:10.1016/S0920-3796(03)00343-0

Boyer, C., Seo, K., & Hamada, K. (2014). Validation of Helium Inlet Design for ITER Toroidal Field Coil. *IEEE Transactions on Dielectrics and Electrical Insulation*, *24*(3).

Champion, J. V., & Dodd, S. J.. (2001). Simulation of Partial Discharges in Conducting and Non-conducting Electrical Tree Structures. *Journal of Physics. D, Applied Physics*, *34*(8), 1235–1242. doi:10.1088/0022-3727/34/8/314

Chen, G., & Tham, C. H. (2009). Electrical Treeing Characteristics in XLPE Power Cable Insulation in Frequency Range Between 20 and 500 Hz. *IEEE Transactions on Dielectrics and Electrical Insulation*, *16*(1), 179–188. doi:10.1109/TDEI.2009.4784566

Chen, Y., Imai, T., Ohki, Y., & Tanaka, T. (2010). Tree Initiation Phenomena in Nanostructured Epoxy Composites. *IEEE Transactions on Dielectrics and Electrical Insulation*, *17*(5), 1509–1515. doi:10.1109/TDEI.2010.5595552

Chowdhuri, P. (1981). Some Characteristics of Dielectric Materials at Cryogenic Temperatures for HVDC Systems. *IEEE Transactions on Electrical Insulation*, *16*(1), 40–51. doi:10.1109/TEI.1981.298402

Chowdhuri, P. (1981). Some Characteristics of Dielectric Materials at Cryogenic Temperatures for HVDC Systems. *IEEE Transactions on Electrical Insulation*, *16*(1), 40–51. doi:10.1109/TEI.1981.298402

Danikas, M. G., & Adamidis, G. (1997). Partial Discharges in Epoxy Resin Voids and the Interpretational Possibilities and Llimitations of Pedersens Model. *Electrical Engineering*, *80*(2), 105–110. doi:10.1007/BF01245959

Dissado, L. A. (2002). Understanding Electrical Trees in Solids: From Experiment to Theory. *IEEE Transactions on Dielectrics and Electrical Insulation*, *9*(4), 483–497. doi:10.1109/TDEI.2002.1024425

Du, B. X., Han, T., & Su, J. G. (2013). Effect of Low Temperature on Treeing Phenomena of Silicone Rubber/SiO2 Nanocomposites. *IEEE Conf. electri. Insul. Dielectr. Phenomena (CEIDP)*, 514-517.

Du, B. X., Han, T., & Su, J. G. (2014). Tree Characteristics in Silicone Rubber/SiO2 Nanocomposites under Low Temperature. *IEEE Transactions on Dielectrics and Electrical Insulation*, *21*(2), 503–510. doi:10.1109/TDEI.2013.004140

Du, B. X., Li, J., Hu, W. J., & Mu, J. (2012). Discharge breakdown characteristics in the gap with polyimide nanocomposite films. *High Voltage Engineering*, *38*, 336–339.

Du, B. X., Liu, Y., & Liu, H. J. (2008). Effects of Low Pressure on Tracking Failure of Printed Circuit Boards. *IEEE Transactions on Dielectrics and Electrical Insulation*, *15*(5), 1379–1384. doi:10.1109/TDEI.2008.4656247

Du, B. X., Ma, Z. L., & Gao, Y. (2009), Phenomena and Mechanism of Electrical Tree in Silicone Rubber. *Proc. IEEE Int. Conf. Prop. Appl. Dielectr. Mater.*, 37-40. doi:10.1109/ICPADM.2009.5252511

Du, B. X., Zhang, M. M., Han, T., & Zhu, L. W. (2016). Effect of pulse frequency on tree characteristics in epoxy resin under low temperature. *IEEE Transactions on Dielectrics and Electrical Insulation*, *23*(1), 104–112. doi:10.1109/TDEI.2015.005342

Du, B. X., & Zhu, L. W. (2015). Electrical tree characteristics of XLPE under repetitive pulse voltage in low temperature. *Dielectrics & Electrical Insulation IEEE Transactions on*, *22*(4), 1801–1808. doi:10.1109/TDEI.2015.005183

Hong, J. G., Jeong, I. H., & Seokho, N. (2014). Analysis on the Dielectric Characteristics of Solid Insulation Materials in LN2 for Development of High Voltage Magnet Applications. *IEEE Transactions on Applied Superconductivity*, *24*(3).

Honjo, S., Shimodate, M., Takahashi, Y., Masuda, T., Yumura, H., Suzawa, C., & Suzuki, H. et al. (2003). Electric properties of a 66 kV 3-core superconducting power cable system. *IEEE Transactions on Applied Superconductivity*, *13*(2), 1680–1683. doi:10.1109/TASC.2003.812975

Ivanov, D. P., Anashkin, D. I. O., & Khvostenko, P. P. (2013). Reliability of Force Cooled Superconducting Magnets for Fusion. *IEEE Transactions on Applied Superconductivity*, *24*(3).

Kai Wu, , & Dissado, L. A.. (2005). Model for Electrical Tree Initiation in Epoxy Resin. *IEEE Transactions on Dielectrics and Electrical Insulation*, *12*(4), 655–668. doi:10.1109/TDEI.2005.1511091

Kai Wu, , & Dissado, L. A.. (2005). Model for Electrical Tree Initiation in Epoxy Resin. *IEEE Transactions on Dielectrics and Electrical Insulation*, *12*(4), 655–668. doi:10.1109/TDEI.2005.1511091

Lee, B. W., Choi, W., Choi, Y. M., Kim, Y. H., & Koo, J. Y. (2013). Comparison between PD inception voltage and BD voltage of PPLP in LN_2 considering HTS cable insulation. *IEEE Transactions on Applied Superconductivity*, *23*(3).

Li, W. W., Li, J. Y., Yin, G. L., Li, S. T., Zhao, J. K., Ouyang, B. H., & Ohki, Y. (2012). Frequency Dependence of Breakdown Performance of XLPE with Different Artificial Defects. *IEEE Transactions on Dielectrics and Electrical Insulation*, *19*(4), 1351–1359. doi:10.1109/TDEI.2012.6260011

Liu, Y., & Cao, X. L. (2013). Electrical Tree Initiation in XLPE Cable Insulation by Application of DC and Impulse Voltage. *IEEE Transactions on Dielectrics and Electrical Insulation*, *20*(5), 1691–1698. doi:10.1109/TDEI.2013.6633699

Rezaeifar, F., Suzuki, Y., Kumada, A., & Hidaka, K. (2010). Characterization of partial discharge in composite insulation system with PPLP® for HTS cable. *IEEE Transactions on Applied Superconductivity*, *17*(6), 1747–1753.

Shimizu, N., & Laurent, C. (1998). Electrical Tree Initiation. *IEEE Transactions on Dielectrics and Electrical Insulation*, *5*(5), 651–659. doi:10.1109/94.729688

Stone, G. C., Van Heeswijk, R. G., & Bartnikas, R. (1992). Electrical Aging and Electroluminescence in Epoxy under Repetitive Voltage Surges. *IEEE Transactions on Electrical Insulation*, *27*(2), 233–244. doi:10.1109/14.135595

Zheng, X. Q., & Chen, G. (2008). Propagation Mechanism of Electrical Tree in XLPE Cable Insulation by Investigating a Double Electrical Tree Structure. *IEEE Transactions on Dielectrics and Electrical Insulation*, *15*(3), 800–807. doi:10.1109/TDEI.2008.4543118

Index

www.ingramcontent.com/pod-product-compliance
Lightning Source LLC
LaVergne TN
LVHW080605300926
849162LV00005B/1053